Power Electronics and Power Systems

Series Editors:
M.A. Pai
Alex M. Stankovic

For further volumes:
http://www.springer.com/series/6403

Aranya Chakrabortty • Marija D. Ilić
Editors

Control and Optimization Methods for Electric Smart Grids

 Springer

Editors

Aranya Chakrabortty
Electrical and Computer Engineering
FREEDM Systems Center
North Carolina State University
Raleigh, NC 27695, USA
aranya.chakrabortty@ncsu.edu

Marija D. Ilić
Electrical and Computer Engineering
Engineering and Public Policy
Carnegie Mellon University
Pittsburgh, PA 15213-3890, USA
milic@ece.cmu.edu

ISBN 978-1-4614-1604-3   e-ISBN 978-1-4614-1605-0
DOI 10.1007/978-1-4614-1605-0
Springer New York Dordrecht Heidelberg London

Library of Congress Control Number: 2011942899

Printed on acid-free paper

Springer is part of Springer Science+Business Media (www.springer.com)

To Joe,
A true pioneer in rethinking sensing,
communications and control of electric
power systems

Preface

In the 21st century, electric power engineering is going green and smart. Triggered by several recent catastrophes such as the major blackout in the Northeastern USA in 2003 and Hurricane Katrina in New Orleans in 2005 together with the Energy Act of 2007, the term *smart grid* has become almost ubiquitous across the world not only as a political concept but also as an entirely new technology requiring a tremendous amount of inter-disciplinary research. The initiatives taken by the smart grid research community in the United States have so far been successful in bringing researchers from power system engineering, signal processing, computer science, communications, business, and finance as well as chemical and wind engineering among other disciplines under the same roof in order to cater to the diverse research needs of this technology. As a part of this enterprise power engineers, for example, are investigating efficient and intelligent ways of energy distribution and load management, computer scientists are researching cyber security issues for reliable sharing of information across the grid, the signals community is looking into advancing instrumentation facilities for detailed grid monitoring, wind engineers are studying renewable energy integration, while business administrators are reframing power system market policies to adapt to these new changes in the system.

Concomitant with these advances, researchers have also come to recognize the urgent need for new systems-level knowledge of power and energy systems for sustaining the advancement of this emerging field. This, in turn, has led to a natural demand for the two lifelines of system theory in smart grid research, namely – control and optimization. Almost every facet of making a power system *smart* or self-regulated boils down to using control theory in some form or other. Relevant examples include modeling, identification, estimation, robustness, optimal control, and decision-making over networks. Over the past two years, for instance, several research workshops, conference tutorial sessions, and national meetings have been organized to discuss the strong potential of control and optimization in smart grid applications starting from small residential-level energy management, smart metering, and power markets to much broader-scale problems such as wide-area monitoring and control. The editors of this book have been involved in the organization of many of these workshops. Several of the contributing authors too have given invited presentations in these gatherings, three of the most notable ones being a special session in the IEEE Conference on Decision and Control (Atlanta, GA, 2010), a tutorial session in the American Control Conference (San Francisco, CA, 2011) and a Workshop on Cyber-Physical Applications in Smart Power Systems (North Carolina State University, Raleigh, NC, 2011). The initial idea for publishing this book arose from these meetings with the objective of consolidating some of the most promising and transformative recent research in smart grid control in hopes of laying the foundation for future advances in this critical field of study.

The book contains eighteen chapters written by leading researchers in power, control, and communication systems. The essays are organized into three broad sections, namely Architectures and Integration, Modeling and Analysis, and Communication and Control. As is apparent from their titles, the main perspective of these sections is to capture in a holistic way how tomorrow's grid will need to be an enormously complex system in order to solve the problems that we are facing today. Literally, with every passing day, our national grid is becoming integrated with new generation in the form of renewable energy resources, new loads in the form of smart vehicles, new sensors such as smart meters and Phasor Measurement Units, and newer mechanisms of decision-making guided by complex power market dynamics. Our goal is to capture the spectrum of this exponential transformation, and at the same time present the plethora of open problems that this transformation poses for our control theory colleagues. Many of these problems may sound like routine questions in control and optimization, but they often lead to challenging, interesting, and ultimately highly rewarding directions for theoretical research. To the best of our knowledge, this is the first comprehensive book on this topic.

The Architectures and Integration section opens the book with visionary ideas on sustainable architectures for power system operation and control under significant penetration of highly variable renewable energy resources presented in Chap. 1. This is followed by a discussion on the economics of electricity markets and their impacts on demand response in Chap. 2. Chapter 3 furthers the demand response concept for enabling random energy integration. Chapter 4 illustrates several practical constraints in smart grid sensing and communications that may destabilize real-time power market operations and proposes new communication topologies that can bypass such problems. Chapter 5 presents a fresh control perspective to demand-side energy management in residential and commercial units using convex optimization-based model predictive control. Chapter 6 highlights the architectural challenges needed for integration of plug-in-hybrid vehicles into the grid focusing on problems related to demand response and communications necessary to accomplish the smart features of these smart vehicles.

The Modeling and Analysis section presents the upcoming research directions on mathematical modeling, data analysis, and information processing in power systems. Chapter 7 opens this section with a modeling framework that can be highly useful for analyzing the impacts of wind power penetration on the dynamics of the conventional grid. Chapter 8 delves into novel data analysis techniques for wide-area oscillation tracking in large-scale power systems and highlights the importance of signal processing as a major tool for wide-area monitoring research. Chapter 9 models the dynamic mechanisms of cascading failures in geographically dispersed grids, while Chap. 10 presents a reliability modeling framework for tomorrow's phasor-integrated power system using ideas of real-time fault diagnosis and Markovian models of measurement networks. The discussion switches gears towards the computational aspects of the smart grid in Chap. 11, which presents a modeling and control strategy for data centers that are becoming essential parts of today's grid operations. Chapter 12 closes this section with a discussion on how various geometrical properties of power system topologies can influence disturbance propagation in the system, thereby highlighting the importance of combinatorics in smart grid research.

The Communication and Control section unites some of the most critical control design challenges for tomorrow's grid with a focus on how communications and security will play integral roles in the execution of such controllers. The section opens with Chap. 13, which presents game-theoretic optimization problems for charging coordination of plug-in electric vehicles, and, thereby addresses the intersection of two emerging trends in the modernization of power systems, namely vehicle electrification and flexible loading. Chapter 14 addresses the seminal problem of cyber-security of smart grids and presents a vulnerability assessment framework to quantify risk due to intelligent coordinated attacks on grid assets. Chapter 15 discusses the applications of wide-area phasor measurement technology in distribution-level power systems, initiating a new line of thinking on monitoring and control. Chapter 16 fuses different ideas of cooperative control theory to develop *distributed* algorithms for economic dispatch, thereby enabling the future grid to be independent of centralized decision-making strategies. Chapters 17 and 18 address wide-area control problems for oscillation damping in power transmission systems. The former presents a new adaptive control approach for delay compensation in Synchrophasor-based feedback, while the latter using model reference control and clustering methods for adding damping to inter-area oscillations.

As can be seen, the chapters in each section maintain their own thematic continuity and at the same time have significant overlaps with chapters in other sections as well. Therefore, one may read the book in its entirety or focus on individual chapters. Due to its broad scope, this will be an ideal resource for students in advanced graduate-level courses and special topics in both power and control systems. It will also interest utility engineers who seek an intuitive understanding of the emerging applications of control and optimization methods in smart grids. Until now there has been very little literature concerning the formulation of a comprehensive control problem for the smart grid enterprise, and on relating different models and approaches to its overall solution methodology. The chapters in this book represent work in progress by the community on the way towards such solutions.

We would like to express our gratitude to all the contributing authors for providing their valuable input toward the development of this book. We also thank our colleagues Murat Arcak, John Wen, Tariq Samad, Ning Lu, Steven Elliot, M. A. Pai, Massoud Amin, Manu Parashar, Sumit Roy, and Yufeng Xin for many interesting discussions that have motivated us to undertake the idea of publishing this book. Sincere thanks also go to Merry Stuber from Springer for proofreading different versions of the manuscript and guiding the editorial work.

Last but not least, this book is dedicated to Joe Chow on the occasion of his 60th birthday. Dr. Chow is one of the most distinguished researchers and educators in the field of power systems and control theory. His research career spans nearly 40 years, and includes pioneering contributions to singular perturbation theory in the late 1970s, multivariable control of power systems in the 1980s, FACTS controller designs in the 1990s, and Wide-area Phasor Measurements over the past two decades. His ground-breaking work has earned him the highest acclaim from both power and control research communities all around the world. We take great pleasure in offering this book as a small token of appreciation in honor of Joe on this very special occasion.

Raleigh, NC, USA   Aranya Chakrabortty
Pittsburgh, PA, USA   Marija D. Ilić

Contents

Contributors

Anuradha Annaswamy Department of Mechanical Engineering, Massachusetts Institute of Technology, Cambridge, MA, USA

I. Safak Bayram Department of Electrical and Computer Engineering, North Carolina State University, Raleigh, NC, USA

Subhashish Bhattacharya Department of Electrical and Computer Engineering, North Carolina State University, Raleigh, NC, USA

Duncan Callaway Energy and Resources Group, University of California, Berkeley, CA, USA

Aranya Chakrabortty Department of Electrical and Computer Engineering, North Carolina State University, Raleigh, NC, USA

Lijun Chen Engineering and Applied Science, California Institute of Technology, Pasadena, CA, USA

Joe H. Chow Electrical, Computer and Systems Engineering, Rensselaer Polytechnic Institute, Troy, NY, USA

Mo-Yuen Chow Department of Electrical and Computer Engineering, North Carolina State University, Raleigh, NC, USA

Michael Devetsikiotis Department of Electrical and Computer Engineering, North Carolina State University, Raleigh, NC, USA

Alejandro D. Domínguez García Department of Electrical Engineering, University of Illinois, Urbana Champaign, IL, USA

S. Ghiocel Electrical, Computer and Systems Engineering, Rensselaer Polytechnic Institute, Troy, NY, USA

Manimaran Govindarasu Department of Electrical and Computer Engineering, Iowa State University, Ames, IA, USA

Fabrizio Granelli Department of Information Engineering and Computer Science, University of Trento, Trento, Italy

Daniel Greene Palo Alto Research Center, Palo Alto, CA, USA

Daniel A. Haughton School of Electrical, Computer and Energy Engineering, Arizona State University, Tempe, AZ, USA

Gerald Heydt School of Electrical, Computer and Energy Engineering, Arizona State University, Tempe, AZ, USA

T. Hikihara Department of Electrical Engineering, Kyoto University, Kyoto, Japan

Haitham Hindi Palo Alto Research Center, Palo Alto, CA, USA

Ian Hiskens Department of Electrical Engineering and Computer Science, University of Michigan, Ann Arbor, MI, USA

Marija D. Ilić Department of Electrical and Computer Engineering, Carnegie Mellon University, Pittsburgh, PA, USA

Libin Jiang Engineering and Applied Science, California Institute of Technology, Pasadena, CA, USA

Arman Kiani Institute of Automatic Control Engineering, Technische Universität München, Munich, Germany

Anupama Kowli CSL and the ECE Department, University of Illinois, Urbana-Champaign, IL, USA

Bruce Krogh Department of Electrical and Computer Engineering, Carnegie Mellon University, Pittsburgh, PA, USA

Caitlin Laventall AOL.com, Palo Alto, CA, USA

Na Li Engineering and Applied Science, California Institute of Technology, Pasadena, CA, USA

Chen-Ching Liu School of Electrical, Electronic and Mechanical Engineering, University College Dublin, Dublin, Ireland

Qixing Liu Department of Electrical and Computer Engineering, Carnegie Mellon University, Pittsburgh, PA, USA

Steven H. Low Engineering and Applied Science, California Institute of Technology, Pasadena, CA, USA

Zhongjing Ma School of Automation, Beijing Institute of Technology, and the Key Laboratory of Complex System Intelligent Control and Decision (Beijing Institute of Technology), Ministry of Education, Beijing, China

Sean Meyn CSL and the ECE Department, University of Illinois, Urbana-Champaign, IL, USA

I. Mezic Department of Mechanical Engineering, University of California, Santa Barbara, CA, USA

George Michailidis Department of Statistics, University of Michigan, Ann Arbor, MI, USA

Matias Negrete-Pincetic CSL and the ECE Department, University of Illinois, Urbana-Champaign, IL, USA

Xueping Pan School of Energy and Electrical Engineering, Hohai University, Nanjing, China

Luca Parolini Department of Electrical and Computer Engineering, Carnegie Mellon University, Pittsburgh, PA, USA

Matthew C. Ruschmann Department of Electrical and Computer Engineering, Binghamton University, Binghamton, NY, USA

Anna Scaglione University of California, Davis, CA, USA

Ehsan Shafieepoorfard CSL and the ECE Department, University of Illinois, Urbana-Champaign, IL, USA

Uday V. Shanbhag Department of Industrial and Enterprise Systems Engineering, University of Illinois, Urbana Champaign, IL, USA

Bruno Sinopoli Department of Electrical and Computer Engineering, Carnegie Mellon University, Pittsburgh, PA, USA

Siddharth Sridhar Department of Electrical and Computer Engineering, Iowa State University, Ames, IA, USA

Y. Susuki Department of Electrical Engineering, Kyoto University, Kyoto, Japan

Robert J. Thomas Cornell University, Ithaca, NY, USA

Vaithianathan "Mani" Venkatasubramanian School of Electrical Engineering and Computer Science, Washington State University, Pullman, WA, USA

Gui Wang CSL and the ECE Department, University of Illinois, Urbana-Champaign, IL, USA

Zhifang Wang University of California, Davis, CA, USA

N. Eva Wu Department of Electrical and Computer Engineering, Binghamton University, Binghamton, NY, USA

Ziang Zhang Department of Electrical and Computer Engineering, North Carolina State University, Raleigh, NC, USA

Part I
Architectures and Integration

Toward Sensing, Communications and Control Architectures for Frequency Regulation in Systems with Highly Variable Resources

Marija D. Ilić and Qixing Liu

Abstract The basic objective of this chapter is to rethink frequency regulation in electric power systems as a problem of cyber system design for a particular class of complex dynamical systems. It is suggested that the measurements, communications, and control architectures must be designed with a clear understanding of the temporal and spatial characteristics of the power grid as well as of its generation and load dynamics. The problem of Automatic Generation Control (AGC) and frequency regulation design lends itself well to supporting this somewhat general observation because its current implementation draws on unique structures and assumptions common to model aggregation in typical large-scale dynamic network systems. We describe how these assumptions are changing as a result of both organizational and technological industry changes. We propose the interactions variable-based modeling framework necessary for deriving models, which relax conventional assumptions when that is needed. Using this framework, we show that the measurements, communications, and control architectures key to ensuring acceptable frequency response depend on the types of disturbances, the electrical characteristics of the interconnected system and the desired technical and economic performance. The simulations illustrate several qualitatively different electric energy systems. This approach is by and large motivated by today's AGC and its measurement, communications and control architectures. It is with this in mind that we refer to our interactions variable-based frequency regulation framework as "enhanced AGC" (E-AGC). The enhancements come from accounting for temporal and spatial characteristics of the system which require a more advanced frequency regulation design than the one presently in place. Our proposed interactions variable-based aggregation modeling could form the basis

M.D. Ilić (✉) • Q. Liu
Carnegie Mellon University, 5000 Forbes Ave., Pittsburgh, PA 15213, USA
e-mail: milic@ece.cmu.edu; lqx@cmu.edu

A. Chakrabortty and M.D. Ilić (eds.), *Control and Optimization Methods for Electric Smart Grids*, Power Electronics and Power Systems 3,
DOI 10.1007/978-1-4614-1605-0_1, © Springer Science+Business Media, LLC 2012

for a coordination of interactions between the smart balancing authorities (SBAs) responsible for frequency regulation in the changing industry. Given the rapid deployment of synchrophasors, the proposed E-AGC can be easily implemented.

1 Introduction

A high quality of electricity service assumes near-ideal nominal frequency, which is the result of almost instantaneous power supply and demand balancing. In actual systems, deviations from this perfect balance always exist and are mainly caused by hard-to-predict demand and by the lack of direct power control exchange between neighboring utilities. Therefore, the resulting frequency must be regulated back to nominal by means of output feedback control responding to the frequency errors caused by the power imbalances. Much is known about Automatic Generation Control (AGC), and this control scheme is considered one of the most ingenious elegant feedback schemes in large-scale man-made network systems. Its effectiveness comes from viewing the problem as quasi-stationary and accounting for the fact that, at equilibrium, the system operates at a single frequency. This, together with the assumption that utilities are weakly connected, forms the basis for using Area Control Error (ACE) as the single output to which several fast-responding power plants respond in each utility (control area) and drive the error to zero within 10 min or so. It can be shown that, at theoretical equilibrium, the entire interconnected system will balance and frequency should return to nominal, provided that each control area (CA) responds to its own ACE without communicating with other CAs.

However, the question of "correct" AGC standards and a utility's ability to meet these standards has been a long-standing subject of many industry committees led by the North American Electric Reliability Organization (NERO). Many of the industry's great thinkers, starting with the late Nathan Cohn, the father of AGC, have brought the scheme to near perfection over the years. Understanding the fundamentals of AGC, including the elusive notions of Inadvertent Energy Exchange (IEE) and Time Error Correction (TEC), requires an in-depth treatment, which is outside the objectives of this chapter; the interested reader should explore the extensive literature on this subject. Ensuring that the standards are met has been difficult up until very recently because of the lack of synchronized measurements. As a result, there have been growing concerns about the worsening of frequency quality. These concerns have become magnified with the prospects of deploying highly variable energy resources, such as wind and solar power farms. This overall situation requires a rethinking of today's AGC industry practice from the viewpoint of the key assumptions underlying its design. In particular, given the recent deployments of synchrophasor technologies capable of providing fast and synchronized measurements across wide areas, the question of an enhanced AGC (E-AGC) design and its purpose presents itself quite forcefully.

In this chapter, we pose the problem of E-AGC as a control design problem based on a carefully designed dynamic model directly relevant for frequency regulation in complex power systems. We derive a model which can capture the effects of disturbances likely to be seen in future electric power systems with many intermittent resources. We observe that today's AGC is used for the fine tuning of frequency deviations caused by small hard-to-predict power imbalances around the generation dispatched to supply forecast demand. As such, it is viewed as regulating steady-state frequency offsets over the time interval of 10–15 min; primary governor control at the same time compensates for local fluctuations very quickly, without considering dynamic interactions with the rest of the system. Simplicity is achieved by having fast local primary control, which is tuned without modeling fast dynamic interactions with the rest of the system. AGC, on the other hand, compensates for steady-state net power imbalances in each utility, assuming that the electrical distances between the generators in the CA are negligible; the effects of other CAs are compensated for by responding to the ACE, which is a measure of the power imbalance created by the internal demand deviations from the forecast and the deviations in net power exchange from the scheduled exchanges at the time of dispatch in each CA.

In emerging electric energy systems, the implied spatial simplifications when designing fast-stabilizing primary control, as well as the implied temporal simplifications assuming near steady-state conditions when performing AGC and accounting for the effects of net imbalances within a large multi-CA interconnected power system, will become hard to justify. Depending on the electrical properties of the system, the nature of the disturbances causing the frequency changes, and the dynamics of the power plants and loads, it is plausible that frequency response may become very different than the historic response has been. Models for analyzing and predicting the likely frequency response and, consequently, designing controls for stabilizing and regulating frequency have become a very difficult problem. A power system driven by continuously varying persistent disturbances does not lend itself well to separating stabilization and regulation objectives.

In order to begin to answer these difficult questions, we pose in Sect. 2 the problem of frequency regulation by first reviewing a general dynamic model of an interconnected power system capable of representing typical system response to these new disturbances. To start with, this general model is very complex and without obvious structure. In Sect. 3, we propose a systematic model reduction which lends itself well to the enhanced frequency regulation design. Of particular interest is the derivation of models capable of capturing dynamic interactions between (groups of) system users given the ultimate objective of designing the minimum coordination architecture across these groups of system users necessary to control power imbalance interactions causing potentially unacceptable frequency response. Notably, the concepts introduced early on by Joe Chow and his collaborators for model reduction in electric power systems are shown to be key to arriving at the models for enhanced frequency regulation of interest here. Model simplification using standard singular perturbation is used to introduce acceptable temporal simplifications of the generator models used. Similarly, the lesser-known nonstandard singular perturbation method is used to prove the existence of the interactions variable, which can capture the dynamics of power imbalances across large power grid interconnections. The question of aggregation within an interconnected system and the implications of the grouping selected on the achievable frequency quality and the complexity of the required sensing, communications and control architectures is discussed in considerable detail. We recognize that aggregation principles could be based on: (1) the pre-defined organizational boundaries of the consumers and producers responsible for balancing supply and demand; the most typical representatives are utilities and/or CAs; (2) the bottom-up created portfolio of users and producers; representatives of such aggregated system users can likely be entities comprising users with their own distributed energy resources; and (3) best technical decomposition of a given dynamical system from the point of view of having coherent dynamic response and being controllable and observable without relying on help from the others. In Sect. 4, the model relevant for frequency regulation of each SBA is simulated to show the qualitatively different interactions variables resulting from the different relative electrical distances internal and external to the SBAs. In Sect. 5, we illustrate the use of coherency-based method introduced by Joe Chow for the four qualitatively different physical systems of seemingly identical design. We show the system aggregation which results from this approach. In Sect. 6, we compare the aggregation obtained using our proposed interactions variable-based modeling of SBAs with coherency-based aggregation. Interestingly, when inquiring whether interconnection-level coordination may be needed, the two methods arrive at the same conclusion via different paths. For SBAs with strong internal interactions, the coherency-based method leads to the conclusion that a meaningful aggregation would be to have one single system. The interactions variable-based approach concludes that, because the interactions are strong, it is essential to coordinate them. Notably, the model proposed here does not neglect the effects of electrical distances; this is in sharp contrast with today's AGC. We show how the interactions variables are affected by the relative strength of electrical interconnections, both internal to the SBA and also in-between the SBAs.

We next show how these qualitatively different cases lend themselves to different measurement, communications, and control architectures. In Sect. 7, a required sensing, communications and control architecture based on the interactions variable-based model is drawn up, and the results of using such a control architecture are illustrated for the four cases of a small system studied. In Sect. 8, a required sensing, communications, and control architecture based on coherency-based aggregation is sketched out. The complexity of the two, and a comparison of the two, are summarized. In Sect. 9, we discuss the relationship between the proposed cyber architecture for ensuring acceptable frequency response and the AGC architecture of today. We highlight and confirm how the complexity of a relevant dynamic model for frequency regulation depends on the locations and the type of disturbances, and on the electrical characteristics of the interconnected system comprising the CAs with the same boundaries. This complexity will determine the most adequate cyber architecture. We stress that as the industry undergoes both technological and organizational changes, it is going to become difficult to break down a complex interconnected system into weakly connected subsystems, which would lend themselves to uncoordinated frequency regulation. For example, a CA with significant wind power capacity and very few fast-responding power plants may need to rely on power sent by other areas which have this type of plants. Provided this is done, the cost of regulating frequency at the interconnection level as a whole will be lower than without coordination. Examples of such sharing of resources across CAs already exist and are known as dynamic scheduling. A hydro power plant in CA_2 may provide regulation to the CA_1, for example. This all leads to the observation that a dynamic model relevant for the effective frequency regulation of a given interconnected system, independent of how it is partitioned, must represent the relative importance of dynamic power imbalance interactions. It is with this in mind that we formally define the notion of a dynamic interactions variable for each SBA. We show that this variable is driven by the disturbances internal to the SBA, and by the disturbances from the rest of the system as seen by the SBA model.

In the closing Sect. 10, we point out that the case of E-AGC described in this chapter is only an illustration of the enormous need to rethink what the role of cyber is and what it might become in future electric energy systems. While much effort has been made toward major breakthroughs in the fundamental science of energy processing, it is important to recognize the huge opportunities for the enhanced utilization of energy resources presented to us by the soft technologies of sensing, communications, computing, and control. These opportunities are clear but not very tangible at present in part because of the major lack of viewing the cyber design of energy systems with a full understanding of the physical characteristics of these systems, and of their temporal, spatial, and contextual structures. We hope that this chapter will illustrate the enormous importance of relating the physical understanding of the systems problem at hand, the models used, and the implications of these on the type of cyber required. Conversely, the models are also cyber-dependent and the recent progress in cyber technologies for power systems has opened the doors wide to progress in this area. We have attempted with great pleasure and honor to discuss the technical problem of interest, keeping in mind the early work of Joe Chow, which paved our way.

2   A Dynamic Model of Electric Energy Systems with Persistent Disturbances

In this section, the general frequency dynamic model for interconnected power systems is reviewed. We assume that only generators contribute to the frequency dynamics, and we use a module-based approach to derive a dynamic model by subjecting the dynamics of generator modules and loads to the transmission network constraints. Persistent disturbances in the system are caused by both variable loads and renewable power sources.

2.1 Modeling Dynamics of Generator Modules

We model the frequency dynamics of generator modules by assuming that the effects of reactive power and voltage change can be ignored and focus only on the governor-turbine model, which captures the real power-frequency dependence. In addition, in the governor-turbine model, we assume that the primary control loop has already been designed and its parameters are known. Recall that our purpose is to propose a secondary level regulation and therefore the control variable is viewed as being predefined.

The continuous-time frequency dynamics of each individual generator module with closed-loop primary control is modeled as follows [1]:

$$
\begin{bmatrix} \Delta\dot{\delta}_G \\ \Delta\dot{f}_G \\ \Delta\dot{P}_T \\ \Delta\dot{a} \end{bmatrix} = \begin{bmatrix} 0 & \omega_0 & 0 & 0 \\ 0 & -\frac{D}{M} & \frac{1}{M} & \frac{e_t}{M} \\ 0 & 0 & -\frac{1}{T_t} & \frac{K_t}{T_t} \\ 0 & -\frac{1}{T_g} & 0 & -\frac{r}{T_g} \end{bmatrix} \begin{bmatrix} \Delta\delta_G \\ \Delta f_G \\ \Delta P_T \\ \Delta a \end{bmatrix} + \begin{bmatrix} 0 \\ 0 \\ 0 \\ \frac{1}{T_g} \end{bmatrix} \Delta f_G^{ref} + \begin{bmatrix} 0 \\ -\frac{1}{M} \\ 0 \\ 0 \end{bmatrix} \Delta P_G, \quad (1)
$$

where the state variables $\Delta\delta_G$, Δf_G, ΔP_T, and Δa correspond to the deviations of generator voltage phase angle, frequency, turbine mechanical power output and the incremental change of the steam valve position, respectively, around the system equilibrium. Δf_G^{ref} is the governor set-point adjustment, which will be used as the control on the secondary level. ΔP_G refers to the deviation of electrical power output of the generator around the equilibrium value. ω_0 which is the rated angular velocity. M, D, T_g, and T_t are used to denote the inertia constant of the generator, its damping coefficient and the time constants of the governor and turbine, respectively. K_t and e_t are the constant parameters of the governor-turbine primary control loop. r is defined so that $\frac{1}{r}$ is the generator speed droop.

For the ith generator module, we denote the state variables and the secondary control input as

$$
x_{G,i} = [\Delta\delta_{G,i} \quad \Delta f_{G,i} \quad \Delta P_{T,i} \quad \Delta a_i],
$$

$$
u_{G,i} = \Delta f_{G,i}^{ref},
$$

and the system, input and coupling matrices of the module as

$$
A_{G,i} = \begin{bmatrix} 0 & \omega_0 & 0 & 0 \\ 0 & -\frac{D_i}{M_i} & \frac{1}{M_i} & \frac{e_{t,i}}{M_i} \\ 0 & 0 & -\frac{1}{T_{t,i}} & \frac{K_{t,i}}{T_{t,i}} \\ 0 & -\frac{1}{T_{g,i}} & 0 & -\frac{r_i}{T_{g,i}} \end{bmatrix}, \quad B_{G,i} = \begin{bmatrix} 0 \\ 0 \\ 0 \\ \frac{1}{T_{g,i}} \end{bmatrix}, \quad F_{G,i} = \begin{bmatrix} 0 \\ -\frac{1}{M_i} \\ 0 \\ 0 \end{bmatrix}.
$$

The generator module can then be represented in the state-space form

$$
\dot{x}_{G,i} = A_{G,i}x_{G,i} + B_{G,i}u_{G,i} + F_{G,i}\Delta P_{G,i}. \quad (2)
$$

2.2 Modeling of the Variable Load

For the purposes of deriving a dynamic model for frequency regulation the load is characterized as a constant forecasted real power $P_L(0)$. Deviations of load power around its forecast $\Delta P_L(t)$ create a disturbance in the dynamics of the interconnected power system. We consider $\Delta P_L(t)$ to be either a hard-to-predict deviation in demand and/or a hard-to-predict deviations in the renewable source located at the load bus. The dynamics of renewable power sources are not modeled here in order to explicitly define them as negative variable loads, which inject randomly disturbed power into the grid. Therefore, the actual load $P_L(t)$ can be represented as

$$P_L(t) = P_L(0) + \Delta P_L(t). \tag{3}$$

2.3 Modeling of the Transmission Network Constraints

Both the dynamics of generators and load deviations are subject to transmission network constraints. The network constraints are typically expressed in terms of nodal algebra equations. When modules get interconnected through a transmission network the basic Kirchhoff's laws have to be satisfied. Let $S = P + jQ$ be the vector of the net complex power injections to all the buses; the algebraic complex power flow equation can be written as [1]

$$S = \mathrm{diag}(V)(Y_{bus}V)^*, \tag{4}$$

where diag($\cdot$) stands for the diagonal matrix with each element of the vector as a diagonal element. V is the vector of all the bus voltage phasors. The kth element of V is given by $V_k e^{j\delta_{G,k}}$. Y_{bus} is the admittance matrix of the power grid.

The net real part of complex power S, in general, is comprised of active power injection of the generator P_G and consumption of the load P_L, which is $P = \begin{bmatrix} P_G & -P_L \end{bmatrix}^T$. Linearizing the real part of the complex power flow equation (4) around the system equilibrium yields

$$\Delta P_G = J_{GG}\Delta\delta_G + J_{GL}\Delta\delta_L, \tag{5a}$$

$$-\Delta P_L = J_{GL}\Delta\delta_G + J_{LL}\Delta\delta_L, \tag{5b}$$

where

$$J_{ij} = \left.\frac{\partial P_i}{\partial \delta_j}\right|_{\delta_j = \delta_j^*}, \quad i,j \in \{G, L\}$$

is the Jacobian matrix evaluated at the system equilibrium. $\Delta \boldsymbol{\delta}_L$ stands for the phase angle deviations on the load buses. Assuming that $\boldsymbol{J}_{LL}$ is invertible in normal operating conditions, we can substitute $\Delta \boldsymbol{\delta}_L$ from (5b) to (5a) and obtain the system-level algebraic network coupling equation:

$$\Delta \boldsymbol{P}_G = \boldsymbol{K}_p \Delta \boldsymbol{\delta}_G + \boldsymbol{D}_p \Delta \boldsymbol{P}_L, \tag{6}$$

where

$$\boldsymbol{K}_p = \boldsymbol{J}_{GG} - \boldsymbol{J}_{GL} \boldsymbol{J}_{LL}^{-1} \boldsymbol{J}_{LG},$$
$$\boldsymbol{D}_p = -\boldsymbol{J}_{GL} \boldsymbol{J}_{LL}^{-1}.$$

2.4 Dynamic Model of the Interconnected System

The dynamic model of the interconnected system with n generators is derived by combing the dynamics of the individual generator modules given in equation 2 whose local states are $\boldsymbol{x}_{G,i}$, $i = 1, 2, \ldots, n$ and the network constraints (6). The system-level state variables, system disturbances and the secondary control inputs are defined as

$$\boldsymbol{x} = [\boldsymbol{x}_{G,1}, \boldsymbol{x}_{G,2}, \ldots \boldsymbol{x}_{G,n}]^T,$$
$$\boldsymbol{W}(t) = \Delta \boldsymbol{P}_L(t),$$
$$\boldsymbol{u} = [\boldsymbol{u}_{G,1}, \boldsymbol{u}_{G,2}, \ldots \boldsymbol{u}_{G,n}]^T,$$

and the system matrices of all generator modules $\boldsymbol{A}_{uc}$, $\boldsymbol{B}_{uc}$ and $\boldsymbol{F}_{uc}$ are defined by combining corresponding matrices of all generator modules the $\boldsymbol{A}_{G,i}$, $\boldsymbol{F}_{G,i}$, and $\boldsymbol{B}_{G,i}$ to obtain

$$\boldsymbol{A}_{uc} = \text{blockdiag}\,(\boldsymbol{A}_{G,1}, \boldsymbol{A}_{G,2}, \ldots \boldsymbol{A}_{G,n}),$$
$$\boldsymbol{B}_{uc} = \text{blockdiag}\,(\boldsymbol{B}_{G,1}, \boldsymbol{B}_{G,2}, \ldots \boldsymbol{B}_{G,n}),$$
$$\boldsymbol{F}_{uc} = \text{blockdiag}\,(\boldsymbol{F}_{G,1}, \boldsymbol{F}_{G,2}, \ldots \boldsymbol{F}_{G,n}).$$

The interconnected system-level dynamic model then becomes

$$\dot{\boldsymbol{x}} = \boldsymbol{A}_{uc}\boldsymbol{x} + \boldsymbol{B}_{uc}\boldsymbol{u} + \boldsymbol{F}_{uc}\Delta \boldsymbol{P}_G. \tag{7}$$

We define selection matrix $\boldsymbol{S}$ such that

$$\Delta \boldsymbol{\delta}_G = \boldsymbol{S}\boldsymbol{x}, \tag{8}$$

and it can be substituted into the network constraint equation (6), which yields

$$\Delta \boldsymbol{P}_G = \boldsymbol{K}_p \boldsymbol{S}\boldsymbol{x} + \boldsymbol{D}_p \Delta \boldsymbol{W}. \tag{9}$$

The full state-space model is then obtained by combining (7) into (9):

$$\dot{x} = Ax + Bu + FW, \tag{10a}$$

$$y = Cx, \tag{10b}$$

where

$$A = A_{uc} + F_{uc}K_pS,$$

$$B = B_{uc},$$

$$F = F_{uc}D_p,$$

$$w = \Delta P_L.$$

Note that since no slack generator bus is specified nor removed from the state space model (10), the system matrix A is structurally singular with $rank(A) = 4n - 1$.

3 A New Interactions Variable-based Dynamic Model for Frequency Regulation in Large Interconnected Electric Energy Systems

The interconnected power system model (10) introduced in Sect. 2 may be overly complex and not necessary when designing a sensing, communication, and control architecture for frequency regulation. Both temporal and spatial simplifications are possible. However, as the system changes and the nature of generation and load changes, one must proceed very carefully with such simplifications. A temporal simplification takes into consideration that, at the generator level, the internal states of the generator module consist of the fast states ΔP_T and Δa and the slow states $\Delta \delta_G$ and Δf_G. The spatial simplification mainly considers that, at the system level, the structurally singular matrix A leads to the slow dynamics, which can represent the entire system's response to persistent disturbances. We address both the component and system-level time-scale separation in this section by, respectively, using the standard and nonstandard singular perturbation form model, which had its origin in the early work of Joe Chow and his collaborators [2, 3].

The existence of an interactions variable, which is crucial to the cyber architecture proposed in this chapter, is a direct consequence of the system-level dynamic model having a structurally non-standard singularly perturbed form. This form reflects the fact that system matrix A is not a full rank matrix. We show that the structural singularity also holds at the CA level. We introduce the definition of SBA, which refers to the CA in the new power system environment.

3.1 Use of Standard Singularly Perturbed Form for Temporal Simplifications of the System Model

The standard singularly perturbed form-based method is applied at each generator module level to reduce the full state x to the reduced state $\hat{x}$ by identifying the time-scale separation between the slow states and fast states of the turbine-generator-governor set. The standard state separable form is described by

$$\dot{x} = f(x, z, u, \varepsilon, t), \quad x(t_0) = x_0, \quad x \in R^n, \tag{11a}$$

$$\varepsilon \dot{v} = g(x, v, u, \varepsilon, t), \quad v(t_0) = v_0, \quad v \in R^m, \tag{11b}$$

where x refers to the slow system states and v to the fast system states. ε is a small positive scalar, which accounts for the small time constant. If the dynamics of the two states are widely separated, ε will become very small and can be approximated as $\varepsilon = 0$. This approximation is equivalent to setting the speed of v as infinitely large and the transient of v as instantaneous. Hence, (11b) reduces to a set of algebra equations:

$$0 = g(\hat{x}, \hat{v}, \hat{u}, 0, t), \tag{12}$$

and the substitution of a root of (12)

$$\hat{v} = \boldsymbol{\phi}(\hat{x}, \hat{u}, t), \tag{13}$$

into (11a) yields a reduced model:

$$\frac{d\hat{x}}{dt} = f(\hat{x}, \boldsymbol{\phi}(\hat{x}, \hat{u}, t), \hat{u}, 0, t), \quad \hat{x}(t_0) = x_0, \quad \hat{x} \in R^n, \tag{14}$$

where the upper hat is used to indicate that the variables belong to a system with $\varepsilon = 0$. The singularly perturbed model of (2) can be obtained as

$$\frac{d\hat{x}_{G,i}}{dt} = \hat{A}_{G,i}\hat{x}_{G,i} + \hat{B}_{G,i}\hat{u}_{G,i} + \hat{F}_{G,i}\Delta\hat{P}_{G,i}, \tag{15}$$

where

$$\hat{x}_{G,i} = [\Delta\delta_{G,i} \quad \Delta f_{G,i}],$$

$$\hat{u}_{G,i} = \Delta f_{G,i}^{ref},$$

and

$$\hat{A}_{G,i} = \begin{bmatrix} 0 & \omega_0 \\ 0 & -\frac{r_i D_i + K_{t,i} + e_{t,i}}{r_i M_i} \end{bmatrix}, \quad \hat{B}_{G,i} = \begin{bmatrix} 0 \\ \frac{K_{t,i} + e_{t,i}}{r_i M_i} \end{bmatrix}, \quad \hat{F}_{G,i} = \begin{bmatrix} 0 \\ -\frac{1}{M_i} \end{bmatrix}.$$

This temporal simplification at the generator module level leads to the reduced order model of the interconnected system as follows:

$$\frac{d\hat{x}}{dt} = \hat{A}\hat{x} + \hat{B}\hat{u} + \hat{F}\hat{W}, \tag{16a}$$

$$\hat{y} = \hat{C}\hat{x}. \tag{16b}$$

3.2 Use of Nonstandard Singularly Perturbed Form for Spatial Simplifications of the System Model

The reduced system dynamic model (16) still has a structural singularity because $rank(\hat{A}) = 2n - 1$. Therefore, the dimension of the null-space $\mathcal{N}$ of $\hat{A}$ is 1. Therefore, a further time-scale separation based on the nonstandard singularly perturbed form is possible. According to [3], the slow variable z can be obtained by deriving a $(1 \times 2n)$ vector $\hat{T}$ transformation, which spans the left null-space of $\hat{A}$, that is, $\hat{T}\hat{A} = 0$. Then in the reduced system model (16a), we multiply by $\hat{T}$ on both sides and obtain

$$\frac{d\hat{z}}{dt} = \hat{T}\frac{d\hat{x}}{dt} = \hat{T}\hat{B}\hat{u} + \hat{T}\hat{F}\hat{W}. \tag{17}$$

This implies that the time response of the slow variable $\hat{z}$ only depends on the control input and the disturbances. When no control is implemented, $\hat{z}$ represents the response of the system to disturbances. The dynamics of $\hat{z}$ are a consequence of the existence of the zero eigenvalue in system matrix $\hat{A}$. They are defined by equation (17) from which it can be seen that $\hat{z}$ can only be controlled by the secondary level control $\hat{u}$ in response to disturbances $\hat{w}$. This points out the necessity for secondary level frequency control: with only the locally designed primary controllers of the generator-turbine sets, the frequency of the power system exposed to persistent disturbances will never return to the nominal frequency. This is because the interactions variable $\hat{z}$ will not settle to zero, which can be shown to be the necessary condition for system frequency to settle to zero.

3.3 The Dynamics of CA-level Interactions Variable

Motivated by today's AGC approach, we consider the temporally and spatially simplified model at the subsystem level. The subsystem in today's AGC approach is denoted as the CA. Later in this subsection, we will introduce the dynamic model and a new notion to the subsystem.

The subsystem is equivalent to a stand-alone power system with external interconnections represented as disturbances. Therefore, each subsystem can be modeled similar to (16) with the subscript a to differentiate it from (16),

$$
\begin{aligned}
\frac{\mathrm{d}\hat{\boldsymbol{x}}_a}{\mathrm{d}t} &= \hat{\boldsymbol{A}}_a\hat{\boldsymbol{x}}_a + \hat{\boldsymbol{B}}_a\hat{\boldsymbol{u}}_a + \hat{\boldsymbol{F}}_{a_{uc}}\hat{\boldsymbol{D}}_{p_a}\Delta\hat{\boldsymbol{P}}_{L_a} + \hat{\boldsymbol{F}}_{a_{uc}}\hat{\boldsymbol{D}}_{p_a}\Delta\hat{\boldsymbol{F}}_{L_a} - \hat{\boldsymbol{F}}_{a_{uc}}\Delta\hat{\boldsymbol{F}}_{G_a} \\
&= \hat{\boldsymbol{A}}_a\hat{\boldsymbol{x}}_a + \hat{\boldsymbol{B}}_a\hat{\boldsymbol{u}}_a + \hat{\boldsymbol{F}}_a\hat{\boldsymbol{W}}_a,
\end{aligned}
\tag{18}
$$

where

$$
\hat{\boldsymbol{F}}_a = \begin{bmatrix} \hat{\boldsymbol{F}}_{a_{uc}}\hat{\boldsymbol{D}}_{p_a} & \hat{\boldsymbol{F}}_{a_{uc}}\hat{\boldsymbol{D}}_{p_a} & -\hat{\boldsymbol{F}}_{a_{uc}} \end{bmatrix},
$$

$$
\hat{\boldsymbol{W}}_a = \begin{bmatrix} \Delta\hat{\boldsymbol{P}}_{L_a} \\ \Delta\hat{\boldsymbol{F}}_{L_a} \\ \Delta\hat{\boldsymbol{F}}_{G_a} \end{bmatrix}.
\tag{19}
$$

The term $\Delta\hat{\boldsymbol{F}}_{G_a}$ stands for the power flow from the neighboring CAs to the generator buses, and $\Delta\hat{\boldsymbol{F}}_{L_a}$ is the real power flow from the neighboring CAs to the load buses. Hence, at the CA level, disturbances to a CA could be caused by its own load fluctuations and/or the tie-line fluctuations connecting to other areas.

Structural singularity also exists in (18), so we can apply the nonstandard form singular perturbation method to the subsystem model as in Sect. 3.2, which leads to

$$
\frac{\mathrm{d}\hat{z}_a}{\mathrm{d}t} = \hat{\boldsymbol{T}}_a\frac{\mathrm{d}\hat{\boldsymbol{x}}_a}{\mathrm{d}t} = \hat{\boldsymbol{T}}_a\hat{\boldsymbol{B}}_a\hat{\boldsymbol{u}}_a + \hat{\boldsymbol{T}}_a\hat{\boldsymbol{F}}_a\hat{\boldsymbol{W}}_a.
\tag{20}
$$

The dynamics of the slow variable $\hat{z}_a$ depend solely on the internal control input and the disturbances from both internal and external energy sources. When no internal control is applied, $\hat{z}_a$ is the interactions variable of the subsystem and it represents power imbalances caused by the internal and external disturbances.

We define the CA interactions variable $\hat{z}_a$ has the unique property given as follows [1]:

Definition 1. The interactions variable $\hat{z}_a$ of a CA is the variable which satisfies

$$
\frac{\mathrm{d}\hat{z}_a}{\mathrm{d}t} \equiv 0,
$$

when no secondary control input is applied, and when there are no internal disturbances and all interconnections with other CAs are removed.

As in Sect. 3.2, the interactions variable $\hat{z}_a$ is an aggregation of the state variables of a CA, namely,

$$
\hat{z}_a = \hat{\boldsymbol{T}}_a\hat{\boldsymbol{x}}_a,
\tag{21}
$$

and $\hat{\boldsymbol{T}}_a$ is the basis of the eigenspace $\mathcal{N}(\hat{\boldsymbol{A}}_a)$ and can be determined by solving

$$
\hat{\boldsymbol{T}}_a\hat{\boldsymbol{A}}_a = 0.
\tag{22}
$$

We further define $\hat{z}_a$ as the output variable of the subsystem and write the subsystem model as

$$\frac{d\hat{x}_a}{dt} = \hat{A}_a\hat{x}_a + \hat{B}_a\hat{u}_a + \hat{F}_a\hat{W}_a, \tag{23a}$$

$$\hat{z}_a = \hat{C}_a\hat{x}_a, \tag{23b}$$

where

$$\hat{C}_a = \hat{T}_a.$$

At the end of this section, we introduce a new notion of SBA as the CA (subsystem), which communicates its interactions variable with the rest of the system in order to balance supply and demand in real time by utilizing on-line adjustments of both the resources internal to the subsystem and the neighboring subsystems. Both the concepts of interactions variable of CA and SBA will be used in the follow up sections for designing the cyber architecture of the frequency regulation system.

4 Interactions Variable-based Dynamical Models for Frequency Regulation: Comparison of Model Complexity for Four Qualitatively Different Cases

In this section, we illustrate the role of the interaction variables in four qualitatively different systems. The five-bus power system shown in Fig. 1 is used to demonstrate the four different scenarios without loss of generality. The pre-defined organizational boundaries divide the system into two interconnected subsystems (SBAs). However, as the internal and external electrical distances and disturbances change qualitatively, the dynamic interaction between these two SBAs can differ dramatically. Tables 1 and 2 list the system parameters and $\hat{A}$ matrices, respectively. In this section, we assume that the magnitude and rate of change of the disturbances remain within a pre-specified range and differentiate among the cases with different electrical distances.

4.1 Cases 1 and 2

In the first case, the system has two weakly interconnected SBAs whose internal states are strongly connected. The system is therefore modeled using the concept of interactions variable proposed in this chapter, for which we have

$$\hat{z}_a^I \triangleq \hat{C}_a^I\hat{x}_a^I, \tag{24a}$$

$$\hat{z}_a^{II} \triangleq \hat{C}_a^{II}\hat{x}_a^{II}, \tag{24b}$$

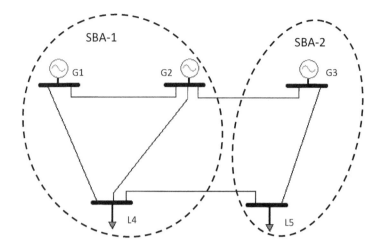

Fig. 1 Five-bus power system

Table 1 Parameters of the five-bus test system ($S_{base} =$ 100 MVA)

Transmission line reactance data (p.u.)

	X_{12}	X_{14}	X_{23}	X_{24}	X_{35}	X_{45}
Case 1	0.01	0.01	20	0.01	0.01	20
Case 2	1	1	20	1	1	20
Case 3	0.01	0.01	0.01	0.01	0.01	0.01
Case 4	1	1	0.01	1	1	0.01

Generator data (p.u.)

	M	D	T_t	T_g	e_t	K_t	r
Gen 1	8	2	0.2	0.25	39.4	250	19
Gen 2	10	2	0.18	0.23	39.4	250	19
Gen 3	9	1.6	0.3	0.3	39	280	21

where $\hat{C}_a^I$ and $\hat{C}_a^{II}$ can be solved from (22). It is shown in Fig. 3a that when the system is driven by fast oscillating disturbances (depicted by Fig. 2), the dynamics of the interactions variables of SBA-2 are significantly affected. By contrast, the interactions variables of SBA-1 have much slower dynamics. Note that the positive direction of the interactions variables can be arbitrarily assigned.

Case 2 represents an overall weakly connected system. The interactions variables for this case are presented in Fig. 3b. The results resemble those of case 1 due to the fact that the dynamic behaviors of $\hat{z}_a^I$ and $\hat{z}_a^{II}$ differ significantly and SBA-1 and SBA-2 interact weakly. These two cases are different with regard to internal dynamics in the CAs.

We conclude that in cases 1 and 2 the dynamics of the interactions variables depend mainly on internal disturbances; and we propose that, because of the insignificant interactions between the SBAs in these cases, there is no need to

Table 2 System $\hat{A}$ matrices in the four cases

Case 1

$$\hat{A} = \begin{bmatrix} 0 & 377 & 0 & 0 & 0 & 0 \\ -18.75 & -2.15 & 18.75 & 0 & 0.0031 & 0 \\ 0 & 0 & 0 & 377 & 0 & 0 \\ 15 & 0 & -15 & -1.72 & 0.0075 & 0 \\ 0 & 0 & 0 & 0 & 0 & 377 \\ 0.0028 & 0 & 0.0083 & 0 & -0.0111 & -1.87 \end{bmatrix}$$

Case 2

$$\hat{A} = \begin{bmatrix} 0 & 377 & 0 & 0 & 0 & 0 \\ -0.19 & -2.15 & 0.19 & 0 & 0.0029 & 0 \\ 0 & 0 & 0 & 377 & 0 & 0 \\ 0.15 & 0 & -0.16 & -1.72 & 0.0073 & 0 \\ 0 & 0 & 0 & 0 & 0 & 377 \\ 0.0026 & 0 & 0.0081 & 0 & -0.0107 & -1.87 \end{bmatrix}$$

Case 3

$$\hat{A} = \begin{bmatrix} 0 & 377 & 0 & 0 & 0 & 0 \\ -20 & -2.15 & 17.5 & 0 & 2.5 & 0 \\ 0 & 0 & 0 & 377 & 0 & 0 \\ 14 & 0 & -26 & -1.72 & 12 & 0 \\ 0 & 0 & 0 & 0 & 0 & 377 \\ 2.22 & 0 & 13.33 & 0 & -15.55 & -1.87 \end{bmatrix}$$

Case 4

$$\hat{A} = \begin{bmatrix} 0 & 377 & 0 & 0 & 0 & 0 \\ -0.17 & -2.15 & 0.14 & 0 & 0.0345 & 0 \\ 0 & 0 & 0 & 377 & 0 & 0 \\ 0.11 & 0 & -10.13 & -1.72 & 10.02 & 0 \\ 0 & 0 & 0 & 0 & 0 & 377 \\ 0.0307 & 0 & 11.14 & 0 & -11.17 & -1.87 \end{bmatrix}$$

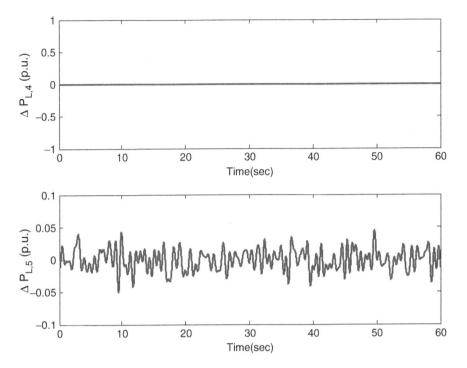

Fig. 2 Continuous real power disturbances

dynamically exchange measurements across the SBAs in order to ensure the desired frequency quality. Instead, one could design a decentralized controller for each SBA based on its respective internal output measurements.

4.2   Cases 3 and 4

In this subsection, we analyze two more cases which are qualitatively different from the previous cases. In case 3, the system is characterized by strong electrical connections both between and within the SBAs. In case 4, the system has strong inter-area electrical connections and weak intra-area connections. The interactions variables of SBA-1 and SBA-2 can be obtained similarly to (24) and are shown in Fig. 3c, d, respectively.

In Fig. 3c, the strong interactions between SBAs 1 and 2 in case 3 are illustrated through the highly similar behavior of the interactions variables of the SBAs. This is because the components in the system are tightly connected and the two SBAs consequently behave like one system. Figure 3d shows the interactions variables of case 4, which are as strong as those of case 3. In case 4, Generator 2 is electrically much closer to SBA-2 but is organizationally grouped into SBA-1.

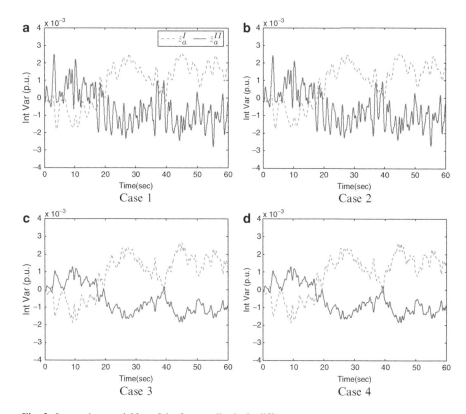

Fig. 3 Interactions variables of the four qualitatively different cases

It is seen from the results that both $\hat{z}_a^I$ and $\hat{z}_a^{II}$ have strong dynamics even though only SBA-2 is subject to continuous disturbances. So we conclude that unless the coordination among SBAs for minimizing interactions variables is carefully designed, a decentralized CA level control will not guarantee good frequency response.

5   Coherency-based Dynamic Models for Frequency Regulation: Comparison of Model Complexity for Four Qualitatively Different Cases

The early work by Joe Chow on slow coherency is briefly summarized in this section [2, 4–6]. His slow coherency-based approach is used to partition the interconnected power system into weakly interconnected subsystems. This partitioning leads to the time-scale separation of slow inter-area dynamics from fast intra-area dynamics. Generators within each resulting subsystem should exhibit similar contribution to the slow eigenmodes to the system [6]. This concept of slow coherency dynamics is similar to the concept of interactions variables in our work proposed in Sect. 3. We briefly review the slow coherency-based approach and illustrate how it applies to the four-case study in the five-bus power system.

In the linear time-invariant (LTI) system $\dot{x} = Ax$, states x_i and x_j are defined as σ_a-coherent if and only if none of these modes are observable from

$$y_j(t) = x_j(t) - x_i(t) = c_j x(t), \tag{25}$$

where the only nonzero entries of row c_j are its ith entry -1 and jth entry 1 [5]. Therefore, according to this definition, if x_i and x_j are coherent with respect to the excited σ_a-modes, then $x_j(t) - x_i(t) = 0$. This is algebraically equivalent to identifying the identical rows of an $(n \times r)$ matrix V:

$$c_j V = 0, \tag{26}$$

where the columns of V are the corresponding eigenvectors of the σ_a-modes. Since the rank of V is r, the smallest number of different subgroups containing identical rows of V is r. Therefore, the smallest possible number of coherent subgroups is r.

By applying the slow coherency-based approach to the five-bus power system shown in Fig. 1 for the four-case study, the resulting weakly connected coherent subgroups for each case are shown in Fig. 4, respectively. To design a frequency regulation architecture for the slow coherency-partitioned system, we propose to use the decentralized state feedback controller for each subsystems based on its own state measurements with no information exchange across the subsystems because they are either weakly connected (Fig. 4a, b, d) or completely stand-alone (Fig. 4c).

6 Comparison of Interactions Variable-based and Coherency-based Dynamic Models Relevant for Frequency Regulation

It is seen from the previous Sects. 4 and 5 that both the interactions variable-based approach and the slow coherency-based approach can interpret the interactions between subareas and could assist in the modeling, sensing, control, and communications system design for frequency regulation. The conclusions, based on the numerical results of Sects. 4 and 5, at which the two approaches arrive via different paths are compared in this section. Their implications for frequency regulation system design are then discussed based on the comparisons.

The interactions variable-based approach determines the strength of interactions among SBAs by examining the interactions variables of the SBAs. For example, in the case when the SBA has weak external but strong internal electrical connections (case 1 in Sects. 4 and 5), the interactions variables presented in Fig. 3a have a noticeable difference in dynamics with $\hat{z}_a^I$ evolving much slower than $\hat{z}_a^{II}$.

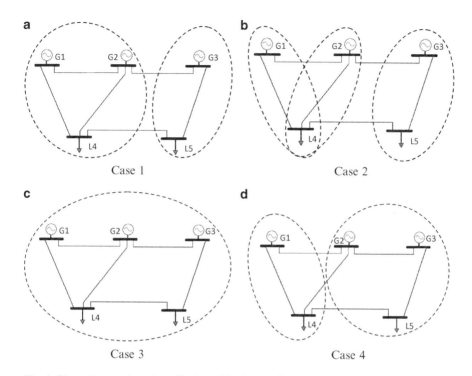

Fig. 4 Slow coherency-based partitioning of the four qualitatively different cases

The different dynamics imply a weak interaction between the two SBAs. In case 3, the interactions variables shown in Fig. 3c have similar dynamics to each other. This high similarity indicates a strong interaction between the two SBAs. In the other two cases, the strength of the interaction between the two SBAs can also be identified by following the same analysis.

In comparison, the slow coherency-based approach decomposes the system into weakly connected coherent subsystems (CSubs). We identify the strength of interaction between SBAs by comparing the boundaries of SBAs and the boundaries of CSubs. If the boundaries of SBAs match the boundaries of CSubs, the interactions among the SBAs will be weak; if not, strong interactions can be identified. In case 1, Figs. 1 and 4a illustrate that the SBAs have the same boundary as the CSubs, so that weak interaction exists between the SBAs. In case 2, as shown in Fig. 4b, although the SBA-1 is divided into two subgroups, the boundary of SBA-1 and SBA-2 still matches the boundary of the other two CSubs, which indicates that weak interactions exist inside SBA-1 and also between SBA-1 and SBA-2. In cases 3 and 4, the slow coherency-based approach tends to either group all generators into one single system (case 3) or group part of the generators in SBA-1 into SBA-2 (case 4); both cases exhibit strong interaction between SBA-1 and SBA-2.

As a result, the two approaches arrive at the same conclusion, via different paths, concerning the strength of the SBA interactions. Both approaches can assist us in the design of cyber architecture for the E-AGC system with respect to the interaction strength. In short, when the interactions among the SBAs are weak, there is no need for coordination of the SBAs. If the interactions are not weak, the SBAs will have to exchange information dynamically and be coordinated in order to ensure the global minimization of the imbalance between power supply and demand and good frequency response at the system level.

7 Two Qualitatively Different Cyber Architectures for Regulating Frequency Using Interactions Variable-based Models

In this section, the cyber architecture of the E-AGC system is designed by employing the interactions variable-based model. According to the concepts in Sects. 4 and 6 different control and communication architectures are needed for systems with strong and weak interactions, respectively. For each system type, we formulate the control design problem, using the Linear Quadratic Regulator (LQR) to ensure that the interactions among all SBAs and the control inputs are limited to acceptable levels.

7.1 Systems Characterized by the Weak Interactions

For systems with interactions, a decentralized linear output feedback control with no information exchange among the SBAs is designed. For the ith SBA modeled by (23), the linear output feedback control is expressed as

$$\hat{u}_a^i = -\hat{K}_a^i \hat{z}_a^i, \tag{27}$$

where $\hat{K}_a^i$ is the feedback control gain vector and $\hat{z}_a^i$ is the interactions variable of the ith SBA.

We determine the proper constant gains $\hat{K}_a^i$ by solving the following LQR-based optimization problem:

$$\underset{\hat{u}_a^i}{\text{minimize}} \quad J = \frac{1}{2} \int_0^\infty \left[(\hat{z}_a^i)^2 q_a^i + (\hat{u}_a^i)^T R_a^i (\hat{u}_a^i) \right] \mathrm{d}t$$

$$\text{subject to} \quad \frac{\mathrm{d}\hat{x}_a^i}{\mathrm{d}t} = \hat{A}_a^i \hat{x}_a^i + \hat{B}_a^i \hat{u}_a^i$$

$$\hat{z}_a^i = \hat{C}_a^i \hat{x}_a^i$$

$$\hat{u}_a^i = -\hat{K}_a^i \hat{z}_a^i. \tag{28}$$

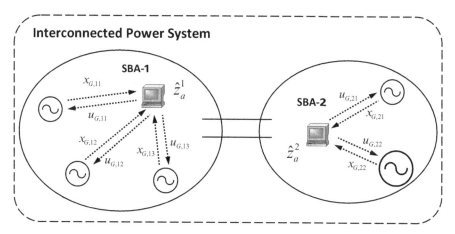

Fig. 5 The communication scheme of the E-AGC system in the weak interaction scenario

The non-negative scalar q_a^i in the quadratic objective function is considered to be an indicator of the willingness of the ith SBA to eliminate its own interactions variable. The positive definite matrix R_a^i, on the other hand, presents how much it will cost the sources to provide the frequency regulation service.

Accordingly, the communication scheme for the entire system can be designed on the assumption that the communication channels only exist inside each SBA, which are between the control center and the generators that belong to this SBA (shown in Fig. 5). Each generator measures its internal states and uploads the information to the control center, which estimates the interactions variable after gathering the measurement of all states in the area. Control signals will then be distributed back to each generator, which participates in the frequency regulation service.

7.2 Systems Characterized by the Strong Interactions

In the strong interaction scenario, a system-level coordinated output feedback control with information dynamically exchanged among the SBAs is designed. Recalling the singularly perturbed dynamic model for the entire system (16), the linear output feedback control can be represented by

$$\hat{u} = -\hat{K}\hat{y}, \tag{29}$$

where $\hat{y}$ is a vector whose entries are the interactions variables of all SBAs; if we let N be the number of SBAs, we have

$$\hat{y} = \hat{C}\hat{x}$$
$$\hat{C} = \text{blockdiag}\left(\hat{C}_a^1, \hat{C}_a^2, \dots \hat{C}_a^N\right). \tag{30}$$

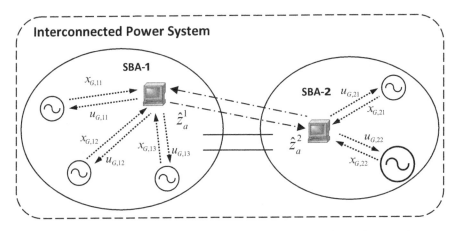

Fig. 6 The communication scheme of the E-AGC system in the strong interaction scenario

For this scenario, we propose an LQR-based optimization problem to obtain the output feedback control gain matrix $\hat{K}$, which is

$$
\underset{\hat{u}}{\text{minimize}} \quad J = \frac{1}{2} \int_0^\infty \left[\hat{y}^T Q \hat{y} + \hat{u}^T R \hat{u} \right] \mathrm{d}t
$$

$$
\text{subject to} \quad \frac{\mathrm{d}\hat{x}}{\mathrm{d}t} = \hat{A}\hat{x} + \hat{B}\hat{u}
$$

$$
\hat{y} = \hat{C}\hat{x}
$$

$$
\hat{u} = -\hat{K}\hat{y}. \tag{31}
$$

The positive semidefinite matrix Q is the quadratic objective function, which indicates all SBAs' willingness to eliminate their own interactions variables. The positive definite matrix R stands for the cost the generators will incur in providing frequency regulation service.

Compared to the weak interaction scenario, the communication scheme in the strong interaction scenario is designed so that communication channels exist not only internally between the SBA control center and the internal generators but also among the control centers (shown in Fig. 6). The SBA control center gathers the internal states information, exchanges its own interactions variable with other SBAs, and distributes the control signals to the generators that contribute to the frequency regulation service.

8 Cyber Architecture for Slow Coherency-based Regulating Frequency and Comparative Simulation Study

8.1 Design of the Cyber-Architecture

This section designs the control and communication architecture of the frequency regulation system by following the slow coherency-based approach, which was introduced and discussed in Sects. 5 and 6. A decentralized state feedback control is proposed for each CSub separately due to the fact that all the electrical interconnections are weak among the CSubs revealed by the slow coherency-based approach, and the size of each CSub is usually relatively small, so that no much measurement and communication is needed even though the control is full state feedback based.

The control and communication scheme is much the same as the one in Sect. 7.1. The only difference is that, in this section, we replace the interactions variable in the objective function with the full states of the CSub.

$$\hat{u}_c^i = -\hat{K}_c^i \hat{x}_c^i. \tag{32}$$

The LQR-based optimization problem can be formulated as

$$\underset{\hat{u}_c^i}{\text{minimize}} \quad J = \frac{1}{2} \int_0^\infty \left[(\hat{x}_c^i)^T Q_c^i \hat{x}_c^i + (\hat{u}_c^i)^T R_c^i \hat{u}_c^i \right] dt$$

$$\text{subject to} \quad \frac{d\hat{x}_c^i}{dt} = \hat{A}_c^i \hat{x}_c^i + \hat{B}_c^i \hat{u}_c^i$$

$$\hat{u}_c^i = -\hat{K}_c^i \hat{x}_c^i. \tag{33}$$

The communication architecture is shown in Fig. 7; it has information exchange only within each CSub and works quite similarly to the interactions variable-based architecture in the weak interaction scenario.

8.2 Comparative Study via Numerical Simulations

The interactions variable-based frequency regulation approach (E-AGC) proposed in Sect. 7 and the slow coherency-based frequency regulation approach are compared, via numerical simulations, on a five-bus power system (Fig. 1), which is exposed to the disturbances shown in Fig. 2. The system performances of the four qualitatively different cases are illustrated respectively.

In Figs. 8–11, the system frequency performances of the four cases are demonstrated. In the first row of each figure, the subplots correspond to the frequency

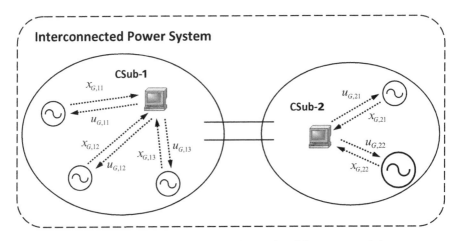

Fig. 7 The communication scheme of the slow coherency-based frequency regulation system

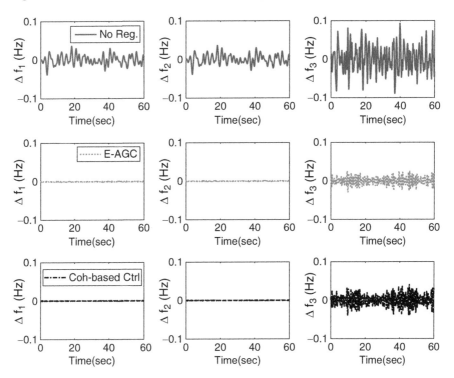

Fig. 8 System frequency performance for case 1

of Generators 1–3 when no regulation exists. In the second row, the subplots are for the E-AGC; in the third row, they are for the slow coherency-based approach. In Figs. 12–15, the responses of the interactions variables are demonstrated for the four

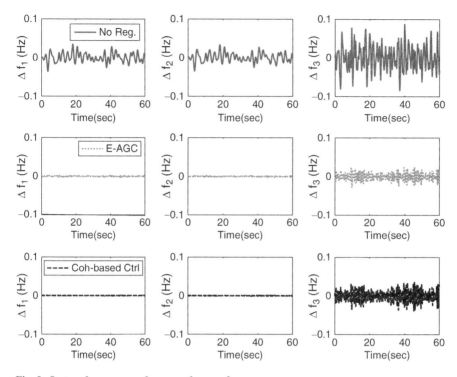

Fig. 9 System frequency performance for case 2

cases. Note that only the E-AGC defines the interactions variable and utilizes it in the feedback control design. Therefore, each figure has only two rows of subplots, in which the first row stands for no regulation and the second row represents the performance of E-AGC.

The results show that both approaches are able to effectively suppress the frequency deviation caused by the persistent disturbances. The E-AGC approach eliminates the imbalances (represented by the interactions variables in Figs. 12–15), and the frequency deviations which are, in fact, a consequence of the imbalances. The slow coherency-based approach performs quite closely to the interactions variable-based E-AGC but it costs much more on communications for the state feedback controller. Besides, the CSubs' boundaries remain unfixed in the slow coherency-based approach and may vary with respect to the changes in system structure or operating equilibrium. In this case, the communication scheme for frequency regulation needs to be dynamically updated in real time, which is hard for large-scale system operations. By contrast, the interactions variable-based E-AGC system uses a predetermined boundary and communication scheme for each SBA. Changes in system structure or operating equilibrium only affect the necessity of information exchange among SBAs. We conclude that it is more applicable to design an interactions variable-based E-AGC for frequency regulation.

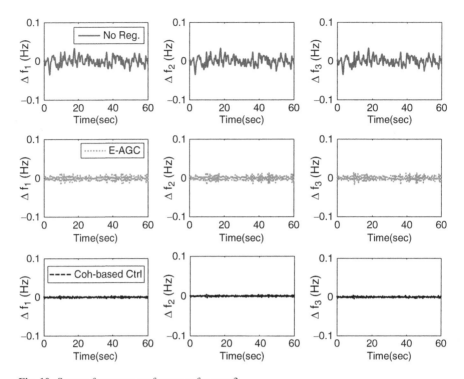

Fig. 10 System frequency performance for case 3

9   Comparison of the E-AGC and Today's AGC

This section compares the proposed E-AGC approach and today's AGC approach. We first recall the current state of the ACE in today's AGC system. The linear combination of area-wide frequency deviation and net tie-line flow is defined as the ACE to represent the area-wide real power imbalance. The ingenious aspect of ACE is that for a CA, which is free of disturbances, the corresponding ACE will be zero. In other words, only the CAs exposed to disturbance sources have nonzero ACEs. Consequently, by using ACE to feed into the PI-based secondary controller, each CA responses only to its own imbalance. And as long as all CAs compensate for their ACEs, the imbalances of the entire system will be eliminated and the frequency and net tie-line flow will be regulated. The ACE-based AGC scheme is fully decentralized with no need for information exchange among the CAs since both the net tie-line flow and frequency deviations are measurable at the CA level.

However, the ACE-based AGC scheme is not optimally designed. First, it assumes a perfect estimation of the frequency bias, which is hard to achieve practically. Second, from the perspective of cost-effectiveness, the implementation of ACE initially blocks the opportunity for systematically utilizing cheap and fast

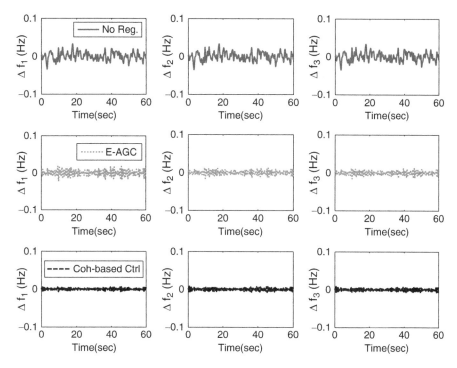

Fig. 11 System frequency performance for case 4

resources in frequency regulation. For example, if one CA is subjected to a large amount of fast disturbances and its resources for frequency regulation are either too expensive or too slow, the ACE-based AGC has no choice but to use these resources and pay much more for them rather than to import service from much cheaper and faster resources belonging to other CAs.

The potential advantages of applying the E-AGC are pointed out and discussed in the following:

- The E-AGC accounts for the SBA's requisition on the magnitude of imbalance and the capability of participating in frequency regulation service. And it provides an opportunity to the SBAs equipped with cheap and fast frequency regulation dedicated generating units to provide service to other SBAs and improve the cost-effectiveness of the frequency regulation service of the entire system. However, in contrast to the E-AGC, the ACE-based AGC of each CA only regulates its own imbalance, which does not translate automatically to a systematic cost-effectiveness of the frequency regulation service.
- The amount of exchanged information is limited. The interactions variable is the only information that each SBA needs to share with the rest of the system. This indicates a significant savings on investment and on maintenance for the communication system. In contrast, for the normally used LQR-based state

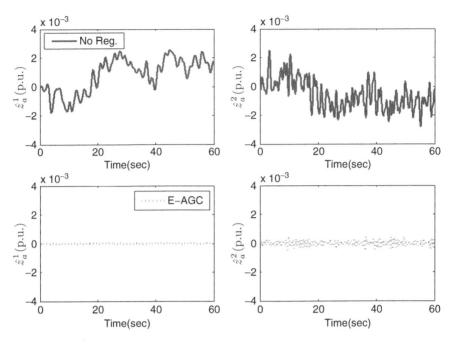

Fig. 12 System interactions variable performance for case 1

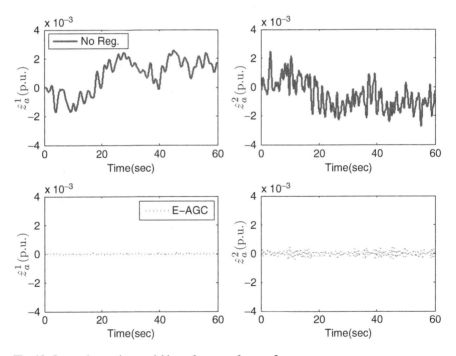

Fig. 13 System interactions variable performance for case 2

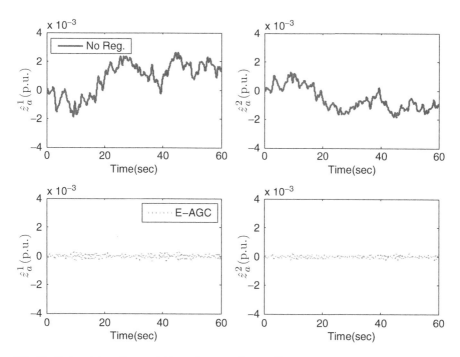

Fig. 14 System interactions variable performance for case 3

feedback control architecture, the full system states need to be measured, and either gathered in a centralized control center or exchanged with all other areas. So state feedback control is impractical in relatively large-scale power systems because too much communication is required. Large amounts of communication will result in both high costs and a high probability of congestion in communication channels, which will result in a deterioration of control performance.

- The role of Phasor Measurement Units (PMUs) is profound. In the proposed E-AGC approach, the deployment of PMUs is necessary because, as in (21), the SBA needs to measure its internal state variables; these include the generator voltage phase angles. Only PMUs can provide fast and highly accurate measurements of the voltage phase angles. In addition, the differences between frequencies on different generator buses inside the SBA can only be distinguished by high-accuracy and high-resolution devices such as PMUs. Therefore, in order to have a correct estimation of the interactions variable for the SBA, sufficient PMUs have to be deployed.

- The use of interactions variables as interchanged information also protects the privacy of each SBA. Since the $\hat{C}_a^i$ for the SBA is not an invertible matrix but a vector, the other SBAs, as long as they have no complete knowledge of the structure and components of this SBA, cannot extract its information about the states from the interactions variable.

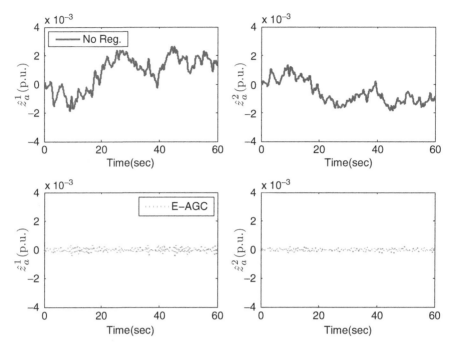

Fig. 15 System interactions variable performance for case 4

10   Open Questions and Future Work

In this chapter, we rethink the balancing of hard-to-predict supply and demand variations around their forecast by using frequency regulation in future electric energy systems. We conclude that the design of sensing, communication, and control architectures requires a clear understanding of the temporal and spatial characteristics of the power grid as well as of its generation and load dynamics. We propose an E-AGC, which would fulfill these requirements. Modeling, sensing, and control are introduced and illustrated on a small two CA system. We explain the relevance of Joe Chow's work to this problem. In particular, we suggest that the early standard and nonstandard form singular perturbation methods should be examined again in light of the changing technologies. The interactions variable, which is a concept motivated by nonstandard form singular perturbation, is proposed to estimate the strength of the interactions among the CAs. The interactions variable-based results and conclusions are verified by the slow coherency-based approach, which had its origin in the early work of Joe Chow. Moreover, the online implementation of the proposed E-AGC would not be possible without the efforts made in the direction of PMUs, concentrators of the type Joe Chow recently worked on. This chapter has been written with the idea of highlighting the long-lasting impact of the methodological thinking nurtured by Joe Chow. The co-author expresses his gratitude for the life-long work which has led to this school of thinking.

Acknowledgments The authors would like to express their appreciation for the partial support for this research provided by US NSF award 0931978, PSERC S-37 and the Semiconductor Research Corporation (SRC) Smart Grid Research Center (SGRC) at Carnegie Mellon University, Research Task 2111.002.

References

1. Ilić MD, Zaborszky J (2000) Dynamics and control of large electric power systems. Wiley, New York
2. Chow JH (1982) Time-scale modeling of dynamic networks with applications to power systems. Springer, New York
3. Kokotović PV, Khalil HK, O'Reilly J (1986) Singular perturbation methods in control analysis and design. Academic Press, London
4. Winkelman JR, Chow JH, Bowler BC, Avramovic B, Kokotovic PV (1981) An analysis of interarea dynamics of multi-machine systems. IEEE Trans Power Apparatus Syst PAS-100(2):754–763
5. Kokotović PV, Avramovic B, Chow JH, Winkelman JR (1982) Coherency based decomposition and aggregation. Automatica 18(1):47–56
6. Chow JH, Winkelman JR, Pai MA, Sauer PW (1990) Singular perturbation analysis of large-scale power systems. Int J Electr Power Energ Syst 12(2):117–126

Dynamic Competitive Equilibria in Electricity Markets

Gui Wang, Matias Negrete-Pincetic, Anupama Kowli, Ehsan Shafieepoorfard, Sean Meyn, and Uday V. Shanbhag

Abstract This chapter addresses the economic theory of electricity markets, viewed from an idealized competitive equilibrium setting, taking into account volatility and the physical and operational constraints inherent to transmission and generation. In a general dynamic setting, we establish many of the standard conclusions of competitive equilibrium theory: Market equilibria are efficient, and average prices coincide with average marginal costs. However, *these conclusions hold only on average*. An important contribution of this chapter is the explanation of the exotic behavior of electricity prices. Through theory and examples, we explain why, in the competitive equilibrium, sample-paths of prices can range from negative values, to values far beyond the "choke-up" price—which is usually considered to be the maximum price consumers are willing to pay. We also find that the variance of prices may be very large, but this variance decreases with increasing demand response.

1 Introduction

Electricity markets are intended to fulfill a range of goals: the pricing of electricity, the prescription of supply and demand-side decisions, and the creation of incentives for enhanced services and better technologies for electricity generation. The earliest efforts at deregulation appear to have been in Latin America in the 1980s, and subsequently in the UK in 1990. In the late 1990s, electricity markets were

G. Wang (✉) • M. Negrete-Pincetic • A. Kowli • E. Shafieepoorfard • S. Meyn
CSL and the ECE Department, University of Illinois, Urbana-Champaign, IL 61820, USA
e-mail: xjtu.wg@gmail.com; mnegret2@illinois.edu; akowli2@illinois.edu; eh.shafiee@gmail.com; spmeyn@gmail.com

U.V. Shanbhag
ISE Department, University of Illinois, Urbana-Champaign, IL 61820, USA
e-mail: udaybag@illinois.edu

A. Chakrabortty and M.D. Ilić (eds.), *Control and Optimization Methods for Electric Smart Grids*, Power Electronics and Power Systems 3,
DOI 10.1007/978-1-4614-1605-0_2, © Springer Science+Business Media, LLC 2012

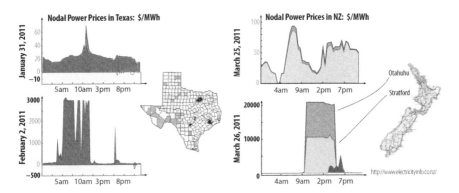

Fig. 1 Electricity prices in Texas and New Zealand in 2011

created in various US zones, such as in California, New York, New England, and the Pennsylvania–Jersey–Maryland (PJM) interchange [33]. The last decade has witnessed many changes in the design and organization of electricity markets both within the United States and beyond.

An electricity market may be viewed as a coupling between two constrained and highly complex dynamical systems, of which the first is purely physical while the second is financial.

- The physical system is a complex network comprising of power flowing through transmission lines, modulated by distributed generation units, Kirchhoff's laws, and operational and security constraints. Loads and generation are each subject to uncertainty. The stability of this network relies on instantaneous balancing between the production and consumption of electricity. Achieving this balance in a constrained environment is challenging due to the combination of uncertainty and a wide range of system constraints.
- The financial system is typically comprised of a coupled set of markets, such as the forward, day-ahead, and real-time auctions. A sequence of market clearings leads to an associated trajectory of electricity prices for a specific type of auction. The financial system is also subject to both constraints and uncertainty: Constraints from the market structure itself, and uncertainty in the form of unpredictable fuel prices, volatility in demand, and supply-side stochasticity (e.g., via wind-based resources). The dynamics of the financial system can be dramatic: prices may vary by two orders of magnitude within short durations—A few examples are shown in Fig. 1.

The presence of such complicating factors in these coupled systems, along with the often orthogonal relationship between societal needs and the economic goals of the market players, makes electricity market design a challenging proposition.

Given the importance of an efficient and reliable grid infrastructure, the modeling and subsequent analysis of electricity markets has been a dominant area of research. Static equilibrium analysis via competitive equilibrium models [14], supply-function models [2, 4, 19] and Nash–Cournot models [13, 23] has provided insights regarding the nature of equilibria. Of these, supply-function equilibrium problems often lead to infinite dimensional variational problems, but conjectured [10] and parameterized [17] variants lead to more tractable problems. Extensions to more representative regimes that incorporate forward and day-ahead markets were examined in [16, 28, 35]. Dynamic generalizations have seen far less study in strategic settings, with recent work on dynamic Nash–Cournot models by Mookherjee et al. [25] being an exception. In general, the implication of dynamics in market analysis has received limited attention, in spite of the remarkable volatility seen in electricity markets around the world. This chapter surveys modest beginnings in the area of dynamic equilibrium analysis, drawing mainly from [6, 24, 30], following Cho and Meyn [7, 27], and also inspired by recent research reported in [18, 31, 36].

In the US, price swings observed in California in 2000–2001 are the most famous examples of price volatility. A few years prior, unexpected and equally dramatic price patterns brought down the Illinois electricity market [12]. More recent examples are illustrated in Fig. 1. Shown on the left are prices in ERCOT (the Texas market) for two days in 2011. January 31 was a typical day, with prices ranging from a high near $80/MWh, and a low that was just below zero. Two days later on February 2, 2011, unusually cold weather in Texas resulted in real-time prices hitting the price cap of $3,000, which is about 100 times the average price of $30/MWh. Shown on the right is a far more dramatic example drawn from New Zealand in March, 2011, where electricity prices exceeded $20,000/MWh in one region of the country, and remained near this extraordinary level for about 6 h. These two incidents resulted in investigations in their respective communities [15, 32]. In New Zealand, the Electricity Authority has recently responded by retroactively reversing the prices to roughly one-tenth of the peak value, since these high prices threatened to "undermine confidence in, and ... damage the integrity and reputation of the wholesale electricity market."

While it is possible that strategic behavior and manipulation resulted in the wild price swings observed in Texas and New Zealand, in this chapter we demonstrate that such price patterns are *not* inconsistent with the *efficient* competitive equilibrium, where such manipulation is ruled out by assumption.

In a broad survey of reliability and market issues in the grid [9], Joe Chow and his coauthors assert that "*System reliability is an integral part of a properly designed deregulated electricity market, even though wholesale energy prices are its most visible piece.*" Models that capture dynamics, constraints, and volatility are needed to gain any understanding of how to create markets that enhance reliability, with stable price profiles, in the face of uncertain generation assets and volatile demand. Understanding the coupled dynamics of the physical grid overlaid by a set of sequentially occurring markets is especially important today: There are incentives from the government to make wider use of "smart meters" to create a "smart grid." Concomitantly, there has been an impetus to install greater levels of intermittent renewable energy from sun, wind, and waves, which will bring greater uncertainty to the grid. Until some intelligence can be introduced to accommodate the increased complexity and uncertainty that will come with these changes, we believe that the term *entropic grid* best describes the power grid envisioned by policy makers and researchers today in 2011 [26]. A basic message to the power system community is that while we recognize the potential benefits of the Smart Grid vision, we must be aware of the potential issues arising from its entropic characteristics. With proper design, this uncertainty can be reduced, and the term *Smart Grid* will be justified.

The focus of this chapter is on developing models to capture the dynamics associated with the market side of the entropic grid. Our goal is to contribute to the understanding of the impact of dynamics, constraints, and uncertainty on dynamic competitive equilibria; factors which we expect to become more acute with the increased deployment of renewable resources. Such sentiment is aptly conveyed by Smith et al. in the recent article [29], where the authors write that, "*little consideration was given to market design and operation under conditions of high penetrations of remote, variable renewable generation, such as wind … and solar energy, which had not yet appeared on the scene in any significant amounts.*" Our approach is the development of models that are able to characterize the competitive equilibria for a power network model that captures these complexities. This may be regarded as a stepping stone toward the creation of reliable markets for a smart grid.

We present a competitive economic equilibrium model that refines standard economic models (e.g., [1]), by including dynamics, uncertainty in supply and demand, and operational constraints associated with generation and transmission. Using a Lagrangian decomposition that is standard in static economic analysis and certain dynamic economic analyses [8, 22], we provide conditions for the existence of a competitive equilibrium in this dynamic setting, and its optimality with regard to suitably defined social planner's problem (SPP). Many of the conclusions obtained may be predicted from classical economic theory. In particular, under general conditions, the average equilibrium price coincides with the average marginal cost. However, in a competitive equilibrium, the sample path behavior can be as volatile as seen in the examples shown in Fig. 1. Moreover, in the presence of transmission constraints, equilibrium prices may become negative or they may exceed beyond the "choke-up" price that was predicted in [7].

The chapter also contains a brief economic analysis of a market with mixed generation sources, in which fast responding, expensive ancillary services are available to improve reliability. We consider a single example based on [6], where the solution to the "SPP" was obtained, but a market analysis was not considered. We illustrate through numerical experiments that volatile price patterns result in very large variance, which can negatively impact ancillary service providers. We explore demand response as a potential solution and demonstrate its effectiveness in reducing price variance, We contend that models and concepts surveyed in this chapter will open new pathways for accommodating uncertainty, dynamics, and strategic behavior in electricity market models.

The remainder of this chapter contains three additional sections and is organized as follows. In Sect. 2, we present the economic and physical models for the electricity market players in a dynamic setting, and illustrate the models through representative test case examples. We devote Sect. 3 to the characterization of the competitive equilibrium of the dynamic electricity market model using a control-theoretic scaffolding. In particular, we derive conditions for the existence of a competitive equilibrium in terms of duality concepts from optimization theory. We also prove the two celebrated theorems of welfare economics, thus establishing the efficiency of market equilibria. In Sect. 4, we characterize the equilibrium prices for the dynamic market, and show that average prices coincide with average marginal cost under some general assumptions. However, the test case examples emphasize the volatility observed in the sample paths of prices as well as the price range, which reaches both negative and extremely large positive values. We provide concluding remarks and final thoughts in Sect. 5.

2 Electricity Market Model

Electricity markets are driven by economic objectives of the market participants as well as the reliability constraints associated with the physical limitations on generation and transmission. Hence, we explicitly consider both the economic objectives and the reliability constraints in our modeling.

The market model described here is a high-level abstraction of the electric industry. It is an extension of the equilibrium models found in standard economic texts, suitably modified to accommodate dynamics, physical constraints on generation and transmission, and uncertainty in supply and demand. The model consists of three "players": As in typical economic analyses, the two main players are the consumers and the suppliers representing the utility companies and generation owners, respectively. For simplicity, we restrict the discussion to a single consumer and a single supplier that, respectively, represent aggregation of all utility companies and generators across the grid. The third player is the *network*. The network is introduced as a player to capture the impacts of transmission constraints and exploit the network structure. We can think of this third player as corresponding to the independent system operator[1] (ISO) that operates the transmission grid in most electricity markets in the world. In what follows, we discuss the modeling of the three players and motivate the modeling aspects through examples.

2.1 The Players

The power grid is represented by a graph in which each node represents a bus (which corresponds to a specific area/location), and each link represents a transmission line. There are N nodes, indexed as $\{1, \dots, N\}$, and L transmission lines, indexed

---

[1]The ISO is an entity independent of the consumers and the suppliers that coordinates, controls and monitors a large electric power transmission grid and its associated electricity markets.

as $\{1, \ldots, L\}$. The network is assumed to be connected. A lossless DC model is used to characterize the relationship between nodal generation and demand, and power flow across the various links. For simplicity, throughout most of this chapter we assume that at each node there is exactly one source of generation, and one exogenous demand.

2.1.1 The Consumer

We denote by $D_n(t)$ the demand at bus n at time t, and by $E_{Dn}(t)$ the energy withdrawn by the consumer at that bus. We assume that there is no free disposal for energy, which requires that $E_{Dn}(t) \leq D_n(t)$ for all t. At time t, if sufficient generation is available at bus n, then $E_{Dn}(t) = D_n(t)$. In the event of insufficient generation, we have $E_{Dn}(t) < D_n(t)$, i.e., the consumer's demand is not met and he experiences a forced blackout.

The consumer must pay for energy consumed: The price[2] at bus n is denoted by $P_n(t)$. We use $D(t)$, $E_D(t)$, and $P(t)$ to denote the associated N-dimensional column vectors, and we use boldface font to denote the entire sample path. For instance, $\boldsymbol{P} := \{P(t) : t \geq 0\}$.

The consumer obtains utility for energy consumption and disutility when demand is not met: This is modeled by the following two functions:

$$\text{Utility of consumption: } v_n(E_{Dn}(t)), \tag{1a}$$

$$\text{Disutility of blackout: } c_n^{bo}(D_n(t) - E_{Dn}(t)). \tag{1b}$$

The welfare of the consumer at time t is the signed sum of benefits and costs:

$$\mathcal{W}_D(t) := \sum_n \left[v_n(E_{Dn}(t)) - c_n^{bo}(D_n(t) - E_{Dn}(t)) - P_n(t)E_{Dn}(t) \right]. \tag{2}$$

2.1.2 The Supplier

We denote by $E_{Sn}(t)$ and $R_{Sn}(t)$ the energy and reserve produced by the supplier at bus n at time t. The generation capacity $\boldsymbol{G}_S$ available online coincides with $\boldsymbol{E}_S + \boldsymbol{R}_S$. The operational and physical constraints on available generation are expressed abstractly as

$$(\boldsymbol{E}_S, \boldsymbol{R}_S) \in \mathbf{X}_S. \tag{3}$$

These constraints include minimum up/down-time constraints, ramping constraints, and capacity constraints imposed by the physical limitations on generation.

---

[2]Observe that prices may vary by the location. In the language of electricity markets, they are *locational prices*.

At each time t, the supplier incurs costs for producing energy as well as maintaining reserves, which are represented as

$$\text{Cost of energy: } c_n^E(E_{Sn}(t)), \tag{4a}$$

$$\text{Cost of reserve: } c_n^R(R_{Sn}(t)). \tag{4b}$$

The supplier at bus n receives the revenue $P_n E_{Sn}(t)$ for producing energy. The welfare of the supplier at time t is the difference between the supplier's revenue and costs,

$$\mathcal{W}_S(t) := \sum_n \left[P_n E_{Sn}(t) - c_n^E(E_{Sn}(t)) - c_n^R(R_{Sn}(t)) \right]. \tag{5}$$

2.1.3 The Network

The network player can be thought of as a *broker*, who buys energy from the supplier and sells it to the consumer. The transactions brokered by the network player are subject to the physical constraints of the transmission grid. The first constraint is based on the assumption that the network is lossless, so it neither generates nor consumes energy. Consequently, the network transactions are subject to the supply-demand balance constraint,

$$1^T E_S(t) = 1^T E_D(t) \quad \text{for } t \geq 0. \tag{6}$$

The next set of constraints are due to the limitations of power flow through transmission lines. We adopt the DC power flow model [34] for the power flows through the lines. Without loss of generality, bus 1 is selected as the reference bus and the *injection shift factor matrix* $H \in [-1, 1]^{N \times L}$ is defined. Note that H_{nl} represents the power that is distributed on line l when 1 MW is injected into bus n and withdrawn at the reference bus. If f_l^{max} denotes the capacity of transmission line l and $H_l \in \mathbb{R}^N$ represents the l-th column of H, the power flow constraint for line l can be expressed as

$$- f_l^{max} \leq [E_S(t) - E_D(t)]^T H_l \leq f_l^{max} \quad \text{for } t \geq 0. \tag{7}$$

We find it convenient to introduce a "network welfare function" to define a competitive equilibrium for the dynamic market model. The welfare of the network at time t represents the "brokerage charges" and is defined by

$$\mathcal{W}_T(t) := \sum_n \left[P_n (E_{Dn}(t) - E_{Sn}(t)) \right]. \tag{8}$$

The introduction of the network welfare function is purely for the sake of analysis. In Sect. 3, we assume that the "network" maximizes its welfare, subject to the supply-demand balance constraint (6), and power flow constraint (7) for each line l, which are collectively summarized by the notation,

$$(E_D, E_S) \in \mathbf{X}_T. \tag{9}$$

2.2 Test Case Examples

The following three examples are intended to illustrate the modeling conventions described in the preceding section. We return to these in Sect. 4 to illustrate the conclusions of Sect. 3.

2.2.1 Example A—Single-Bus Model

In this simple model, there is a single generator, a single consumer, and no transmission network. The exogenous demand D is scalar valued, and the supply-demand balance constraints amount to $E := E_S = E_D$. This example is similar to the model of [7].

We impose the simplest constraints on generation—those imposed by limitations on ramping capabilities. The positive constants ζ^+ and ζ^- represent the maximum rates for ramping up and down the generators: For each $t_1 > t_0 \geq 0$,

$$-\zeta^- \leq \frac{E_S(t_1) - E_S(t_0)}{t_1 - t_0} + \frac{R_S(t_1) - R_S(t_0)}{t_1 - t_0} \leq \zeta^+. \tag{10}$$

The constraint set $\mathbf{X}_S$ consists of all pairs (E_S, R_S), which satisfy (10).

Piecewise-linear utility, disutility, and cost functions take the following form:

$$\text{Utility of consumption: } vE(t),$$

$$\text{Disutility of blackout: } c^{bo} \max(D(t) - E(t), 0),$$

$$\text{Cost of generation: } cE(t) + cR(t), \tag{11}$$

where v, c^{bo}, and c are positive constants. The parameter c^{bo} is the cost of outage incurred by the consumers for each unit of demand that is not satisfied. This may be tens of thousands of dollars per megawatt hour, depending on the location and time of day. In typical static analysis, the quantity $(v + c^{bo})$ represents the maximum price the consumer is willing to pay for electricity, and is known as the *choke-up* price [7].

2.2.2 Example B—Texas Model

The network topology is shown in Fig. 2. There are sources of demand and supply at each of the three nodes shown in the figure, and each node is connected to the other nodes via transmission lines.

The supply-demand balance constraints (6) result in the single equation:

$$E_{S1}(t) + E_{S2}(t) + E_{S3}(t) = E_{D1}(t) + E_{D2}(t) + E_{D3}(t) \quad \text{for } t \geq 0. \tag{12}$$

Fig. 2 Texas model:
a three-bus network

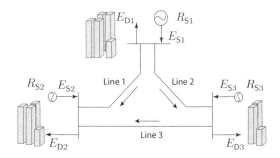

Suppose that the impedances of all three transmission lines are identical. Further, the arrows indicate the direction of power flow, which is assumed to be positive. Then with bus 1 chosen as the reference bus, the injection shift factor matrix is given by

$$H = \frac{1}{3}\begin{bmatrix} 0 & 0 & 0 \\ -2 & -1 & -1 \\ -1 & -2 & 1 \end{bmatrix}. \tag{13}$$

The power flow constraints for the three transmission lines are expressed as

$$-F^{max} \leq \begin{bmatrix} E_{S1}(t) - E_{D1}(t) \\ E_{S2}(t) - E_{D2}(t) \\ E_{S3}(t) - E_{D3}(t) \end{bmatrix}^{T} H \leq F^{max} \quad \text{for } t \geq 0, \tag{14}$$

where $F^{max} = [f_{12}^{max}, f_{13}^{max}, f_{23}^{max}]$ with f_{mn}^{max} representing the capacity limit of each line $l = 1, 2, 3$ connected between nodes m and n. Then the network constraints set $\mathbf{X}_T$ consists of all sample paths (E_D, E_S), which satisfy (12) and (14).

2.2.3 Example C—Primary and Ancillary Services

Although we have restricted the analysis to a single source of generation at each node, the physical characteristics of generation may vary by node and also for generators located at one node. In the case of two generation sources, and with network constraints relaxed, we arrive at a model similar to the dynamic newsboy model for generation introduced in [6].

Example C is used to illustrate the role of *ancillary service* in power systems operation. In power markets operating in the world today, the forecast demand is often met by a primary service—the cheapest source of service capacity—through a prior contract. Nuclear and coal generators are typical primary service providers. The deviations from forecast demand can be met through primary service, as well as ancillary service. Ancillary service may be costly, but it can be ramped up/down at a

Fig. 3 Primary and Ancillary
Service

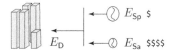

much faster rate than the primary service. Gas turbine generators are an example of
the more expensive, yet more responsive sources of ancillary service. The simplified
model for a system with primary and ancillary generators is depicted in Fig. 3.

We denote by $G_p(t)$ and $G_a(t)$ the available online capacity of primary and
ancillary generators at time t. Note that $G_p(t)$ represents the deviations in primary
service from the day-ahead schedules and, hence, can be less than zero while
$G_a(t) \geq 0$. The demand at time t is given as $D(t)$ and the reserve at that time is,

$$R(t) = G_p(t) + G_a(t) - D(t), \quad t \geq 0. \tag{15}$$

The two sources of generation are distinguished by their *ramping* capabilities:
ζ_p^+, ζ_a^+, ζ_p^- and ζ_a^- represent the maximum rates for ramping up and down the
primary and ancillary services, respectively. The ramping limits imposed on $G_p(t)$
and $G_a(t)$ take the form similar to (10). We assume that $\zeta_a^+ > \zeta_p^+$ and $\zeta_a^- > \zeta_p^-$
to reflect the ability of ancillary service to ramp up faster than primary service.
Similarly, with c^p and c^a representing the per unit production costs of primary and
ancillary services, respectively, we assume $c^a \gg c^p$ to emphasize that ancillary
service is more expensive than primary service.

3 Competitive Equilibria and Efficiency

The competitive equilibrium is used in economic analysis as a vehicle to study
the outcomes of a market under a set of idealized assumptions [1, 3, 22]. It is
characterized by the allocation of goods and their prices: for the given prices, each
player maximizes its welfare subject to constraints on production/consumption of
the goods. That is, the competitive equilibrium characterizes an allocation that is
optimal from the point of view of individual players. In this theory, the solution
to the so-called SPP serves as a benchmark and is an allocation of goods that
maximizes the economic well-being of the aggregation of all players, and is optimal
from the point of view of the entire system. This optimal solution is known as an
efficient allocation.

The fundamental theorems of welfare economics provide a strong link between
competitive equilibria and the SPP. The first theorem states that any competitive
equilibrium leads to an efficient allocation, while the second states the converse. In
this section, we revisit the concepts of competitive equilibrium and efficiency for the
dynamic electricity market model presented in Sect. 2, and establish the fundamental
welfare theorems for such a market.

3.1 Preliminaries

The usual definition of a competitive equilibrium with two players is based on the respective optimization problems of the consumer and the supplier. We adopt the same convention in the dynamic setting, but we extend the equilibrium definition to accommodate the third player—the network. In the dynamic setting, we adopt the long-run discounted expected surplus as the consumer's objective function. With discount rate γ, the long-run discounted expected consumer welfare is

$$K_D := \mathsf{E}\left[\int e^{-\gamma t} \mathcal{W}_D(t)\,dt\right].$$

The long-run discounted welfare of the supplier and the network are defined similarly, denoted K_S, K_T, respectively.

The consumer, supplier, and network each aims to optimize its respective mean discounted welfare K_D, K_S, and K_T. In general, these quantities depend on the initial condition of the system; we suppress this dependency whenever possible.

The assumptions imposed in this dynamic setting are intended to mirror those used in equilibrium theory for static economic models. To emphasize the similarities, we adopt the following Hilbert-space notation: For two real-valued stochastic processes F and G, we introduce the inner product,

$$\langle F, G\rangle := \mathsf{E}\left[\int e^{-\gamma t} F(t)G(t)\,dt\right], \tag{16}$$

and write $F \in L_2$ if $\langle F, F\rangle < \infty$. For example, using this notation, we have $K_D = \langle \mathcal{W}_D, 1\rangle$, where 1 denotes the process that is identically unity. It is assumed throughout that the components of the vector-valued processes $\{E_D, E_S, R_S, P\}$ belong to L_2.

In the formal definition of a competitive equilibrium given in Definition 1 below, the optimization problems of the supplier and of the network are each subject to physical constraints. However, as in [7], the set of feasible strategies for the consumer are *not* subject to the constraints on generation or transmission specified in (3, 9).

We impose several additional assumptions for the dynamic market model, each of which is an extension of what is typically assumed in the competitive equilibrium analysis of a static model.

(A1) Consumers and suppliers share equal information. This is modeled using a common *filtration*: an increasing family of σ-algebras, denoted $\mathcal{H} = \{\mathcal{H}_t : t \geq 0\}$. The demand process D, the price process P, and the decisions of the consumer E_D and the supplier (E_S, R_S) are adapted to this filtration.

(A2) The components of these vector-valued processes lie in L_2: Consumer decisions E_D, supplier decisions (E_S, R_S), and the price process P. Moreover, utility and cost functions are subject to quadratic growth: These functions are non-negative, and for each n, there exists $k_n < \infty$ such that for each scalar r, and each $e \geq 0$,

$$v_n(e) \leq k_n(1 + e^2) \quad c_n^{\mathrm{E}}(e) \leq k_n(1 + e^2),$$

$$c_n^{\mathrm{bo}}(r) \leq k_n(1 + r^2) \quad c_n^{\mathrm{R}}(r) \leq k_n(1 + r^2).$$

(A3) Prices are *exogenous* in the following sense: for each $t_0 > 0$, the future prices $\{P(t) : t > t_0\}$ are conditionally independent of $\{E_{\mathrm{D}}(t), E_{\mathrm{S}}(t) : t \leq t_0\}$, given current and past prices $\{P(t) : t \leq t_0\}$.

Under assumption (A3), also known as the *price-taking assumption*, no market player is large enough to influence prices and, hence, market manipulation is eliminated.

A competitive equilibrium for the dynamic market model is defined as follows.

Definition 1. A *competitive equilibrium* is a quadruple of process vectors: consumed energy, supplied energy, supplied reserve, and energy price; denoted as $\{E_{\mathrm{D}}^e, E_{\mathrm{S}}^e, R_{\mathrm{S}}^e, P^e\}$, which satisfies the following conditions:

E_{D}^e solves the consumer's optimization problem:

$$E_{\mathrm{D}}^e \in \arg\max_{E_{\mathrm{D}}} \langle \mathcal{W}_{\mathrm{D}}, 1 \rangle. \tag{17}$$

$(E_{\mathrm{S}}^e, R_{\mathrm{S}}^e)$ solves the supplier's optimization problem:

$$(E_{\mathrm{S}}^e, R_{\mathrm{S}}^e) \in \arg\max_{E_{\mathrm{S}}, R_{\mathrm{S}}} \langle \mathcal{W}_{\mathrm{S}}, 1 \rangle$$

$$\text{subject to} \quad \textit{generation constraints (3).} \tag{18}$$

$(E_{\mathrm{D}}^e, E_{\mathrm{S}}^e)$ solves the network's optimization problem with P^e being the Lagrange multiplier associated with (9),

$$(E_{\mathrm{D}}^e, E_{\mathrm{S}}^e) \in \arg\max_{E_{\mathrm{D}}, E_{\mathrm{S}}} \langle \mathcal{W}_{\mathrm{T}}, 1 \rangle$$

$$\text{subject to} \quad \textit{network constraints (9).} \tag{19}$$

The consumer, supplier, and network are also subject to the measurability constraint outlined in assumption (A1) in their respective optimization problems.

As discussed at the start of this section, to evaluate the market, we introduce a *social planner* who aims to maximize the economic well-being of aggregate representation of the players in the system. We stress that there is no actual planner—this is another analytical device. The social planner uses the total welfare, denoted by $\mathcal{W}_{\mathrm{tot}}(t)$, to measure the economic well-being of the system with

$$\mathcal{W}_{\mathrm{tot}}(t) := \mathcal{W}_{\mathrm{S}}(t) + \mathcal{W}_{\mathrm{D}}(t) + \mathcal{W}_{\mathrm{T}}(t). \tag{20}$$

The total welfare can be expressed as

$$\mathcal{W}_{tot}(t) = \sum_n \left[v_n(E_{Dn}(t)) - c_n^{bo}(D_n(t) - E_{Dn}(t)) - c_n^E(E_{Sn}(t)) - c_n^R(R_{Sn}(t)) \right],$$

which is independent of the price process.

The efficient allocation is derived by solving the SPP described at the start of this section. It is formally defined as follows:

Definition 2. The SPP is given by

$$\max_{E_D, E_S, R_S} \langle \mathcal{W}_{tot}, 1 \rangle$$

$$\text{subject to} \quad \textit{generation constraints (3)},$$

$$\text{and} \quad \textit{network constraints (9)}. \tag{21}$$

Its solution is called an *efficient allocation*.

We assume that the SPP (21) has a solution, denoted by (E_D^*, E_S^*, R_S^*).

Given these preliminaries, we have set the stage for the two fundamental theorems of welfare economics, which link the above two definitions.

Theorem 1 (First Welfare Theorem). *Any competitive equilibrium, if it exists, is efficient.*

Theorem 2 (Second Welfare Theorem). *If a market admits a competitive equilibrium, then any efficient allocation can be sustained by a competitive equilibrium.*

The proofs of these results are contained in Sect. 3.2 that follows.

3.2 Analysis

The main result here is Theorem 3 that characterizes the existence of a competitive equilibrium. At the end of this section, we establish the first and second fundamental theorems as corollaries to this result.

The analysis in this section is based on the Lagrangian relaxation framework presented in [30]. Lagrange multipliers are scalar processes, denoted $(\lambda, \mu_l^+, \mu_l^-)$, with λ unconstrained, and with $\mu_l^+(t) \geq 0$ and $\mu_l^-(t) \geq 0$ for all t and l. These processes are assumed to lie in L_2, and adapted to $\mathcal{H}$ (see (A1)). The Lagrangian of the SPP is denoted as

$$\mathcal{L} = -\langle \mathcal{W}_{tot}, 1 \rangle + \langle \lambda, (1^T E_D - 1^T E_S) \rangle + \sum_l \langle \mu_l^+, (E_S - E_D)^T H_l - f_l^{max} \rangle$$

$$+ \sum_l \langle \mu_l^-, -(E_S - E_D)^T H_l - f_l^{max} \rangle. \tag{22}$$

A key step is to define the candidate price process P as

$$P_n(t) := \lambda(t) + \sum_l (\mu_l^-(t) - \mu_l^+(t)) H_{ln}, \quad t \geq 0, \, n \geq 1. \tag{23}$$

The Lagrangian is then expressed as the sum of three terms, each of which is a sum over the N nodes,

$$\mathcal{L} = -\sum_n \{\langle v_n(E_{Dn}) - c_n^{bo}(D_n - E_{Dn}), 1\rangle - \langle P_n, E_{Dn}\rangle\}$$

$$-\sum_n \{\langle P_n, E_{Sn}\rangle - \langle c_n^E(E_{Sn}) + c_n^R(R_{Sn}), 1\rangle\} - \sum_l \langle \mu_l^+ + \mu_l^-, f_l^{max}\rangle.$$

The first two terms correspond to $-\mathcal{W}_D$ and $-\mathcal{W}_S$, respectively (the negative consumer and supplier welfare functions), defined using the price P given in (23).

The dual functional for the SPP is defined as the minimum,

$$h(\lambda, \mu^+, \mu^-) = \min_{E_D, E_S, R_S} \mathcal{L}. \tag{24}$$

The following weak duality bound follows since the minimization in (24) amounts to a relaxation of the SPP (21).

Lemma 1 (Weak Duality). *For any allocation* $\{E_D, E_S, R_S\}$ *and Lagrangian multiplier* (λ, μ^+, μ^-) *with* $\mu^+, \mu^- \geq 0$, *we have*

$$-\langle \mathcal{W}_{tot}, 1\rangle \geq h(\lambda, \mu^+, \mu^-). \tag{25}$$

We say that *strong duality* holds if we have equality in (25).

Theorem 3 characterizes the existence of a competitive equilibrium in terms of strong duality.

Theorem 3 (Existence of Competitive Equilibria). *The market admits a competitive equilibrium if and only if the SPP satisfies strong duality.*

Proof. We first prove the *sufficient condition*: strong duality implies existence of competitive equilibrium. Since strong duality holds, we have

$$-\langle \mathcal{W}_{tot}, 1\rangle = h(\lambda, \mu^+, \mu^-). \tag{26}$$

Suppose that the allocation $\{E_D, E_S, R_S\}$ is feasible for the SPP. We then construct a competitive equilibrium with price as given in (23).

The feasibility of the triple $\{E_D, E_S, R_S\}$ for SPP implies $1^T E_S(t) = 1^T E_D(t)$ for all t, and hence

$$\mathcal{L} = -\langle \mathcal{W}_{\text{tot}}, 1\rangle + \sum_{l}\langle \mu_l^+, (E_S - E_D)^T H_l - f_l^{\max}\rangle$$

$$+ \sum_{l}\langle \mu_l^-, -(E_S - E_D)^T H_l - f_l^{\max}\rangle.$$

Furthermore, feasibility also implies $-f_l^{\max} \leq (E_S(t) - E_D(t))^T H_l \leq f_l^{\max}$, and given the non-negativity of μ^+, μ^-, we have

$$\langle \mu^+, (E_S - E_D)^T H_l - f_l^{\max}\rangle \leq 0, \text{ and } \langle \mu^-, -(E_S - E_D)^T H_l - f_l^{\max}\rangle \leq 0.$$

Therefore, using (26), we have $\mathcal{L} \leq h(\lambda, \mu^+, \mu^-)$. But, by the definition (24), we have $\mathcal{L} \geq h(\lambda, \mu^+, \mu^-)$, so that we obtain the identity,

$$h(\lambda, \mu^+, \mu^-) = \mathcal{L}. \tag{27}$$

This identity implies that E_D maximizes the consumer's welfare, $\{E_S, R_S\}$ maximizes the supplier's welfare, and

$$\langle \mu^+, (E_S - E_D)^T H_l - f_l^{\max}\rangle = 0, \quad \text{and} \quad \langle \mu^-, -(E_S - E_D)^T H_l - f_l^{\max}\rangle = 0.$$

Using the prices defined in (23), we compute the network's optimization objective as follows:

$$\left\langle \sum_n [P_n(E_{Dn} - E_{Sn})], 1\right\rangle = \left\langle \sum_n \left[\sum_l (\mu_l^- - \mu_l^+)H_{ln}(E_{Dn} - E_{Sn})\right], 1\right\rangle$$

$$= \left\langle \sum_l (\mu_l^+ + \mu_l^-)f_l^{\max}, 1\right\rangle.$$

Note that the network's objective function is independent of E_D and $\{E_S, R_S\}$, which implies that the network welfare is maximized under the prices $\{P_n\}$. Thus, we conclude that P as defined in (23) is the equilibrium price as claimed and that $\{E_D, E_S, R_S, P\}$ constitute a competitive equilibrium.

Next we establish the *necessary condition*: existence of a competitive equilibrium implies strong duality. Suppose that $\{E_D^e, E_S^e, R_S^e, P^e\}$ is a competitive equilibrium. Then we know that $\{E_D^e, E_S^e\}$ maximizes the network welfare when the price is P^e. The Lagrangian associated with the network's optimization problem is expressed as follows: For any $\mu^+, \mu^- \geq 0$,

$$\mathcal{L}_T = -\sum_n \langle P_n^e, (E_{Dn}^e - E_{Sn}^e)\rangle + \langle \lambda, (1^T E_D^e - 1^T E_S^e)\rangle + \sum_l \langle \mu_l^+, (E_S^e - E_D^e)^T$$

$$\times H_l - f_l^{\max}\rangle + \sum_l \langle \mu_l^-, -(E_S^e - E_D^e)^T H_l - f_l^{\max}\rangle. \tag{28}$$

The network's optimization problem is a linear program and, hence, the optimum $\{E_D^e, E_S^e\}$ satisfies the Karush–Kuhn–Tucker (KKT) conditions. As a consequence, associated with the constraints $\frac{\partial \mathcal{L}_T}{\partial E_{Dn}} = \frac{\partial \mathcal{L}_T}{\partial E_{Sn}} = 0$ at the optimum, there exist $\{\lambda, \mu^+, \mu^-\}$ such that

$$P_n^e = \lambda + \sum_l (\mu_l^- - \mu_l^+) H_{ln};$$

that is, equilibrium price process P_n^e satisfies (23). Moreover, by complementary-slackness, we have

$$\langle \lambda, (1^T E_D^e - 1^T E_S^e) \rangle = 0,$$

$$\langle \mu^+, (E_S^e - E_D^e)^T H_l - f_l^{\max} \rangle = 0,$$

$$\langle \mu^-, -(E_S^e - E_D^e)^T H_l - f_l^{\max} \rangle = 0. \tag{29}$$

Suppose that the dual function $h(\cdot)$ for the SPP is formulated using the multipliers λ, μ^+, and μ^- from the network's optimization problem. Since $\{E_D^e, E_S^e, R_S^e, P^e\}$ is a competitive equilibrium, E_D^e maximizes the consumer's welfare, and $\{E_S^e, R_S^e\}$ maximizes the supplier's welfare. Based on the form (23) for price process P^e, we conclude that

$$\{E_D^e, E_S^e, R_S^e\} \in \underset{E_D, E_S, R_S}{\arg \min} \; \mathcal{L}.$$

Substituting $\{E_D^e, E_S^e, R_S^e\}$ into (22), and applying the complementary slackness conditions from (29), we have

$$-\langle \mathcal{W}_{\text{tot}}, 1 \rangle = h(\lambda, \mu^+, \mu^-).$$

That is, strong duality holds. □

We stress that Theorem 3 characterizes prices in any competitive equilibrium.

Corollary 1. *The only candidates for prices in a competitive equilibrium are given by* (23), *based on the optimal Lagrange multipliers.*

Proof. This is because only the optimal Lagrange multipliers could possibly lead to strong duality. □

Thus, Theorem 3 tells us that computation of prices and quantities can be decoupled: The quantities $\{E_D, E_S, R_S\}$ are obtained through the solution of the SPP, and the price process P^e is obtained as a solution to its dual. The following corollary underlines the fact that if P^e supports one competitive equilibrium, then it supports any other competitive equilibrium.

Corollary 2. *If* $\{E_D^1, E_S^1, R_S^1, P^1\}$ *and* $\{E_D^2, E_S^2, R_S^2, P^2\}$ *are two competitive equilibria, then* $\{E_D^2, E_S^2, R_S^2, P^1\}$ *is also a competitive equilibrium.*

Proof. Due to the necessary condition of Theorem 3, there exist Lagrange multipliers $(\lambda_1, \mu_1^+, \mu_1^-)$ and $(\lambda_2, \mu_2^+, \mu_2^-)$ corresponding to the two equilibria, such that

$$h(\lambda_1, \mu_1^+, \mu_1^-) = -\langle \mathcal{W}_{\text{tot1}}, \mathbf{1} \rangle = -\langle \mathcal{W}_{\text{tot2}}, \mathbf{1} \rangle.$$

By the sufficient condition of Theorem 3, any of these price and quantity pairs satisfying strong duality will constitute a competitive equilibrium.   □

Proof of Theorem 1. From the proof of necessary condition of Theorem 3, for any competitive equilibrium $\{E_D^e, E_S^e, R_S^e, P^e\}$, there exist dual variables (λ, μ^+, μ^-) such that

$$h(\lambda, \mu^+, \mu^-) = -\langle \mathcal{W}_{\text{tot}}, \mathbf{1} \rangle.$$

From weak duality (25), we know that both the SPP and its dual problem are optimized. Therefore, $\{E_D^e, E_S^e, R_S^e\}$ is an efficient allocation.   □

Proof of Theorem 2. We assume that a competitive equilibrium exists. Then applying Theorem 1, there exists at least one equilibrium price process P^e that supports one of the efficient allocations. By Theorem 3, the existence of competitive equilibrium implies strong duality. Consequently, from the sufficient condition of Theorem 3, any of these price and quantity pairs satisfying strong duality will constitute a competitive equilibrium. In other words, the price P^e supports all efficient allocations.   □

4   Equilibrium Prices for the Test Cases

The dynamic electricity market model of Sect. 2 is an appropriate representation of a competitive market as analyzed in a standard economics text. The dynamic nature of the constraints on generation may lead to volatile prices that negatively impact the consumers, the suppliers, or both. The sample path behavior of prices in a competitive equilibrium can look as erratic as the worst days during the crises in Illinois or California in the 1990s, or in Texas and New Zealand during the early months of 2011.

   In this section, we characterize the equilibrium prices for the test case examples presented in Sect. 2.2 using the Lagrangian relaxation framework of Sect. 3. We find that even in the dynamic setting, market outcomes reflect the standard conclusions for efficient markets: Prices equal marginal costs, but only *on average*. We also illustrate how binding transmission constraints can result in prices that are negative or that exceed the choke-up price. Finally, we investigate the characteristics of prices for fast responding, expensive ancillary services introduced in Example C and show, through numerical experiments, that volatile price patterns result in very large variance of price.

4.1   Single-Bus Model

Here, we investigate sample path prices and their mean value for Example A—
the single-bus model. Recall that the generation constraint set $\mathbf{X}_S$ is defined by the
ramp constraints (10), and piecewise-linear forms are assumed for the utility and
cost functions. Also, supply-demand balance necessitates that $E := E_S = E_D$.

A special case of Example A is the model of [7], wherein the equilibrium price
is obtained as

$$P^e(t) = (v + c^{bo})\mathbb{I}\{R^*(t) < 0\}, \qquad (30)$$

with R^* denoting the reserve process in the solution of the SPP. The quantity $(v + c^{bo})$ is the choke-up price, which can be extremely large in a real power system.
Consequently, equilibrium prices show tremendous volatility. However, when initial
reserves are sufficiently large, the *average* price coincides with marginal cost c for
generation, in the sense that

$$\gamma \mathsf{E}\left[\int e^{-\gamma t} P^e(t)\,dt\right] = c. \qquad (31)$$

These conclusions were first obtained in [7] through direct calculation, based on
specific statistical assumptions on demand. We show here that the same conclusions
can be derived in far greater generality based on a Lagrangian relaxation technique.

We first establish the formula for P^e.

Proposition 1. *Suppose that* (E^*, R^*) *is a solution to the SPP that defines a
competitive equilibrium with price process* P^e. *Then*

$$P^e(t) = \nabla v(E^*(t)) + \nabla c^{bo}(D(t) - E^*(t)), \quad t \geq 0. \qquad (32)$$

Proof. In the single-bus model, we have $\mathcal{W}_D(t) := v(E_D(t)) - c^{bo}(D(t) - E_D(t)) - P^e(t)E_D(t)$. The result (32) follows because $E^* = E_D$ in the competitive
equilibrium, and the consumer is myopic (recall that the consumer does not consider
ramp constraints).   □

To find the average price, we consider the supplier's optimization problem. We
then consider a Lagrangian relaxation, in which the constraint $E_S(0) + R_S(0) = g_0$
is captured by the Lagrange multiplier v. The Lagrangian is denoted,

$$\mathcal{L}_S(E_S, R_S, v) = \mathsf{E}\left[\int e^{-\gamma t} \mathcal{W}_S(t)\,dt\right] - v[E_S(0) + R_S(0) - g_0]. \qquad (33)$$

The following result is a consequence of the local Lagrange multiplier theorem [21,
Theorem 1 of Sect. 9.3].

Lemma 2. *Let* $\mathbf{X}_S^{g_0} \subset \mathbf{X}_S$ *represent the set of feasible* (E_S, R_S), *subject to the
given initial condition* g_0. *Suppose* $(E_S^{g_0}, R_S^{g_0})$ *maximizes* $\mathcal{L}_S(E_S, R_S, 0)$ *over the*

set $\mathbf{X}_S^{g_0}$. Then there exist $\nu^* \in \mathbb{R}$ such that $(E_S^{g_0}, R_S^{g_0})$ maximizes $\mathcal{L}_S(E_S, R_S, \nu^*)$ over the larger set $\mathbf{X}_S$.

The next result is a construction required in the application of the local Lagrange multiplier result of Lemma 2.

Lemma 3. *Suppose that* $(E_S, R_S) \in \mathbf{X}_S$. *Then there exists a family of solutions* $\{(E_S^\alpha, R_S) : |\alpha| \leq 1\} \subset \mathbf{X}_S$ *satisfying* $E_S^\alpha(0) = \max(E_S(0) + \alpha, 0)$; $|E_S^\alpha(t) - E_S(t)| \leq |\alpha|$ *for all* t; *and,* $\lim_{\alpha \to 0} \frac{1}{\alpha}(E_S^\alpha(t) - E_S(t)) = 1_S^+(t) := \mathbb{I}\{E_S(t) > 0\}$ *for* $t > 0$.

The extension of the average-price formula (31) is obtained on combining these results:

Theorem 4. *Suppose that* (E^*, R^*) *is a solution to the SPP that defines a competitive equilibrium with price process* P^e. *Suppose that* $E^*(0) > 0$. *Moreover, assume that the following strengthening of (A2) holds: The cost function satisfies* $c^E(e) + |\nabla c^E(e)| \leq k_0(1 + e^2)$ *for some* $k_0 > 0$ *and for all* $e \geq 0$. *Then,*

$$\mathsf{E}\left[\int_0^\infty e^{-\gamma t} 1_S^+(t) P^e(t) \, \mathrm{d}t\right] = \mathsf{E}\left[\int_0^\infty e^{-\gamma t} 1_S^+(t) \nabla c^E(E^*(t)) \, \mathrm{d}t\right] + \nu^*, \quad (34)$$

where ν^* *is the sensitivity term from Lemma 2.*

Proof. Under the assumptions of the theorem, the Lagrangian $\mathcal{L}_S(E^\alpha, R^*, \nu)$ can be differentiated by α so that

$$\frac{\mathrm{d}}{\mathrm{d}\alpha}\mathcal{L}_S(E^\alpha, R^*, \nu^*) = \mathsf{E}\left[\int e^{-\gamma t} \frac{\mathrm{d}}{\mathrm{d}\alpha} \mathcal{W}_S^\alpha(t) \, \mathrm{d}t\right] - \frac{\mathrm{d}}{\mathrm{d}\alpha}\nu^*[E^\alpha(0) + R^*(0) - g_0],$$
$$(35)$$

where $\mathcal{W}_S^\alpha(t) := P^e(t)E_S^\alpha(t) - c^E(E_S^\alpha(t)) - c^R(R^*(t))$. In this calculation, the square integrability assumption and bounds on c^E justify taking the derivative under the expectation and integral in (33).

Then the result follows from two facts. First, the optimality of the Lagrangian at $\alpha = 0$ gives $\frac{\mathrm{d}}{\mathrm{d}\alpha}\mathcal{L}_S(E^\alpha, R^*, \nu) = 0$ for $\alpha = 0$. Second, application of Lemma 3 followed by the chain rule gives,

$$\frac{\mathrm{d}}{\mathrm{d}\alpha}\mathcal{W}_S^\alpha(t)\Big|_{\alpha=0} = 1_S^+(t)\big(P^e(t)E^*(t) - \nabla c^E(E^*(t))\big).$$

Evaluating (35) at $\alpha = 0$ then gives the desired result. $\square$

Thus, we can establish an extended version of (31). The average price depends on the initial value g_0, and this dependence is captured through the sensitivity term ν^*. When ramping down is unconstrained, $\nu^* \geq 0$.

4.2 Network Model

We now study the network-based dynamic market model introduced in Sect. 2. In general, the analysis of this model is complicated by the constraints imposed on the transmission lines, in addition to the dynamic constraints on generation. Therefore, we look for appropriate relaxations to proceed with the analysis.

The relaxation introduced here is based on the assumption that the network and the consumer are not subject to dynamic constraints. Consequently, these two players are myopic in their respective optimization problems. On the other hand, the dynamic nature of the generation constraints restricts the supplier and, consequently, the social planner from adopting myopic strategies in their optimization problems.

To characterize the equilibrium prices, we introduce a market without any suppliers. It is a fictitious model introduced solely for analysis:

Definition 3. The $\mathcal{S}$-market is defined as follows:

(i) The models for the consumer and network players are unchanged. The supplier model is modified by relaxing the operational/physical constraints (3) on (E_S, R_S).
(ii) The welfare functions of the consumer and the network are unchanged. The welfare function of the supplier is *identically zero*. This is achieved by overriding the production cost functions as follows:

$$c_n^{E\mathcal{S}}(E_{Sn}(t)) = P_n^e(t)E_{Sn}(t), \quad c_n^{R\mathcal{S}}(R_{Sn}(t)) = 0. \tag{36}$$

Since the welfare function $\mathcal{W}_S^{\mathcal{S}}$ for the supplier is identically zero, the $\mathcal{S}$-market essentially reduces to a model consisting of two players: the consumer and the network. We find that the equilibrium for the original market provides an equilibrium for this two-player market.

Lemma 4. *A competitive equilibrium for the original three-player market is a competitive equilibrium for the two-player $\mathcal{S}$-market.*

Proof. Let $\{E_D^e, E_S^e, R_S^e, P^e\}$ be a competitive equilibrium of the original three-player market. Clearly, the triple $\{E_D^e, E_S^e, R_S^e\}$ satisfies the supply-demand balance and the network constraints, and maximizes the consumer and the network welfare functions under price P^e. Since the supplier's welfare $\mathcal{W}_S^{\mathcal{S}}$ is always zero by assumption, we can view the pair $\{E_S^e, R_S^e\}$ as maximizing the supplier's welfare in the $\mathcal{S}$-market. Thus, the lemma holds.   □

Therefore, *all* equilibrium prices P^e in the original market model support a competitive equilibrium in the $\mathcal{S}$-market. Hence, we can hope to extract properties of P^e by analyzing the simpler $\mathcal{S}$-market model. This is the main motivation behind the introduction of the $\mathcal{S}$-market.

Recall that in the single-bus model, the characterization of the price in (32) was based on the assumption that consumers are not subject to dynamic constraints and are hence myopic. Lemma 5, which follows from the definition of $\$$-market, justifies the same approach to analyze the network model.

Lemma 5. *All players, as well as the social planner, are myopic in the $\$$-market.*

Hence the optimization problems of the consumer and the network in the $\$$-market are reduced to a "snapshot model" in which we can fix a time t to obtain properties of $P^e(t)$, exactly similar to the derivation of (32).

With time t fixed, we can suppress the time notation so that the snapshot version of the SPP is given as the maximization of the total welfare $\mathcal{W}_{\text{tot}}^{\$}$ with

$$\mathcal{W}_{\text{tot}}^{\$} = \sum_n \left[v_n(E_{Dn}) - c_n^{bo}(D_n - E_{Dn}) - P_n^e E_{Sn} \right], \tag{37}$$

subject to the following constraints:

$$
\begin{aligned}
\mathbf{1}^T E_S = \mathbf{1}^T E_D && \leftrightarrow \lambda, \\
-f_l^{\max} \le (E_S - E_D)^T H_l \le f_l^{\max} && \leftrightarrow \mu_l^-, \mu_l^+ \ge 0 && \text{for all } l, \\
0 \le E_{Dn} \le D_n && \leftrightarrow \eta_n^-, \eta_n^+ \ge 0 && \text{for all } n.
\end{aligned}
$$

$\lambda, \mu_l^-, \mu_l^+, \eta_n^-, \eta_n^+$ are Lagrange multipliers associated with the corresponding constraints. The Lagrangian of the SPP for the $\$$-market is the function of static variables:

$$
\mathcal{L}^{\$} = -\mathcal{W}_{\text{tot}}^{\$} + \lambda(\mathbf{1}^T(E_D - E_S)) + \sum_l \mu_l^+[(E_S - E_D)^T H_l - f_l^{\max}]
$$

$$
+ \sum_l \mu_l^-[-(E_S - E_D)^T H_l - f_l^{\max}] + \sum_n \eta_n^+(E_{Dn} - D_n) - \sum_n \eta_n^+ E_{Dn}. \tag{38}
$$

For the fixed time t, we characterize the nodal price P_n^e as follows.

Proposition 2. *Consider the SPP for the $\$$-market with welfare function defined in (37). Suppose that $\mu_l^-, \mu_l^+, \eta_n^-, \eta_n^+$ are the non-negative, optimal solutions to the dual with Lagrangian (38). Then the equilibrium price has entries given as follows: For $n = 1, \ldots, N$,*

$$P_n^e = \nabla v_n(E_{Dn}^e) + \nabla c_n^{bo}(D_n - E_{Dn}^e) + \Lambda, \tag{39}$$

where $\Lambda = \begin{cases} 0, & 0 < E_{Dn}^e < D_n, \\ -\eta_n^+, & E_{Dn}^e = D_n, \\ \eta_n^-, & E_{Dn}^e = 0, \end{cases}$

and E_{Dn}^e is the energy consumed in the equilibrium.

Proof. Since $\{E_D^e, E_S^e, R_S^e, P^e\}$ is a competitive equilibrium for the $\$$-market, $\{E_D^e, E_S^e\}$ maximizes the SPP for the $\$$-market. By the KKT conditions, we obtain

$$0 = \frac{\partial \mathcal{L}^{\$}}{\partial E_{Dn}} = -\nabla v_n(E_{Dn}^e) - \nabla c_n^{bo}(D_n - E_{Dn}^e) + \lambda + \sum_l (\mu_{ln}^- - \mu_{ln}^+) H_{ln}$$
$$+ \eta_n^+ - \eta_n^-,$$

$$0 = \frac{\partial \mathcal{L}^{\$}}{\partial E_{sn}} = P_n^e - \lambda - \sum_l (\mu_{ln}^- - \mu_{ln}^+) H_{ln}.$$

On summing these two equations, we obtain

$$P_n^e = \nabla v_n(E_{Dn}^e) + \nabla c_n^{bo}(D_n - E_{Dn}^e) - \eta_n^+ + \eta_n^-.$$

The proposition follows from the complementary slackness conditions. □

Note that result (39) holds for all equilibria in the two-player $\$$-market. Consequently, from Lemma 4, it holds for all equilibria in the original three-player market. Although it appears from (39) that prices depend upon the actions of the players, this is not the case. Similar to the Proposition 1, we can write the price as

$$P_n^e(t) = \nabla v_n(E_{Dn}^*(t)) + \nabla c_n^{bo}(D_n(t) - E_{Dn}^*(t)) + \Lambda^*(t), \tag{40}$$

where $\{E_{Dn}^*(t)\}$ and, consequently, $\Lambda^*(t)$ are obtained from the solution of the SPP.

To apply Proposition 2, we must identify the parameters $\{\eta_n^-, \eta_n^+\}$, which represent the sensitivities of the SPP solution with respect to the constraints $E_{Dn} \geq 0$ and $E_{Dn} \leq D_n$ respectively. Next, we motivate the price computations through numerical examples.

4.3 Texas Model: Range of Prices

We revisit Example B to illustrate the application of Proposition 2. We show how prices in a network with binding transmission constraints can go below zero or well above the choke-up price.

For the Texas model shown in Fig. 2, we assume that a single supplier is located at bus 1 while buses 2 and 3 are load buses. The buyer has linear utility of consumption and disutility of blackout (1b) at buses 2 and 3, with identical parameters v and c^{bo}. With identical impedances for the three transmission lines and bus 1 chosen as the reference bus, the injection shift factor matrix H is given by the expression (13).

We assume that the supporting price P_1^e at bus 1 is *zero*, and that this is true not only for the snapshot values $\{E_{Di}, E_{Si}, R_{Si}\}$ and parameters $\{f_{ij}^{max}\}$, but also for all values in a neighborhood of these nominal values. This is not unreasonable based on the results of Sect. 4.1 if the reserves are strictly positive at bus 1. Under these assumptions, we can then compute the prices P_2^e and P_3^e at buses 2 and 3, respectively, using the $\$$-market introduced in Sect. 4.2.

Recall that the supplier's welfare function is identically zero in the $-market. Suppose that $D_2 = 170$ MW and $D_3 = 30$ MW. Then the snapshot SPP for the $-market is

$$\min -\left[v(E_{D2} + E_{D3}) - c^{bo}(200 - E_{D2} - E_{D3})\right]$$

subject to

$$E_{S1} = E_{D2} + E_{D3}$$

$$-f_{12}^{max} \leq \frac{2}{3}E_{D2} + \frac{1}{3}E_{D3} \leq f_{12}^{max}$$

$$-f_{23}^{max} \leq \frac{1}{3}E_{D2} - \frac{1}{3}E_{D3} \leq f_{23}^{max}$$

$$-f_{13}^{max} \leq \frac{1}{3}E_{D2} + \frac{2}{3}E_{D3} \leq f_{13}^{max}$$

$$0 \leq E_{D2} \leq 170, \ 0 \leq E_{D3} \leq 30.$$

4.3.1 Negative Prices

Assume that $f_{23}^{max} = 40$ MW, while the other two lines are unconstrained. Solving the SPP for the $-market we obtain $E_{D2} = 150$ MW and $E_{D3} = 30$ MW, and we find that the limit $f_{23}^{max} = 40$ MW is reached. Since $0 < E_{D2} < D_2$, we have $P_2^e = v + c^{bo}$ by Proposition 2.

For a given $\epsilon \in \mathbb{R}$, we perturb the constraint on E_{D3} to obtain $0 \leq E_{D3} \leq 30 + \epsilon$. On re-solving the SPP, we obtain $E_{D2} = 150 + \epsilon$ MW and $E_{D3} = 30 + \epsilon$ MW. Applying Proposition 2, P_3^e is given by the limit

$$P_3^e := v + c^{bo} + \lim_{\epsilon \to 0} \frac{-(180 + 2\epsilon)v + (20 - 2\epsilon)c^{bo} + 180v - 20c^{bo}}{\epsilon}.$$

That is, $P_3^e = -(v + c^{bo})$, which is clearly *negative*.

4.3.2 Prices Exceeding the Choke Up Price

Assume that $f_{13}^{max} = 50$ MW, while the other two lines are unconstrained. Again, solving the SPP for the $-market gives $E_{D2} = 150$ MW and $E_{D3} = 0$ MW with the limit $f_{13}^{max} = 50$ MW being reached. Proposition 2 gives $P_2^e = v + c^{bo}$ since $0 < E_{D2} < D_2$.

For a given $\epsilon \in \mathbb{R}$, we perturb the constraint on E_{D3} to obtain $0 + \epsilon \leq E_{D3} \leq 30$. On re-solving the SPP, we obtain $E_{D2} = 150 - 2\epsilon$ MW and $E_{D3} = \epsilon$ MW. Again, applying Proposition 2, P_3^e is expressed as a limit

$$P_3^e := v + c^{bo} + \lim_{\epsilon \to 0} \frac{-(180 - \epsilon)v + (20 + \epsilon)c^{bo} + 180v - 20c^{bo}}{\epsilon}.$$

That is, $P_3^e = 2(v + c^{bo})$, which is twice the choke-up price.

Thus, when the transmission constraints come into play, the equilibrium prices are spread over a wide range, encompassing values well below zero as well as values far exceeding the choke-up price.

4.4 Ancillary Service Prices

Given the volatile nature of prices in the dynamic setting, we are interested in understanding the implications of volatility on the providers of ancillary service. The results presented in the preceding sections can be adapted to a market setting in which a number of services can be used to meet the demand.

Here, we restrict to the model of [6] presented in Example C of Sect. 2.2. We consider two sources of generation—primary and ancillary—whose instantaneous output at time t is denoted by $G_p(t)$ and $G_a(t)$, respectively. For simplicity, we focus on the case in which ramping down is unconstrained, i.e., $\zeta_p^-, \zeta_a^- = \infty$.

We assume that the consumers exhibit *demand-response* capabilities: certain loads can be turned off to maintain supply-demand balance in the event of reserves shortfall. When the demand exceeds the available supply, loads with demand response capabilities are the first ones to be turned off. If the load of the responsive consumers can sufficiently cover the reserve shortfall, the price is set by the cost of demand response. Otherwise, forced blackout occurs and the price equals the choke-up price. Since the cost of demand response is typically lower than the cost of blackout, the price when the supply deficiency can be covered by demand response is lower than that if demand response was unavailable. Thus, demand response acts as a cushion between the normal secure operations and the blackout. We refer the reader to [20] and the references therein for more details on demand response.

In our model, we use $\bar{r}_{max}^{dr}$ to denote total load of the consumers with demand response capability. That is, we have forced blackouts only if $R(t) \leq -\bar{r}_{max}^{dr}$. In this example, we convert the SPP into a cost minimization problem, in which the cost function on (G^p, G^a, R) has the following form: For $t \geq 0$,

$$C(t) := c^p G_p(t) + c^a G_a(t) + (c^{bo} - c^{dr})\mathbb{I}(R(t) < -\bar{r}_{max}^{dr}) + c^{dr}\mathbb{I}(R(t) < 0),$$

where c^p, c^a represent the per unit production costs of primary and ancillary services, respectively, c^{dr} is the cost of demand response-based load shedding and c^{bo} is the cost of blackout [6].

To investigate the impact of prices on the ancillary service providers, we simulate the market based on a controlled random-walk (CRW) model for demand, where $D(t)$ has the form,

$$D(k + 1) = D(k) + \mathcal{E}(k + 1), \quad k \geq 0, \quad D(0) = 0,$$

in which the increment process $\mathcal{E}$ is a discrete-valued, bounded, i.i.d. sequence, with marginal distribution symmetric on $\{\pm 1\}$. The reserve R is modeled in discrete

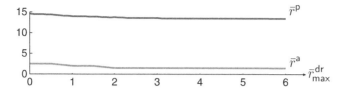

Fig. 4 Optimal thresholds for primary and ancillary service

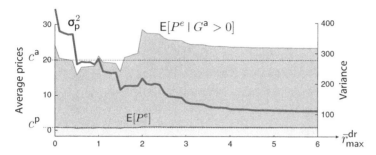

Fig. 5 Average prices and variance of P^e

time using the expression (15). The other parameters used are as follows: $c^p = 1$, $c^a = 20$, $c^{dr} = 100$ and $c^{bo} = 400$. The ramp-up rates are taken as $\zeta_p^+ = 1/10$ and $\zeta_a^+ = 2/5$.

We describe results from experiments using threshold policies, based on two thresholds $(\bar{r}^p, \bar{r}^a)$. Under the policy considered, primary service is ramped up whenever $R(t) \leq \bar{r}^p$. Similarly, ancillary service is ramped up whenever $R(t) \leq \bar{r}^a$. The simulation experiments from [5] were conducted in discrete-time: The discount factor $\beta = 0.995$ used there corresponds approximately to the discount rate $\gamma = 0.005$ in (21). We refer the reader to [5] further details about the threshold policy.

We find the "best-threshold" by approximating the discounted cost by the standard Monte–Carlo estimate. We use c^{dr} and $\bar{r}_{max}^{dr}$ as simulation parameters to study sensitivity of average prices to demand response capabilities.

In Fig. 4, we plot the optimal thresholds for primary and ancillary service against the demand response capacity $\bar{r}_{max}^{dr}$. Note that the optimal threshold for primary service is much higher than that of ancillary service, since primary service ramps up slower than ancillary service and is less expensive. The low sensitivity is consistent with the conclusions of [6].

The average price for primary service, and the conditional average price for ancillary service are shown in Fig. 5, for different values of $\bar{r}_{max}^{dr}$. The average price $E[P^e] = c^p$ is consistent with the conclusions of [7] and Theorem 4 (recall that G_p is not sign-constrained). The conditional average ancillary service price is given by

$$E[P^e|G_a > 0] = \left[\int_0^\infty e^{-\gamma t} P^e(t) \mathbb{I}(G_a(t) > 0)\, dt \right] \left[\int_0^\infty e^{-\gamma t} \mathbb{I}(G_a(t) > 0)\, dt \right]^{-1}.$$

The results shown are consistent with (34): It appears that when $\bar{r}_{max}^{dr} \geq 3$, $\nu^* \approx$ $5\left(\int_0^\infty e^{-\gamma t} \mathbb{I}(G_a(t) > 0)\, dt\right)$.

Also shown in Fig. 5 is a plot of the variance of the equilibrium price with respect to $\bar{r}_{max}^{dr}$. We see that the price variance drops dramatically with an increase in the demand response capacity, even though the optimal reserve thresholds are virtually unchanged.

Recall that the prices for primary and ancillary service are identical. However, the bulk of primary service is allocated in the day ahead market. Consequently, ancillary service, such as provided by gas turbines, will be exposed to much greater variability in the efficient market outcome. We hypothesize that high risk and low average prices may drive generators out of business. Market operators and regulators should consider such consequences when designing electricity markets.

5 Conclusions

In this chapter, we have described a framework for constructing dynamic models for electricity markets, and methods for characterizing the resulting competitive equilibria. The dynamic model is constructed using techniques well-known in the control community and can effectively handle the underlying physics of the power system while taking into account the economic aspects of electricity trading.

Competitive equilibrium theory has elegant mathematical underpinnings, and gives some insight into market outcomes. For example, volatile electricity prices seen throughout the world today are not surprising, given the theory surveyed in Sect. 4.

However, just as in equilibrium theory for control (as taught in senior-level undergraduate electrical engineering courses), an equilibrium is an extreme idealization, intended merely as a *starting point* for understanding market behavior. In particular, (A1) is rarely strictly valid: consumers and suppliers do not share the same information. The price taking assumption in (A3) does not hold in real-life energy markets. The quantitative formulation of "cost" in the SPP requires more careful consideration to capture a broader range of issues: In our discussions, we have ignored environmental impact and other long-term costs [11].

We urgently need to think about market design as engineers think about control design: How can we formulate market rules that assure reliability on a range of time-scales, provide incentives for new technology that is inexpensive in terms of cost and environmental impact, and adapt to an evolving environment and populace?

We hope that the ideas described in this chapter will form building blocks for addressing these questions; our ambition is to create a richer science for the design of smarter energy networks.

Acknowledgments Research supported in part by the Grainger Endowments to the University of Illinois, DOE awards DE-OE0000097 (TCIPG) and DE-SC0003879, and AFOSR Grant FA9550-09-1-0190.

References

1. Aliprantis CD, Cornet B, Tourky R (2002) Economic equilibrium: optimality and price decentralization. Positivity 6:205–241
2. Anderson EJ, Philpott AB (2002) Using supply functions for offering generation into an electricity market. Oper Res 50(3):477–489 MR 1910284 (2003f:91069)
3. Arrow KJ, Hahn F (1971) General competitive analysis. Holden-Day, San Francisco
4. Baldick R, Grant R, Kahn E (2004) Theory and application of linear supply function equilibrium in electricity markets. J Regul Econ 25(2):143–167
5. Chen M, Cho I-K, Meyn SP (2006) Reliability by design in a distributed power transmission network. Automatica 42:1267–1281 (invited)
6. Cho I-K, Meyn SP (2009) A dynamic newsboy model for optimal reserve management in electricity markets. SIAM J Contr Optim (Submitted for publication)
7. Cho I-K, Meyn SP (2010) Efficiency and marginal cost pricing in dynamic competitive markets with friction. Theoret Econ 5(2):215–239
8. Chow GC (1997) Dynamic economics: optimization by the lagrange method. Oxford University Press, USA
9. Chow JH, De Mello W, Cheung KW (2005) Electricity market design: an integrated approach to reliability assurance. Proc IEEE 93(11):1956 –1969
10. Day CJ, Hobbs BF, Pang J-S (2002) Oligopolistic competition in power networks: a conjectured supply function approach. IEEE Trans Power Syst 17:597–607
11. Epstein PR, Buonocore JJ, Eckerle K, Hendryx M, Stout III BM, Heinberg R, Clapp RW, May B, Reinhart NL, Ahern MM, Doshi SK, Glustrom L (2011) Full cost accounting for the life cycle of coal. Ann NY Acad Sci 1219(1):73–98
12. FERC (1998) Staff Report to the Federal Energy Regulatory Commission on the Causes of Wholesale Electric Pricing Abnormalities in the Midwest During June 1998, FERC study, September 22, 1998
13. Hobbs BF (2001) Linear complementarity models of Nash–Cournot competition in bilateral and POOLCO power markets. IEEE Trans Power Syst 16(2):194–202
14. Hobbs BF, Helman U (2004) Complementarity-Based Equilibrium Modeling for Electric Power Markets. Modeling Prices in Competitive Electricity Markets, Wiley Series in Financial Economics. Wiley, West Sussex, England, pp 69–98
15. Houston Chronicle Editorial (2011) Rolling blunder statewide blackouts raise the question: who really runs the lone star power grid? www.chron.com/disp/story.mpl/editorial/7424733.html, Feb. 12, 2011
16. Hu X, Ralph D (2007) Using EPECs to model bilevel games in restructured electricity markets with locational prices. Oper Res 55(5):809–827. MR MR2360950
17. Johari R, Tsitsiklis JN (2009) Efficiency of scalar-parameterized mechanisms. Oper Res 57:823–839
18. Kizilkale AC, Mannor S (2010) Volatility and efficiency in markets with friction. In: 48th Annual Allerton Conference on Communication, Control, and Computing, Monticello, Illinois, pp 50–57
19. Klemperer PD, Meyer MA (1989) Supply function equilibria in oligopoly under uncertainty. Econometrica 57:1243–1277
20. Kowli A, Meyn SP (2011) Supporting wind generation deployment with demand response. In: IEEE PES 11: power energy society general meeting, Detroit, Michigan
21. Luenberger DG (1969) Optimization by vector space methods. Wiley, New York

22. Mas-Colell A, Whinston MD, Green JR (1995) Microeconomic theory. Oxford University Press, New York
23. Metzler C, Hobbs B, Pang J-S (2003) Nash–Cournot equilibria in power markets on a linearized DC network with arbitrage: formulations and properties. Network Spatial Econ 3(2):123–150
24. Meyn S, Negrete-Pincetic M, Wang G, Kowli A, Shafieepoorfard E (2010) The value of volatile resources in electricity markets. In: Proceedings of the 49th IEEE Conference on Decision and Control, Atlanta, Georgia, pp 1029–1036
25. Mookherjee R, Friesz T, Hobbs BF, Ringdon M (2008) Dynamic oligopolistic competition on an electric power network with ramping costs and joint sales constraints. J Ind Manage Org 4(3):425–452
26. Negrete-Pincetic M, Meyn SP (2011) Intelligence by design for the entropic grid. In: IEEE PES 11: Power energy society general meeting, Invited lecture for PES panel: deploying tomorrow's electric power systems: low carbon, efficiency and security, Detroit, Michigan
27. Robinson S (2005) Math model explains volatile prices in power markets. SIAM News
28. Shanbhag UV, Infanger G, Glynn PW (2011) A complementarity framework for forward contracting under uncertainty. Oper Res 59(4):810–834
29. Smith JC, Beuning S, Durrwachter H, Ela E, Hawkins D, Kirby B, Lasher W, Lowell J, Porter K, Schuyler K, Sotkiewicz P (2010) The wind at our backs. IEEE Power Energ Mag 8(5):63–71
30. Wang G, Kowli A, Negrete-Pincetic M, Shafieepoorfard E, Meyn S (2011) A control theorist's perspective on dynamic competitive equilibria in electricity markets. In: Proceedings of 18th world congress of the international federation of automatic control (IFAC), Milano, Italy
31. Wang G, Negrete-Pincetic M, Kowli A, Shafieepoorfard E, Meyn S, Shanbhag UV (2012) Real-time prices in an entropic grid. IEEE Conference on Decision and Control. To appear in the Proceedings of the 2012 PES Innovative Smart Grid Technologies Conference. (Invited), Jan 2012
32. Electricity Authority strikes down Genesis price spike. Business Desk Report, New Zealand Herald, May 6, 2011. www.nzherald.co.nz/business/news/article.cfm?c_id=3&objectid=10723856
33. Wilson R (2002) Architecture of power markets. Econometrica 70(4):1299–1340
34. Wood AJ, Wollenberg BF (1996) Power generation, operation, and control. Wiley, New York
35. Yao J, Adler I, Oren S (2008) Modeling and computing two-settlement oligopolistic equilibrium in a congested electricity network. Oper Res 56(1):34–47. MR MR2402216
36. Zavala VM, Anitescu M (2010) On the dynamic stability of electricity markets. Math Program (submitted for publication)

Optimal Demand Response: Problem Formulation and Deterministic Case

Lijun Chen, Na Li, Libin Jiang, and Steven H. Low

Abstract We consider a set of users served by a single load-serving entity (LSE). The LSE procures capacity a day ahead. When random renewable energy is realized at delivery time, it manages user load through real-time demand response and purchases balancing power on the spot market to meet the aggregate demand. Hence, optimal supply procurement by the LSE and the consumption decisions by the users must be coordinated over two timescales, a day ahead and in real time, in the presence of supply uncertainty. Moreover, they must be computed jointly by the LSE and the users since the necessary information is distributed among them. In this chapter, we present a simple yet versatile user model and formulate the problem as a dynamic program that maximizes expected social welfare. When random renewable generation is absent, optimal demand response reduces to joint scheduling of the procurement and consumption decisions. In this case, we show that optimal prices exist that coordinate individual user decisions to maximize social welfare, and present a decentralized algorithm to optimally schedule a day in advance the LSE's procurement and the users' consumptions. The case with uncertain supply is reported in a companion paper.

1 Introduction

1.1 Motivation

There is a large literature on various forms of load side management from the classical direct load control to the more recent real-time pricing [1, 2]. Direct load control, in particular, has been practised for a long time and optimization methods

L. Chen (✉) • N. Li • L. Jiang • S.H. Low
Engineering and Applied Science, California Institute of Technology, Pasadena, CA, USA
e-mail: chenlj@caltech.edu; nali@caltech.edu; libinj@caltech.edu; slow@caltech.edu

A. Chakrabortty and M.D. Ilić (eds.), *Control and Optimization Methods for Electric Smart Grids*, Power Electronics and Power Systems 3,
DOI 10.1007/978-1-4614-1605-0_3, © Springer Science+Business Media, LLC 2012

have been proposed to minimize generation cost e.g. [3–6], maximize utility's profit e.g. [7], or minimize deviation from users' desired consumptions e.g. [8, 9], sometimes integrated with unit commitment and economic dispatch e.g. [4, 10]. Almost all demand response programs today target large industrial or commercial users, or, in the case of residential users, a small number of them, for two, among other, important reasons. First, demand side management is invoked rarely to mostly cope with a large correlated demand spike due to weather or a supply shortfall due to faults, e.g., during a few hottest days in summer. Second, the lack of ubiquitous two-way communication in the cur rent infrastructure prevents the participation of a large number of diverse users with heterogeneous and time-varying consumption requirements. Both reasons favor a simple and static mechanism involving a few large users that is sufficient to deal with the occasional need for load control, but both reasons are changing.

Renewable sources can fluctuate rapidly and by large amounts. As their penetration continues to grow, the need for regulation services and operating reserves will increase, e.g., [11, 12]. This can be provided by additional peaker units, at a higher cost, or supplemented by real-time demand response [12–16]. We believe that demand response will not only be invoked to shave peaks and shift load for economic benefits, but will also increasingly be called upon to improve security and reduce reserves by adapting elastic loads to intermittent and random renewable generation [17]. Indeed, the authors of [12, 18, 19] advocate the creation of a distribution/retail market to encourage greater load side participation as an alternative source for fast reserves. Such application, however, will require a much faster and more dynamic demand response than practised today. This will be enabled in the coming decades by the large-scale deployment of a sensing, control, and two-way communication infrastructure, including the flexible AC transmission systems, the GPS-synchronized phasor measurement units, and the advanced metering infrastructure, that is currently underway around the world [20].

Demand response in such context must allow the participation of a large number of users, and be dynamic and distributed. Dynamic adaptation by hundreds of millions of end users on a sub-second control timescale, each contributing a tiny fraction of the overall traffic, is being practised everyday on the Internet in the form of congestion control. Even though both the grid and the Internet are massive distributed nonlinear feedback control systems, there are important differences in their engineering, economic, and regulatory structures. Nonetheless, the precedence on the Internet lends hope to a much bigger scale and more dynamic and distributed demand response architecture and its benefit to grid operation. *Ultimately, it will be cheaper to use photons than electrons to deal with a power shortage.* Our goal is to design algorithms for such a system.

1.2 Summary

Specifically, we consider a set of users that are served by a single load-serving entity (LSE). The LSE may represent a regulated monopoly like most utility companies in the United States today, or a nonprofit cooperative that serves a community of end users. Its purpose is (possibly regulated) to promote the overall system welfare. The LSE purchases electricity on the wholesale electricity markets (e.g., day-ahead, real-time balancing, and ancillary services) and sells it on the retail market to end users. It provides two important values: it aggregates loads so that the wholesale markets can operate efficiently, and it hides the complexity and uncertainty from the users, in terms of both power reliability and prices. Our model captures three important features:

- *Uncertainty*: Part of the electricity supply is from renewable sources such as wind and solar, and thus uncertain.
- *Supply and demand*: LSE's supply decisions and the users' consumption decisions must be jointly optimized.
- *Two timescale*: The LSE must procure capacity on the day-ahead wholesale market while user consumptions should be adapted in real time to mitigate supply uncertainty.

Hence, the key is the coordination of day-ahead procurement and real-time demand response over two timescales in the presence of supply uncertainty. Moreover, the optimal decisions must be computed jointly by the LSE and the users as the necessary information is distributed among them. The goal of this chapter is to formulate this problem precisely. Due to space limitation, we can only fully treat the case without supply uncertainty. Results for the case with supply uncertainty are summarized here, but fully developed in a companion paper [21].

Suppose each user has a set of appliances (electric vehicle, air conditioner, lighting, battery, etc.). She (or her energy management system) is to decide how much power she should consume in each period $t = 1, \ldots, T$ of a day. The LSE needs to decide how much capacity it should procure a day ahead and, when the random renewable energy is realized at real time, how much balancing power to purchase on the spot market to meet the aggregate demand. In Sect. 2, we present our user and supply models, and formulate the overall problem as an $(1 + T)$-period dynamic program to maximize expected social welfare. The key idea is to regard the LSE's day-ahead decision as the control in period 0 and the users' consumption decisions as controls in the subsequent periods $t = 1, \ldots, T$. By unifying several models in the literature, our user model incorporates a large class of appliances. Yet, it is simple, thus analytically tractable, where each appliance is characterized by a utility function and a set of linear consumption constraints.

In Sect. 3, we consider the case without renewable generation. In the absence of uncertainty, it becomes unnecessary to adapt user consumptions in real-time and hence supply and consumptions can be optimally scheduled at once instead of over two days. We show that optimal prices exist that coordinate individual users' decisions in a distributed manner, i.e., when users selfishly maximize their own surplus under the optimal prices, their consumption decisions turn out to also maximize the social welfare. We develop an offline distributed algorithm that jointly schedules the LSE's procurement decisions and the users' consumption decisions for each period in the following day. The algorithm is decentralized where the LSE only knows the aggregate demand but not user utility functions or consumption constraints, and the users do not need to coordinate among themselves but only respond to the common prices from the LSE.

With renewable generation, the uncertainty precludes pure scheduling and calls for real-time consumptions decisions that adapt to the realization of the random renewable generation. Moreover, this must be coordinated with procurement decisions over two timescales to maximize the expected welfare. Distributed algorithms for optimal demand response in this case and the impact of uncertainty on the optimal welfare are developed in the companion paper [21].

Finally, we conclude in Sect. 4 with some limitations of this chapter.

We make two remarks. First, the effectiveness of real-time pricing for demand response is still in active research. On the one hand, empirical studies have shown consistently that price elasticity is low and heterogeneous; see [22–24] and references therein. On the other hand, there are strong economic arguments that real-time retail prices improve the efficiency of the overall system by allowing users to dynamically adapt their loads to shortages, with potential benefits far exceeding the cost of implementation [18]. Moreover, the long-run efficiency gain is likely to be significant even if demand elasticity is small, but unfortunately, the popular open-loop time-of-use pricing may capture a very small share of the efficiency gain of real-time pricing [25]. We neither argue for nor are against real-time pricing. Indeed, we do not consider in this chapter the economic issues associated with such a system, such as locational marginal prices, revenue-adequacy. What we refer to as "prices" are simply control signals that provide the necessary information for users to adapt their consumption in a distributed, yet optimal, manner. Whether this control signal should be linked to monetary payments to provide the right incentive for demand response is beyond the scope of this chapter, i.e., we do not address the important issue of how to incentivize users to respond to supply and demand fluctuations.[1]

Second, unlike many current systems, the kind of large-scale distributed demand response system envisioned here must be fully automated. Human users set parameters that specify utility functions and consumption constraints and may change them on a slow timescale, but the algorithms proposed here will execute automatically and transparently to optimize social welfare. The traditional direct load control approach assumes that the controller (e.g., a utility company) knows the user consumption requirements, in the form of payback characteristics of the deferred load, and can optimally schedule deferred consumptions and their paybacks centrally. This is reasonable for the current system where the participating users are few and their requirements are relatively static. We take the view that the utilities and requirements of user consumptions are diverse and private. It is not practical, *nor necessary*, to have direct access to such information in order to optimally coordinate their consumptions in a large, distributed, and dynamic system of the future. The algorithm presented here is an example that can achieve optimality without requiring users to disclose their private information.

---

[1]See, however, [19] for a discussion on some implementation issues of real-time pricing for retail markets and a proposal for the Italian market.

1.3 Other Related Work

A large literature exists on demand response. Besides those cited above, more recent works include, e.g., [26, 27] on load control of thermal mass in buildings, [28–30] on residential load control through coordinated scheduling of different appliances, [31–33] on the scheduling of plug-in electric vehicle charging, and [34] on the optimal allocation of a supply deficit (rationing) among users using their supply functions. Load side management in the presence of uncertain supply has also been considered in [10, 12, 16, 35–37]. Unlike the conventional approach that compensates for the uncertainty to create reliable power, the authors of [16] advocate selling interruptible power and designs service contracts, based on [38], that can achieve greater efficiency than the conventional approach. In [10], various optimization problems are formulated that integrate demand response with economic dispatch with ramping constraints and forecasts of renewable power and load. Both centralized dispatch using model predictive control and decentralized dispatch using prices, or supply and demand functions, are considered. A two-period stochastic dispatch model is studied in [35] and a settlement scheme is proposed that is revenue-adequate even in the presence of uncertain supply and demand. A queueing model is analyzed in [36], where the queue holds deferrable loads that arise from random supply and demand processes. Conventional generation can be purchased to keep the queue small and strategies are studied to minimize the time-average cost. The models that are closest to ours, developed independently, are [12, 37]. All our models include random renewable generation, consider both day-ahead and real-time markets, and allow demand response, but our objectives and system operations are quite different. The authors of [12] advocate the establishment of a retail market, where users (e.g., PHEVs) can buy power from or sell reserves, in the form of demand response capability, to their LSE. The paper formulates the LSE's and users' problems as dynamic programs that minimize their expected costs over their bids, which can be either simple, uncorrelated (price, quantity) pairs for each period, or complex, (price, quantity) pairs with temporal correlations. The model in [37] includes nonelastic users that are price nonresponsive, and elastic users that can either leave the system or defer their consumptions when the electricity price is high. The goal is to maximize LSE's profit over day-ahead procurement, day-ahead prices for nonelastic users, and real-time prices for elastic users.

1.4 Notations

Given quantities such as the demands $q_{ia}(t)$ from appliance a of user i in period t, $q_{ia} := (q_{ia}(t), t \in \mathcal{T})$ denotes the vector of demands at different times, $q_i(t) := (q_{ia}(t), a \in \mathscr{A}_i)$ the vector of demands of different appliances, $q_i := (q_{ia}, a \in \mathscr{A}_i)$ the vector of demands of i's appliances at different times, and $q := (q_i, \forall i)$ the vector of all demands. Similarly, for the aggregate demands $Q_i(t) = \sum_{a \in \mathscr{A}_i} q_{ia}(t)$,

$Q_{ia} := \sum_t q_{ia}(t)$, Q_i, Q. Script letters denote sets, e.g., $\mathcal{N}, \mathcal{A}_i, \mathcal{T}$. Small letters denote individual quantities, e.g., $q_{ia}(t)$, q_{ia}, $q_i(t)$, q_i, q, etc. Capital letters denote aggregate quantities, e.g., $Q_i(t)$, Q_{ia}, $P_d(t)$, $P_r(t)$, $P_o(t)$, $P_b(t)$, etc. We use $q_{ia}(t), q_{ia}, Q_i(t)$, etc. for loads and $P_d(t)$, $P_r(t)$, etc. for supplies. We sometimes write $\sum_i \sum_{a \in \mathcal{A}_i} q_{ia}(t)$ as $\sum_{i,a} q_{ia}(t)$. For any real a, b, c, $[a]_+ := \max\{a, 0\}$ and $[a]_b^c := \max\{b, \min\{a, c\}\}$. Finally, we write a vector as $x = (x_i, \forall i)$ without specifying whether it is a column or row vector so we can ignore the transpose sign to simplify the notation; the meaning should be clear from the context.

2 Model and Problem Formulation

Consider a set $\mathcal{N}$ of N users that are served by a single LSE. We use a discrete-time model with a finite horizon that models a day. Each day is divided into T periods of equal duration, indexed by $t \in \mathcal{T} := \{1, 2, \ldots, T\}$. The duration of a period can be 5, 15, or 60 min, corresponding to the time resolution at which energy dispatch or demand response decisions are made.

2.1 User Model

Each user $i \in \mathcal{N}$ operates a set $\mathcal{A}_i$ of appliances such as HVAC (heat, ventilation, air conditioner), refrigerator, and plug-in hybrid electric vehicle. User i may also possess a battery, which provides further flexibility for optimizing its electricity consumption across time.

Appliance model: For each appliance $a \in \mathcal{A}_i$ of user i, $q_{ia}(t)$ denotes its energy consumption in period $t \in \mathcal{T}$, and q_{ia} the vector $(q_{ia}(t), \forall t)$ over the whole day. An appliance a is characterized by:

- A utility function $U_{ia}(q_{ia})$ that quantifies the utility user i obtains from using appliance a.
- A $K_{ia} \times T$ matrix A_{ia} and a K_{ia}-vector η_{ia} such that the vector of power q_{ia} satisfies the linear inequality

$$A_{ia} q_{ia} \leq \eta_{ia}. \tag{1}$$

In general, U_{ia} depends on the vector q_{ia}. In this chapter, however, we consider four types of appliances whose utility functions take one of the three simple forms. These models are summarized in Table 1 and justified in detail in Appendix A. The utility of a type 1 or type 2 appliance is additive in t^2:

$$U_{ia}(q_{ia}) := \sum_t U_{ia}(q_{ia}(t), t). \tag{2}$$

---

[2]We abuse notation to use U_{ia} to denote both a function of vector q_{ia} and that of a scalar $q_{ia}(t)$; the meaning should be clear from the context.

Table 1 Structure of utility functions and consumption constraints for appliances

Appliances	Utility function	Consumption constraints	Examples
Type 1	(2)	(6)	Lightings
Type 2	(2)	(6), (7)	TV, video game, computer
Type 3	(3)	(6), (7)	PHEV, washers
Type 4	(4)	(6), (8)	HVAC, refrigerator
Battery	$-D_i(r_i)$	(6), (7)	$r_i = q_{ia}$ for battery a

The utility of a type 3 appliance depends only on the aggregate consumption:

$$U_{ia}(q_{ia}) := U_{ia}\left(\sum_t q_{ia}(t)\right). \tag{3}$$

The utility of a type 4 appliance depends on the internal temperature and power consumptions in the past. It is of the form

$$U_{iq}(q_{ia}) := \sum_t U_{ia}\left(T_{ia}(t) + \beta \sum_{\tau=1}^{t}(1-\alpha)^{t-\tau}q_{ia}(\tau)\right), \tag{4}$$

where $T_{ia}(t)$ is a given sequence of temperatures defined in (30) in Appendix A and α, β are given thermal constants. All utility functions are assumed to be continuously differentiable and concave functions for each t.

For example, some of our simulations in [21, 39] use the following time independent and additive utility function of form (2): let $y_{ia}(t)$ be a desired energy consumption by appliance a in period t; then the function

$$U_{ia}(q_{ia}(t), t) := U_{ia}(q_{ia}(t)) := -(q_{ia}(t) - y_{ia}(t))^2 \tag{5}$$

measures the utility of following the desired consumption profile $y_{ia}(t)$. Such utility functions minimize user discomfort as advocated in [8,9].

The consumption constraints (1) for these appliances take three particular forms. First, for all appliances, the (real) power consumption must lie between a lower and an upper bound, possibly time dependent:

$$\underline{q}_{ia}(t) \leq q_{ia}(t) \leq \overline{q}_{ia}(t). \tag{6}$$

An important character of an appliance is its allowable time of operation; e.g., an EV can be charged only between 9 pm and 6 am, TV may be on only between 7–9 am and 6–12 pm. If an appliance operates only in a subset $\mathscr{T}_{ia} \subseteq \mathscr{T}$ of periods, we require that $\underline{q}_{ia}(t) = \overline{q}_{ia}(t) = 0$ for $t \notin \mathscr{T}_{ia}$ and $U_{ia}(0) = 0$. We therefore do not specify $\mathscr{T}_{ia}$ explicitly in the description of utility functions and always sum over all $t \in \mathscr{T}$. The second kind of constraint specifies the range in which the aggregate consumption must lie

$$\underline{Q}_{ia} \leq \sum_t q_{ia}(t) \leq \overline{Q}_{ia}. \tag{7}$$

The last kind of constraint is slightly more general (see derivation in Appendix A)

$$\underline{\eta}_{ia} \leq A_{ia} q_{ia} \leq \bar{\eta}_{ia}. \tag{8}$$

Battery model: We denote by B_i the battery capacity, by $b_i(t)$ the state of charge in period t, and by $r_i(t)$ the power (energy per period) charged to (when $r_i(t) \geq 0$) or discharged from (when $r_i(t) < 0$) the battery in period t. We use a simplified model of battery that ignores power leakage and other inefficiencies, where the state of charge is given by

$$b_i(t) = \sum_{\tau=1}^{t} r_i(\tau) + b_i(0). \tag{9}$$

The battery has an upper bound on charge rate, denoted by $\bar{r}_i$, and an upper bound on discharge rate, denoted by $-\underline{r}_i$. We thus have the following constraints on $b_i(t)$ and $r_i(t)$:

$$0 \leq b_i(t) \leq B_i, \quad \underline{r}_i \leq r_i(t) \leq \bar{r}_i. \tag{10}$$

We assume any battery discharge is consumed by other appliances (zero leakage), and hence it cannot be more than what the appliances need

$$-r_i(t) \leq \sum_{a \in \mathcal{A}_i} q_{ia}(t). \tag{11}$$

Finally, we impose a minimum on the energy level at the end of the control horizon: $b(T) \geq \gamma_i B_i$ where $\gamma_i \in [0, 1]$.

The cost of operating the battery is modeled by a function $D_i(r_i)$ that depends on the vector of charged/discharged power $r_i := (r_i(t), \forall t)$. This cost may correspond to the amortized purchase and maintenance cost of the battery over its lifetime, and depends on how fast/much/often it is charged and discharged; see an example $D_i(r_i)$ in [39]. The cost function D_i is assumed to be a convex function of the vector r_i.

Note that in this model, a battery is equivalent to an appliance: its utility function is $-D_i(r_i)$ and its consumption constraints (9), (10), and $b(T) \geq \gamma_i B_i$ are of the same form as (6) and (7) with $q_{ia} = r_i$. Therefore, a battery can be specified simply as another appliance, in which case the constraint (11) requires that i's aggregate demand be non-negative, $\sum_{a \in \mathcal{A}_i} q_{ia}(t) + r_i(t) \geq 0$. This is summarized in Table 1. Henceforth, we will often use appliances to also include battery and may not refer to battery explicitly when this does not cause confusion.

2.2 Supply Model

We now describe a simple model of the electricity markets. The LSE procures power for delivery in each period t, in two steps. First, it procures day-ahead capacities

$P_d(t)$ for each period t a day in advance and pays for the capacity costs $c_d(P_d(t);t)$. The renewable power in each period t is a non-negative random variable $P_r(t)$ and it costs $c_r(P_r(t);t)$. It is desirable to use as much renewable power as possible, for instance, if the renewable generation is owned by the LSE. For notational simplicity only, we assume $c_r(P;t) \equiv 0$ for all $P \geq 0$ and all t. Then at time t^- (real time), the random variable $P_r(t)$ is realized and used to satisfy demand. The LSE satisfies any excess demand by some or all of the day-ahead capacity $P_d(t)$ procured in advance and/or by purchasing balancing power on the real-time market. Let $P_o(t)$ denote the amount of the day-ahead power that the LSE actually uses and $c_o(P_o(t);t)$ its cost. Let $P_b(t)$ be the real-time balancing power and $c_b(P_b(t);t)$ its cost.

These real-time decisions $(P_o(t), P_b(t))$ are made by the LSE so as to minimize its total cost, as follows. Given the demand vector $q(t) := (q_{ia}(t), a \in \mathscr{A}_i, \forall i)$, let $Q(t) := \sum_{i,a} q_{ia}(t)$ be the total demand and $\Delta(Q(t)) := Q(t) - P_r(t)$ the excess demand, in excess of the renewable generation $P_r(t)$. Note that $\Delta(Q(t))$ is a random variable in and before period $t - 1$, but its realization is known to the LSE at time t^-. Given excess demand $\Delta(Q(t))$ and day-ahead capacity $P_d(t)$, the LSE chooses $(P_o(t), P_b(t))$ that minimizes its total real-time cost, i.e., it chooses $(P_o^*(t), P_b^*(t))$ that solves the problem:

$$c_s(\Delta(Q(t)), P_d(t);t) := \min_{P_o(t), P_b(t)} \{ c_o(P_o(t);t) + c_b(P_b(t);t) \mid P_b(t) \geq 0,$$

$$P_o(t) + P_b(t) \geq \Delta(Q(t)), \ P_d(t) \geq P_o(t) \geq 0\}. \quad (12)$$

Clearly, $P_o^*(t) + P_b^*(t) = \Delta(Q(t))$ unless $\Delta(Q(t) < 0$. The total cost is

$$c(Q(t), P_d(t); P_r(t), t) := c_d(P_d(t);t) + c_s(\Delta(Q(t)), P_d(t);t) \quad (13)$$

with $\Delta(Q(t)) := Q(t) - P_r(t)$. We assume that, for each t, $c_d(\cdot;t), c_o(\cdot;t)$ and $c_b(\cdot;t)$ are increasing, convex, and continuously differentiable with $c_d(0;t) = c_o(0;t) = c_b(0;t) = 0$.

Example. supply cost: Suppose $c_b'(0) > c_o'(P), \forall P \geq 0$, i.e., the marginal cost of balancing power is strictly higher than the marginal cost of day-ahead power, the LSE will use the balancing power only after the day-ahead power is exhausted, i.e., $P_b(t) > 0$ only if $\Delta(Q(t)) > P_d(t)$. The solution $c_s(\Delta(Q(t)), P_d(t);t)$ of (12) in this case is particularly simple and (13) can be written explicitly in terms of c_b, c_o, c_b:

$$c(Q(t), P_d(t); P_r(t), t) = c_d(P_d(t);t) + c_o\left([\Delta(Q(t))]_0^{P_d(t)};t\right)$$

$$+ c_b\left([\Delta(Q(t)) - P_d(t)]_+;t\right), \quad (14)$$

i.e., the total cost consists of the capacity cost c_d and the energy cost c_o of day-ahead power, and the cost c_b of the real-time balancing power.

2.3 Problem Formulation: Welfare Maximization

Recall that $q := (q(t), t \in \mathscr{T})$ and $Q(t) := \sum_{i,a} q_{ia}(t)$. The social welfare is the standard user utility minus supply cost:

$$W(q, P_d; P_r) := \sum_{i,a} U_{ia}(q_{ia}) - \sum_{t=1}^{T} c(Q(t), P_d(t); P_r(t), t). \qquad (15)$$

As mentioned above, the LSE's objective is not to maximize its profit through selling electricity, but rather to maximize the expected social welfare. Given the day-ahead decision P_d, the real-time procurement $(P_o(t), P_b(t))$ is determined by the simple optimization (13). This is most transparent in (14) for the special case: the optimal decision is to use day-ahead power $P_o^*(t)$ to satisfy any excess demand $\Delta(Q(t))$ up to $P_d(t)$, and then purchase real-time balancing power $P_b^*(t) = [\Delta(Q(t)) - P_d(t)]_+$ if necessary. Hence, the maximization of (15) reduces to optimizing over day-ahead procurement P_d and real-time consumption q in the presence of random renewable generation $P_r(t)$. It is therefore critical that, in the presence of uncertainty, $q(t)$ should be decided after $P_r(t)$ have been realized at times t^-. P_d, however, must be decided a day ahead before $P_r(t)$ are realized.

The traditional dynamic programming model requires that the objective function be separable in time t. The welfare function in (15) is not as the first term $U_{ia}(q_{ia})$ depends on the entire control sequence $q_{ia} = (q_{ia}(t), \forall t)$. So does the consumption constraint (1). We now introduce an equivalent state space formulation of that will allow us to state precisely the overall optimization problem as an $(1 + T)$-period dynamic program.

Consider a dynamical system over an extended time horizon $t = 0, 1, \ldots, T$. The control inputs are the LSE's day-ahead decision $P_d := (P_d(t), \forall t)$ in period 0 and the user's decisions $q(t)$ in each subsequent period. Let $v(t)$ denote the inputs, i.e., $v(0) = P_d$ and $v(t) = q(t)$, $t = 1, \ldots, T$. Note that $v(0) \in \Re_+^T$ whereas $q(t) \in \Re^M$, where $M := \sum_{i=1}^{N} |\mathscr{A}_i|$. The system state $x(t) := (x^1(t), x_{ia}^2(t), x^3(t), x_{ia}^4(t), a \in \mathscr{A}_i, \forall i)$ has four components, defined as follows:

- Without loss of generality, $x(0)$ starts from the origin.
- $x^1(t) \in \Re^T$ keeps track of the day-ahead decisions P_d: for each $t = 1, \ldots, T$, $x^1(t) = P_d = (P_d(\tau), \tau = 1, \ldots, T)$.
- $x_{ia}^2(t) \in \Re^{k_{ia}}$ of appropriate dimension k_{ia} for each (i, a) pair keeps track of the consumption constraint (1). The state definition and its transition are problem specific; see a concrete example in Sect. 2.4.
- $x^3(t) \in \Re_+$ keeps track of the random renewable power $x^3(0) = 0$, $x^3(t) = P_r(t), t = 1, \ldots, T$. The purpose of this state definition is merely notational, so that the control policy can depend on the *realization* of the random renewable power $P_r(t)$ through its dependence on state $x^3(t)$.

- $x_{ia}^4(t) \in \mathfrak{R}^{T-1}$ for each (i,a) pair tracks the user decisions $v_{ia}(t-1) = q_{ia}(t-1)$ in the previous period: $x_{ia}^4(1) = 0_{T-1}$, the $T-1$ dimensional zero vector; for each $t = 2,\ldots,T$, the $(t-1)$th component $[x_{ia}^4(t)]_{t-1}$ of $x_{ia}^4(t)$ is set to be the input $v_{ia}(t-1)$ and all the other components $[x_{ia}^4(t)]_\tau$ of $x_{ia}^4(t)$ remain the same as those of $x_{ia}^4(t-1)$, so that the final state $x_{ia}^4(T)$ is the vector $(q_{ia}(t), t = 1,\ldots,T-1)$ of inputs up to period $T-1$. The first term in (15) is then a function of the state and input in period T, $U_{ia}(q_{ia}) = U_{ia}(x_{ia}^4(T), v_{ia}(T))$. This allows us to rewrite the welfare function in (15) in a form that is separable in t; see below.

The above discussion is summarized by a time-varying state transition function f_t:

$$x(t+1) = f_t(x(t), v(t), P_r(t+1)), \quad t = 0,\ldots,T,$$

i.e., the new state $x(t+1)$ depends on the current state $x(t)$, the input $v(t)$, and the new random variable $P_r(t)$, and is therefore random. The consumption constraints (1), which may include the battery constraints, generally translate into constraints on the state $x^2(t)$ and input $v(t)$ and we represent this by $x(t) \in \mathcal{X}(t)$ and $v(t) \in \mathcal{V}(t) \subseteq \mathfrak{R}^M$, $M := \sum_{i=1}^N |\mathcal{A}_i|$. Sometimes, these constraints also give rise to a terminal reward that we denote by $W_{T+1}(x(T+1))$.

Consider the class of feedback control laws $v(t) = \phi_t(x(t))$, where $\phi_0 : \mathcal{X}(0) \to \mathfrak{R}_+^T$ specifies the day-ahead decision P_d and $\phi_t : \mathcal{X}(t) \to \mathcal{V}(t)$ specifies the user decisions $q(t)$ for each period $t = 1,\ldots,T$. Hence, the control $v(t)$ depends only on the current state $x(t)$. Under the control law $\phi := (\phi_t, t = 0,\ldots,T)$, the state evolves (stochastically) according to

$$x(t+1) = f_t(x(t), \phi_t(x(t)), P_r(t+1)). \tag{16}$$

We emphasize that $x(t)$ is obtained under policy ϕ even though this may not be explicit in the notation.

To make the welfare function in (15) separable in t, use (13) to define the welfare in each period t, under the control law ϕ, as a function of the current state $x(t)$ and the current input $v(t) = \phi_t(x(t))$:

$$W_t^\phi := W_t^\phi(x(t), v(t))$$

$$:= \begin{cases} -\sum_{\tau=1}^T c_d([v(0)]_\tau; \tau), & t = 0, \\ -c_s(\Delta(Q^\phi(t)), [x^1(t)]_t; t)), & 1 \le t < T, \\ \sum_{i,a} U_{ia}((x_{ia}^4(T), v_{ia}(T))) - c_s(\Delta(Q^\phi(T)), [x^1(T)]_T; T), & t = T, \end{cases} \tag{17}$$

where $Q^\phi(t) = \sum_{i,a}[v(t)]_{ia}$ is the aggregate demand in period t under ϕ, and $v_{ia}(T) = q_{ia}(T)$ are the real-time consumption decisions in the last control period T. Then the welfare function in (15) is equivalent to

$$J^\phi := \sum_{t=0}^T W_t^\phi(x(t), v(t)) + W_{T+1}^\phi(x(T+1)),$$

where the definition of the terminal reward $W^{\phi}_{T+1}(x(T+1))$ is problem specific. We can now state precisely our objective as the constrained maximization of the expected welfare over the control law ϕ:

$$\max_{\phi} E J^{\phi} = E\left(\sum_{t=0}^{T} W^{\phi}_t + W^{\phi}_{T+1}\right) \quad \text{s.t.} \quad x^{\phi}(t) \in \mathscr{X}(t), \tag{18}$$

where the expectation is taken over $P_r(t), t = 1, \ldots, T$.

Remark. An important assumption in this formulation is that the consumption constraints (1) can be modeled by an appropriate definition of states $x^2_{ia}(t)$, their transitions f_t, the constraint sets $\mathscr{X}(t), \mathscr{V}(t)$, and possibly a terminal reward $W_{T+1}(x(T+1))$.

We now illustrate the problem formulation using a concrete example.

2.4 Example

To simplify the notation, we make two assumptions that do not cause any loss of generality. First, we use the total cost function c in (14) in the definition of the welfare function (15). Second, we assume each user i has a single type-2 appliance and no battery (so we drop the subscript a). From Table 1, user utility functions are additive in time, $U_i(q_i) = \sum_t U_i(q_i(t); t)$ and the consumption constraints are

$$\underline{q}_i(t) \le q_i(t) \le \overline{q}_i(t), \quad \forall i, \tag{19}$$

$$\overline{Q}_i \le \sum_{t=1}^{T} q_i(t). \tag{20}$$

Since the utility functions are separable in t, we do not need to define $x^4(t)$. We now describe the $(1 + T)$-period dynamic program by specifying the definition of $x^2(t)$, the state transition function f_t, and the constraint sets $\mathscr{X}(t), \mathscr{V}(t)$.

The system state $x(t) := (x^1(t), x^2(t), x^3(t))$ consists of three components of appropriate dimensions with

$$x(t) = (P_d, x^2(t), P_r(t)), \quad t = 1, \ldots, T,$$

where $x^2(t)$ is determined by the constraint (20). To simplify exposition, we make the important assumption that $P_r(t)$ are independent for different t; see [21] for a model without this independence assumption. Define $x^2_i(t)$ to be the remaining demand of user i at the beginning of each period t: $x^2_i(1) = \overline{Q}_i$, and for each $t = 1, \ldots, T$, $x^2_i(t + 1) = x^2_i(t) - v_i(t)$, where $v_i(t) = q_i(t)$. To enforce that $x^2(T+1) \le 0$, we define the terminal cost $c_{T+1}(x(T+1)) = 0$ if $x^2(T+1) \le 0_N$

and $c_{T+1}(x(T+1)) = \infty$ otherwise, where 0_n is the n-dimensional zero vector. Let the initial state be $x(0) = 0_{T+N+1}$. Denote $\overline{Q} := (\overline{Q}_i, \forall i)$. The system dynamics is then linear time-varying

$$x(1) = x(0) + \begin{pmatrix} I_T \\ 0_{(N+1)\times T} \end{pmatrix} v(0) + \begin{pmatrix} 0_T \\ \overline{Q} \\ P_r(1) \end{pmatrix}$$

$$x(t+1) = \begin{pmatrix} I_{T+N} & 0_{T+N} \\ 0_{T+N} & 0 \end{pmatrix} x(t) - \begin{pmatrix} 0_{T\times N} \\ I_N \\ 0 \end{pmatrix} v(t) + \begin{pmatrix} 0_{T+N} \\ 1 \end{pmatrix}$$

$$\times P_r(t+1), \quad \forall 1 \le t \le T,$$

where I_n is the $n \times n$ identify matrix, $0_{m \times n}$ the $m \times n$ zero matrix, and $P_r(T+1) := 0$.

The welfare in each period, under input sequence v, is (using (14))

$$W_0^v(x(0), v(0)) := -\sum_{\tau=1}^{T} c_d(P_d(\tau); \tau) = -\sum_{\tau=1}^{T} c_d([v(0)]_\tau; \tau),$$

and for $t = 1, \ldots, T$,

$$W_t^v(x(t), v(t)) := \sum_i U_i(q_i(t); t) - c_o\left([Q(t) - P_r(t)]_0^{P_d(t)}; t\right)$$

$$- c_b\left([Q(t) - P_r(t) - P_d(t)]_+; t\right),$$

$$= \sum_i U_i(v_i(t); t) - c_o\left([\mathbf{1}v(t) - x^3(t)]_0^{[x^1(t)]_t}; t\right)$$

$$- c_b\left([\mathbf{1}v(t) - x^3(t) - [x^1(t)]_t]_+; t\right),$$

where $\mathbf{1}$ is the (row) vector of 1's.

The constraint (19) yields the input constraint sets $\mathcal{V}(0) := \Re_+^T$ and, for $t = 1, \ldots, T$, $\mathcal{V}(t) := \{q(t) \in \Re^N | \underline{q}(t) \le q(t) \le \overline{q}(t)\}$. There is no constraint on the state, i.e., $\mathcal{X}(t) = \Re^{T+N+1}$. Let $\phi := \{\phi_0 : \Re^{T+N+1} \to \Re_+^T, \phi_t : \Re^{T+N+1} \to \mathcal{V}(t), t = 1, \ldots, T\}$ be the control policy so that $v(t) = \phi_t(x(t)), 0 \le t \le T$. Then the welfare maximization problem (18) is

$$\max_\phi E\left(W_0^\phi(x(0), v(0)) + \sum_{t=1}^{T} W_t^\phi(x(t), v(t)) - c_{T+1}(x(T+1))\right), \quad (21)$$

where the state $x(t)$ and the input $v(t)$ are obtained under policy ϕ.

In [21], we study the case with supply uncertainty in detail. We propose a distributed heuristic algorithm to solve the $(1 + T)$-period dynamic program.

We prove that the algorithm is optimal when the welfare is quadratic and the LSEs procurement decisions are strictly positive. Otherwise, we bound the gap between the welfare achieved by the heuristic algorithm and the maximum. Simulation results suggest that the performance of the heuristic algorithm is very close to optimal. As we scale up the size of a renewable generation plant, both its mean production and its variance will likely increase. As expected, the maximum welfare increases with the mean production, when the variance is fixed, and decreases with the variance, when the mean is fixed. More interesting, we prove that as we scale the size of the plant up, the maximum welfare increases.

3 Optimal Scheduling Without Supply Uncertainty

In this chapter, we only fully treat the case where there is no supply uncertainty, i.e., $P_r(t) \equiv 0$. Our goal is to optimally coordinate supply and demand to maximize social welfare. In the absence of uncertainty (our model also ignores demand uncertainty), it becomes unnecessary to adapt user consumptions in real-time and hence supply and consumptions can be optimally scheduled at once instead of over two days. Welfare maximization (18) then takes a simpler form and we develop an offline distributed algorithm that jointly optimizes the LSE's procurements and the users' consumptions for each period in the following day.

3.1 Optimal Procurements and Consumptions

We first consider LSE's procurement decisions. Recall that $Q_i(t) := \sum_{a \in \mathscr{A}_i} q_{ia}(t)$ and $\sum_i Q_i(t)$ is the aggregate demand in period t. With supply uncertainty, while P_d is decided a day ahead, the optimization (12) must be carried out in real time after $P_r(t)$ has been realized to obtain optimal $P_o(t)$, $P_b(t)$. Here, on the other hand, all three decisions $(P_d(t), P_o(t), P_b(t))$ can be computed in advance in the absence of uncertainty. Hence, given an aggregate demand $\sum_i Q_i(t)$, the LSE solves (instead of (12) and (13)):

$$c\left(\sum_i Q_i(t); t\right) := \min_{P_d(t), P_o(t), P_b(t)} c_d(P_d(t); t) + c_o(P_o(t); t) + c_b(P_b(t); t)$$

$$\text{s.t.} \quad P_o(t) + P_b(t) \geq \sum_i Q_i(t), \quad P_d(t) \geq P_o(t) \geq 0,$$

$$P_b(t) \geq 0 \tag{22}$$

to obtain the total cost. The solution of (22) specifies the optimal decisions $(P_d^*(t), P_o^*(t), P_b^*(t))$ to satisfy the aggregate demand $\sum_i Q_i(t)$ for each period t in the following day.

It is not difficult to show that $c(\cdot, t)$ is a nondecreasing, convex, and continuously differentiable function for each t, so the problem (22) is convex. Since $c_d'(P; t) > 0$, the KKT condition implies that $P_d^*(t) = P_o^*(t)$ at optimality, i.e., it is optimal to exhaust all the day-ahead capacity. This is always possible because all procurement decisions are computed jointly without uncertainty. If we further assume that the marginal cost of the balancing power is higher than that of the day-ahead power, i.e., $c_b'(0; t) > c_d'(P; t) + c_o'(P; t)$ for all $P \geq 0$, then KKT implies that it will never pay to use balancing power, i.e., $P_b^*(t) = 0$ at optimality. In this case, $P_d^*(t) = P_o^*(t) = \sum_i Q_i(t)$.

Hence, welfare maximization reduces to the computation of the user consumptions $q_{ia}(t)$; the corresponding procurement decisions are then given by (22). The optimization of the social welfare in (15) then becomes

$$\max_q \sum_{i,a} U_{ia}(q_{ia}) - \sum_t c\left(\sum_i Q_i(t); t\right) \tag{23}$$

$$\text{s.t.} \quad A_{ia}q_{ia} \leq \eta_{ia}, \quad a \in \mathcal{A}_i, \forall i, \tag{24}$$

$$0 \leq Q_i(t) \leq \overline{Q}_i, \quad \forall i. \tag{25}$$

The inequalities in (24) are the consumption constraints (1) of user i's appliances and battery. The lower inequality in (25) is the same as (11); see the discussion at the end of Sect. 2.1 on battery constraints. The upper inequality in (25) imposes a bound on the total power drawn by user i. By assumption, the objective function is concave and the feasible set is convex. Hence, an optimal point can in principle be computed offline centrally by the LSE. This however will require that the LSE know all the users' utility and battery cost functions and all the constraints, which is impractical for technical or privacy reasons. The goal of this section is to derive a distributed algorithm to solve (23)–(25) by decomposing it into subproblems that are solvable in a decentralized manner, where the LSE only needs to know the aggregate demand but not the individual private information.

The key idea is for the LSE to set prices $\pi := (\pi(t), \forall t)$ to induce the users to individually choose socially optimal consumptions $q_i := (q_{ia}(t), \forall t)$ in response. Indeed, given prices π, we assume that each user i chooses its own demand q_i so as to maximize its net benefit, her total utility minus the electricity cost, i.e., each user i solves

$$\max_{q_i} \sum_{a \in \mathcal{A}_i} U_{ia}(q_{ia}) - \sum_t \pi(t)Q_i(t) \quad \text{s.t.} \quad (24)–(25). \tag{26}$$

Given prices π, we denote an *individually* optimal solution of (26) and the corresponding aggregate demand by

$$q_i(\pi) := (q_{ia}(t; \pi), \forall t, \forall a \in \mathcal{A}_i), \quad Q_i(\pi) := (Q_i(t; \pi), \forall t)$$

$$:= \left(\sum_{a \in \mathcal{A}_i} q_{i,a}(t; \pi), \forall t \right).$$

Recall $q(\pi) := (q_i(\pi), \forall i)$. It is a remarkable fact in the competitive equilibrium theory in economics that there exist prices π that align the individual optimality with the social optimality, i.e., there are prices π^* such that if $q_i(\pi^*)$ optimize i's objectives for all users i then they also optimize the social welfare.

Definition 4. A consumption vector q^* is called (socially) *optimal* if it solves (23)–(25). A price vector π^* is called *optimal* if $q(\pi^*)$ is optimal, i.e., any solution $q(\pi^*)$ of (26) also solves (23)–(25).

The following result follows from the welfare theorem in economics. It implies that setting the prices to the marginal costs of power is optimal.

Theorem 5. *The prices that satisfy $\pi^*(t) := c' \left(\sum_i Q_i(t; \pi^*); t \right) \geq 0$ exist and are optimal.*

Proof. Write the welfare maximization problem as

$$\max_{q_i \in \mathcal{Q}_i, Y_i} \sum_{i,a} U_{ia}(q_{ia}) - \sum_t c \left(\sum_i Y_i(t); t \right) \quad \text{s.t.} \quad Y_i(t) = \sum_{a \in \mathcal{A}_i} q_{ia}(t), \quad \forall i, t,$$

where the feasible set $\mathcal{Q}_i$ is defined by the constraints (24) and (25). Clearly, an optimal solution q^* exists. Moreover, there exist Lagrange multipliers $\pi_i^*(t), \forall i, t$, such that (taking derivative with respect to $Y_i(t)$)

$$\pi_i^*(t) = c' \left(\sum_i Y_i^*(t); t \right) = c' \left(\sum_i \sum_{a \in \mathcal{A}_i} q_{ia}^*(t); t \right) \geq 0.$$

Since the right-hand side is independent of i, the LSE can set the prices as $\pi^*(t) := \pi_i^*(t) \geq 0$ for all i. One can check that the KKT condition for the welfare maximization problem is identical to the KKT conditions for the collection of users' problems. Since all these problems are convex, the KKT conditions are both necessary and sufficient for optimality. This proves the theorem.   □

3.2   Offline Distributed Scheduling Algorithm

Theorem 5 motivates a distributed algorithm to compute the optimal prices π^* and user decisions $q(\pi^*)$. The LSE sets prices to be the marginal costs of power and each user solves its own maximization problem (26) in response. The model is that at the beginning of each day the LSE and (the energy management systems of) the users iteratively compute the electricity prices $\pi(t)$ and consumptions $q_i(t)$ for each period t of the following day. These decisions are then carried out for that day. This is an offline algorithm since all decisions are made at once before the day starts. It is decentralized where the LSE only knows the aggregate demand but not user utility functions or consumption constraints and the users do not need to coordinate among themselves but only respond to common prices.

Algorithm 1: Optimal scheduling without supply uncertainty

For each iteration $k = 1, 2, \ldots$, after initialization:

1. The LSE collects aggregate demand forecasts, denoted by $(Q_i^k(t), \forall t)$, from all users i over a communication network. It updates the prices to the marginal costs $\pi^{k+1}(t) := c'\left(\sum_i Q_i^k(t); t\right)$ and broadcasts $\pi^{k+1} := (\pi^{k+1}(t), \forall t)$ to all users.
2. Each user i updates its demands q_i^{k+1} after receiving π^{k+1} according to

$$\tilde{q}_{ia}^{k+1}(t) = q_{ia}^k(t) + \gamma \left(\frac{\partial U_{ia}(q_i^k)}{\partial q_{ia}^k(t)} - \pi^{k+1}(t) \right)$$

$$q_{ia}^{k+1} = \left[\tilde{q}_{ia}^{k+1} \right]_{\mathcal{Q}_i}$$

where $\gamma > 0$ is a constant stepsize, $\tilde{q}_{ia}^{k+1} := (\tilde{q}_{ia}^{k+1}(t), \forall t)$ is the new consumption vector before being projected onto the feasible set $\mathcal{Q}_i$ specified by constraints (24)–(25), and $[\cdot]_{\mathcal{Q}_i}$ denotes the projection. User i's aggregate demand forecast in period t is updated to $Q_i^{k+1}(t) = \sum_{a \in \mathcal{A}_i} q_{ia}^{k+1}(t)$.
3. Increment iteration index to $k + 1$ and goto Step 1.

Algorithm 1 converges asymptotically to optimal prices π^* and optimal consumptions $q(\pi^*)$, provided the stepsize $\gamma > 0$ is small enough. More precisely, suppose:

- A1: The utility functions $U_{ia}(q_{ia})$ are strictly concave in the vector $q_{ia} := (q_{ia}(t), \forall t)$ for all i, a.
- A2: The feasible set of q defined by the consumption constraints (24) and (25) is compact. All our user models in Sect. 2.1 satisfy this condition because of (6).
- A3: Suppose the spectral radius of the Hessian matrix $\nabla^2 U_{ia}$ and the second derivative $c''(\cdot; t)$ are both uniformly bounded: $\|\nabla^2 U_{ia}(q_{ia})\|_2 < \rho$ for all q_{ia} for all i, a, and $c''(Q; t) < \alpha$ for all Q, t.

Theorem 6. *Under the assumptions A1–A3, the sequence (π^k, q^k) generated by Algorithm 1 converges to the optimal price and consumption vectors $(\pi^*, q(\pi^*))$, provided $\gamma < 2/\left(\rho + \alpha \sum_i |\mathcal{A}_i|\right)$.*

Proof. Let the welfare function be

$$h(q) := \sum_{i,a} U_{ia}(q_{ia}) - \sum_t c\left(\sum_i Q_i(t); t\right).$$

Then $h(q)$ is strictly concave since $U_{ia}(q_{ia})$ are strictly concave. The gradient $\nabla h(q)$ has components

$$[\nabla h(q)]_{ia}(t) = \frac{\partial U_{ia}(q_i)}{\partial q_{ia}(t)} - c'\left(\sum_i Q_i(t); t\right). \tag{27}$$

Hence Algorithm 1 is a gradient projection algorithm where in each iteration k, the variable q^k is updated to q^{k+1} according to

$$q^{k+1} = \left[q^k + \gamma \nabla h(q^k)\right]_{\mathcal{Q}},$$

where $\mathcal{Q} := \mathcal{Q}_1 \times \cdots \times \mathcal{Q}_N$. Moreover, assumption A3 implies the following lemma, proved in Appendix B.   □

Lemma 6. $\nabla h(q)$ is Lipschitz with $\|\nabla h(q) - \nabla h(\tilde{q})\|_2 < \left(\rho + \alpha \sum_i |\mathscr{A}_i|\right) \|q - \tilde{q}\|_2$ for all $q, \tilde{q}$.

Lemma 6 implies that, provided $\gamma < 2/\left(\rho + \alpha \sum_i |\mathscr{A}_i|\right)$, any accumulation point q^* of the sequence q^k generated by Algorithm 1 is optimal, i.e., maximizes welfare $h(q)$ [40, p. 214]. Assumption A2 implies that the sequence q^k lies in a compact set and hence must have a convergent subsequence. But assumption A1 implies that the optimal q^* is unique. Therefore all convergent subsequences, hence the original sequence q^k, must converge to q^*. By continuity of c', $\pi^k(t) = c'(\sum_i Q_i^k(t); t)$ converges to the unique price $c'(\sum_i Q_i^*(t); t)$ with $Q_i^*(t) := \sum_{a \in \mathscr{A}_i} q_{ia}^*(t)$ which, by Theorem 5, is optimal.

The rate of convergence of Algorithm 1 depends on the stepsize γ: a larger γ generally leads to faster convergence, but a large γ can also risk instability. The bound on the stepsize γ in Theorem 6 is conservative; in practice a much larger stepsize can usually be used without losing stability. We simulate this algorithm in [39] with realistic system parameters. The simulation results show that, as expected, the prices are capable of coordinating the decisions of different appliances in a decentralized manner, to reduce peak aggregate demand and flatten its profile, greatly increasing the load factor. Furthermore, battery amplifies the benefits of demand response.

4 Conclusion

We have presented a simple yet versatile user model and formulated the optimal demand response problem as an $(1 + T)$-period dynamic program to maximize the expected social welfare. In this chapter, we have focused on the case where there is no uncertainty. In this case, demand response reduces to the deterministic welfare maximization in (23)–(25) that has a natural decentralized and incentive-compatible structure. We have proposed an offline distributed scheduling algorithm where the LSE sets the day-ahead prices to be their marginal costs based on forecast demands and, in response, the users forecast their demands to maximize their own surplus. As long as the stepsize is small enough, this procedure will converge to the unique optimal prices and consumptions. The algorithm is decentralized where the LSE only knows the aggregate demand but not user utility functions or consumption constraints, and the users do not need to coordinate with other users but only respond to the common prices from the LSE.

The current work has several limitations. First, our model does not include the distribution system, implicitly assuming that the underlying network has enough capacity to distribute the power demanded by the users without causing congestion. Second, we only consider power balance in steady-state and ignore fast timescale dynamics such as frequency and voltage fluctuations due to random supply and demand. Third, we do not model power market dynamics; for example, our model assumes that the cost functions faced by the LSE are independent of the demands and we ignore economic issues such as revenue-adequacy for the LSE. Finally, our results are only for the case without uncertainty. When there is random renewable generation, offline scheduling alone will be insufficient and real-time demand response should be employed to match fluctuating supply. This is considered in [21].

Appendix A: Detailed Appliance Models

We describe detailed models of common electric appliances summarized in Sect. 2.1.

Type 1: This category of appliances includes lighting that must be on for a certain period of time. The consumption constraint is (6), with the understanding that $\underline{q}_{ia}(t) = \overline{q}_{ia}(t) = 0$ for periods t that are outside its time of operation. User i attains a utility $U_{ia}(q_{ia}(t), t)$ from consuming power $q_{ia}(t)$ independent of its consumption in other periods, and the overall utility (2) is therefore separable in t.

Type 2: This category includes TV, video games, and computers. For these appliances, a user's utility depends on her consumption in each period she wishes to use it as well as the total amount of consumption in a day. Hence, the consumption constraints are (6) and (7). For example, a user may have a favorite TV program that she wishes to watch everyday. With DVR, she can watch the program at any time. However, the total power demand of TV should at least cover the program. Type 2 appliances have the same kind of utility functions (2) as Type 1 appliances. The time dependent utility function models the fact that a user may get different benefits from consuming the same amount of power at different times, e.g., she may enjoy a TV program to different levels at different times.

Type 3: This category includes PHEV, dish washer, clothes washer. For these appliances, a user only cares about whether the task is completed by a certain time. This means that the aggregate power consumption by such an appliance must exceed a threshold within its time of operation [28, 29, 33]. Hence, the consumption constraints are (6) and (7). The utility depends only on the total power consumed, hence (3).

Type 4: This category includes HVAC (heating, ventilation, air conditioning) and refrigerator that control the temperature of a user's environment. Let $T_{ia}^{in}(t)$ and $T_{ia}^{out}(t)$ denote the temperatures at time t inside and outside the place that appliance (i, a) is in charge of, and $\mathcal{T}_{ia}$ denotes the set of times when user i cares about the temperature. For instance, for air conditioner, $T_{ia}^{in}(t)$ is the temperature inside the house, $T_{ia}^{out}(t)$ is the temperature outside the house, and $\mathcal{T}_{ia}$ is the set of times when she is at home.

The inside temperature evolves according to the following linear dynamics [9, 26, 27]:

$$T_{ia}^{in}(t) = T_{ia}^{in}(t-1) + \alpha(T_{ia}^{out}(t) - T_{ia}^{in}(t-1)) + \beta q_{ia}(t), \tag{28}$$

where α and β are parameters that specify thermal characteristics of the appliance and the environment in which it operates. The second term in (28) models heat transfer. The third term models the thermal efficiency of the system; $\beta > 0$ if appliance a is a heater and $\beta < 0$ if it is a cooler. Here, we define $T_{ia}^{in}(0)$ as the temperature $T_{ia}^{in}(T)$ from the previous day. Let $[\underline{T}_{ia}, \overline{T}_{ia}]$ be a range of preferred temperature, leading to the constraint:

$$\underline{T}_{ia} \leq T_{ia}^{in}(t) \leq \overline{T}_{ia}, \quad \forall t \in \mathcal{T}_{ia}. \tag{29}$$

Using (28), we can write $T_{ia}^{in}(t)$ in terms of $(q_{ia}(\tau), \tau = 1, \dots, t)$:

$$T_{ia}^{in}(t) = (1-\alpha)^t T_{ia}^{in}(0) + \sum_{\tau=1}^{t}(1-\alpha)^{t-\tau}\alpha T_{ia}^{out}(\tau) + \beta \sum_{\tau=1}^{t}(1-\alpha)^{t-\tau}q_{ia}(\tau).$$

Define

$$T_{ia}(t) := (1-\alpha)^t T_{ia}^{in}(0) + \sum_{\tau=1}^{t}(1-\alpha)^{t-\tau}\alpha T_{ia}^{out}(\tau). \tag{30}$$

Then

$$T_{ia}^{in}(t) = T_{ia}(t) + \beta \sum_{\tau=1}^{t}(1-\alpha)^{t-\tau}q_{ia}(\tau). \tag{31}$$

With (31), the constraint (29) becomes a linear constraint on the load vector q_{ia}: for any $t \in \mathcal{T}_{ia}$,

$$\underline{T}_{ia} \leq T_{ia}(t) + \beta \sum_{\tau=1}^{t}(1-\alpha)^{t-\tau}q_{ia}(\tau) \leq \overline{T}_{ia}.$$

This is the constraint (8), in addition to (6). Assume user i attains a utility $U_{ia}(T_{ia}^{in}(t))$ when the temperature is $T_{i,a}^{in}(t)$. Then (31) gives the utility function (4).

Appendix B: Proof of Lemma 6

We first describe the Hessian $\nabla^2 h(q)$. Let $N := |\mathcal{N}|$ be the number of users and $A := |\cup_{i \in \mathcal{N}} \mathcal{A}_i|$ the total number of appliances. Let k take value (i, a) for $i = 1, \ldots, N, a = 1, \ldots, A$. For $k = (i, a)$, let 1_k be 1 if $a \in \mathcal{A}_i$ and 0 otherwise. From (27), $\nabla^2 h(q)$ is given by

$$\frac{\partial^2 h}{\partial q_k^2(t)} = \frac{\partial^2 U_k}{\partial q_k^2(t)}(q_k) - c'' \left(\sum_j Q_j(t); t \right) 1_k,$$

$$\frac{\partial^2 h}{\partial q_k(s) \, \partial q_k(t)} = \frac{\partial^2 U_k}{\partial q_k(s) \, \partial q_k(t)}(q_k), \quad s \neq t,$$

$$\frac{\partial^2 h}{\partial q_{\tilde{k}}(t) \, \partial q_k(t)} = -c'' \left(\sum_j Q_j(t); t \right) 1_k 1_{\tilde{k}}, \quad k \neq \tilde{k},$$

$$\frac{\partial^2 h}{\partial q_{\tilde{k}}(s) \, \partial q_k(t)} = 0, \quad k \neq \tilde{k} \text{ and } s \neq t.$$

To express $\nabla^2 h(q)$ in matrix form, let $H_k(q_k)$ denote the $T \times T$ matrix $\frac{\partial^2 U_k}{\partial q_k^2}(q_k)$, for $k = 1, \ldots, NA := K$. Let $H(q)$ denote the block-diagonal matrix

$$H(q) := \text{diag} \, (H_1(q_1), \ldots, H_K(q_K)).$$

Let C be the $TNA \times TNA$ matrix with $C_{kt,\tilde{k}\tilde{t}} := c'' \left(\sum_j Q_j(t); t \right) 1_k 1_{\tilde{k}}$ if $t = \tilde{t}$ and 0 otherwise. Then $\nabla^2 h(q) = H(q) - C$. Hence, $\|\nabla^2 h(q)\|_2 \leq \|H(q)\|_2 + \|C\|_2$.

Now assumption A3 implies

$$\|H(q)\|_2 \leq \max_k \|H_k(q_k)\|_2 \leq \rho$$

and (with $\tilde{k} = (\tilde{i}, \tilde{a})$)

$$\|C\|_2 = \rho(C) \leq \|C\|_\infty = \max_{kt} \sum_{\tilde{k}\tilde{t}} C_{kt,\tilde{k}\tilde{t}} \leq \alpha \max_k 1_k \sum_{\tilde{k}} 1_{\tilde{k}} = \alpha \sum_i |\mathcal{A}_i|,$$

where $\rho(C)$ is the spectral radius of matrix C and the first equality holds because C is symmetric. Therefore, $\|\nabla^2 h(q)\|_2 \leq \rho + \alpha \sum_i |\mathscr{A}_i|$. Theorem 9.19 of [41] implies that $\|\nabla h(q) - \nabla h(\tilde{q})\|_2 < \left(\rho + \alpha \sum_i |\mathscr{A}_i|\right) \|q - \tilde{q}\|_2$ for all $q, \tilde{q}$.

References

1. Gellings CW, Chamberlin JH (1988) Demand-side management: concepts and methods. The Fairmont Press, Lilburn
2. Albadi MH, El-Saadany EF (2007) Demand response in electricity markets: an overview. In: Proceedings of the IEEE power engineering society general meeting, Tampa, Florida, USA, June 2007
3. Cohen AI, Wang CC (1988) An optimization method for load management scheduling. IEEE Trans Power Syst 3(2):612–618
4. Hsu YY, Su CC (1991) Dispatch of direct load control using dynamic programming. IEEE Trans Power Syst 6(3):1056–1061
5. Wei DC, Chen N (1995) Air conditioner direct load control by multi-pass dynamic programming. IEEE Trans Power Syst 10(1):307–313
6. Chen J, Lee FN, Breipohl AM, Adapa R (1995) Scheduling direct load control to minimize system operation cost. IEEE Trans Power Syst 10(4):1994–2001
7. Ng KH, Sheble GB (1998) Direct load control—a profit-based load management using linear programming. IEEE Trans Power Syst 13(2):688–695
8. Chu W-C, Chen B-K, Fu C-K (1993) Scheduling of direct load control to minimize load reduction for a utility suffering from generation shortage. IEEE Trans Power Syst 8(4):1525–1530
9. Ramanathan B, Vittal V (2008) A framework for evaluation of advanced direct load control with minimum disruption. IEEE Trans Power Syst 23(4):1681–1688
10. Ilic MD, Xie L, Joo J-Y (2011) Efficient coordination of wind power and price-responsive demand part I: theoretical foundations; part II: case studies. IEEE Trans Power Syst 99:1
11. Makarov YV, Loutan C, Ma J, de Mello P (2009) Operational impacts of wind generation on California power systems. IEEE Trans Power Syst 24(2):1039–1050
12. Caramanis MC, Foster JM (2010) Coupling of day ahead and real-time power markets for energy and reserves incorporating local distribution network costs and congestion. In: Proceedings of the 48th annual allerton conference, Monticello, Illinois, USA, September–October 2010
13. Kirschen D (2003) Demand-side view of electricity market. IEEE Trans Power Syst 18(2):520–527
14. Smith JC, Milligan MR, DeMeo EA, Parsons B (2007) Utility wind integration and operating impact: State of the art. IEEE Trans Power Syst 22(3):900–908
15. Ruiz N, Cobelo I, Oyarzabal J (2009) A direct load control model for virtual power plant management. IEEE Trans Power Syst 24(2):959–966
16. Varaiya PP, Wu FF, Bialek JW (2011) Smart operation of smart grid: Risk-limiting dispatch. Proc IEEE 99(1):40 –57
17. Department of Energy (2006) Benefits of demand response in electricity markets and recommendations for achieving them. Technical Report
18. Borenstein S (2005) Time-varying retail electricity prices: theory and practice. In: Griffin JM, Puller SL (eds) Electricity deregulation: choices and challenges. University of Chicago Press, Chicago
19. Triki C, Violi A (2009) Dynamic pricing of electricity in retail markets. Quart J Oper Res 7(1):21–36

20. Ilic MD (2011) Dynamic monitoring and decision systems for enabling sustainable energy services. Proc IEEE 99(1):58–79
21. Jiang L, Low SH (2011) Optimal demand response: with uncertain supply. Technical Report
22. Schwarz PM, Taylor TN, Birmingham M, Dardan SL (2002) Industrial response to electricity real-time prices: short run and long run. Econ Inq 40(4):597–610
23. Goldman C, Hopper N, Bharvirkar R, Neenan B, Boisvert R, Cappers P, Pratt D, Butkins K (2005) Customer strategies for responding to day-ahead market hourly electricity pricing. Technical Report, Lawrence Berkeley National Lab, LBNL-57128, Report for CA Energy Commission.
24. Taylor TN, Schwarz PM, Cochell JE (2005) 24-7 hourly response to real-time pricing with up to eight summers of experience. Journal of Regulatory Economics 27(3):235–262
25. Borenstein S (2005) The long-run efficiency of real-time electricity pricing. Energy J 26(3):93–116
26. Braun JE (2003) Load control using building thermal mass. J Sol Energy Eng 125(3):292–301
27. Xu P, Haves P, Piette MA, Zagreus L (2006) Demand shifting with thermal mass in large commercial buildings: field tests, simulation and audits. Technical Report, Lawrence Berkeley National Lab, LBNL-58815
28. Mohsenian-Rad A, Leon-Garcia A (2010) Optimal residential load control with price prediction in real-time electricity pricing environments. IEEE Trans Smart Grid 1(2):120–133
29. Pedrasa M, Spooner T, MacGill I (2010) Coordinated scheduling of residential distributed energy resources to optimize smart home energy services. IEEE Trans Smart Grid 1(2):134–143
30. Caron S, Kesidis G (2010) Incentive-based energy consumption scheduling algorithms for the smart grid. In Proceedings of the IEEE international conference on smart grid communications, October 2010
31. Caramanis MC, Foster JM (2009) Management of electric vehicle charging to mitigate renewable generation intermittency and distribution network congestion. In: Proceedings of the 48th IEEE conference on decision and control (CDC), December 2009
32. Ma Z, Callaway D, Hiskens I (2010) Decentralized charging control for large populations of plug-in electric vehicles. In: Proceedings of the 49th IEEE conference on decision and control (CDC), December 2010
33. Clement-Nyns K, Haesen E, Driesen J. (2010) The impact of charging plug-in hybrid electric vehicles on a residential distribution grid. IEEE Trans Power Syst 25(1):371–380
34. Chen L, Li N, Low SH (2010) Two market models for demand response in power networks. In: Proceedings of the IEEE international conference on smart grid communications, Gaithersburg, Maryland, USA, October 2010
35. Pritchard G, Zakeri G, Philpott A (2010) A single-settlement, energy-only electric power market for unpredictable and intermittent participants. Oper Res 58(4):1210–1219
36. Neely MJ, Tehrani AS, Dimakis AG (2010) Efficient algorithms for renewable energy allocation to delay tolerant consumers. In: Proceedings of the IEEE international conference on smart grid communications, Gaithersburg, Maryland, USA, October 2010
37. He M, Murugesan S, Zhang J (2010) Multiple timescale dispatch and scheduling for stochastic reliability in smart grids with wind generation integration. preprint, CoRR, abs/1008.3932
38. Tan C-W, Varaiya PP (1993) Interruptible electric power service contracts. J Econ Dynam Contr 17(3):495–517
39. Li N, Chen L, Low SH (2011) Optimal demand response based on utility maximization in power networks. In: Proceedings of IEEE power engineering society general meeting, Detroit, Michigan, USA, July 2011
40. Bertsekas DP, Tsitsiklis JN (1989) Parallel and distributed computation. Prentice Hall Inc., Old Tappan
41. Rudin W (1976) Principles of mathematical analysis. 3rd edn. McGraw-Hill Inc., New York

Wholesale Energy Market in a Smart Grid: A Discrete-Time Model and the Impact of Delays

Arman Kiani and Anuradha Annaswamy

Abstract The main foundations of the emerging Smart Grid are (1) Distributed Energy Resources (DER) enabled primarily by intermittent, nondispatchable renewable energy sources such as wind and solar, and independent microgrids and (2) Demand Response (DR), the concept of controlling loads via cyber-based communication and control and economic signals. While smart grid communication technologies offer dynamic information provide real-time signals to utilities, they inevitably introduce delays in the energy real-time market. In this article, a dynamic, discrete-time model of the wholesale energy market that captures these interactions is derived. Beginning with a framework that includes optimal power flow and real-time pricing, this model is shown to capture the dynamic interactions between generation, demand, and locational marginal price near the equilibrium of the optimal dispatch. It is shown that the resulting dynamic real-time market has stability properties that are dependent on the delay due to the measurement and communication. Numerical studies are reported to illustrate the dynamic model, and a suitable communication topology is suggested.

1 Introduction

The Smart Grid paradigm has been firmly set into motion driven by the need to reduce energy imports from foreign sources, environmental concerns, and the need to maintain energy efficiency in various economic sectors even in the

A. Kiani (✉)
Institute of Automatic Control Engineering, Technische Universität München,
D-80290 Munich, Germany
e-mail: arman.kiani@tum.de

A. Annaswamy
Department of Mechanical Engineering, Massachusetts Institute of Technology, Cambridge,
MA 02319, USA
e-mail: aanna@mit.edu

A. Chakrabortty and M.D. Ilić (eds.), *Control and Optimization Methods for Electric Smart Grids*, Power Electronics and Power Systems 3,
DOI 10.1007/978-1-4614-1605-0_4, © Springer Science+Business Media, LLC 2012

presence of rapidly increasing energy demand. The cyber-enabled and cyber-secure communication systems, which are key components of a Smart Grid and link Distributed Energy Resources (DERs) and Demand Response (DR) together, should therefore be designed so that the security of supply and reliability of the electricity transmission, distribution systems meet the desired performance targets.

In a Smart Grid, real-time balancing and decentralized decision making are likely to be implemented via day-ahead and real-time markets [11]. In a real-time market, an idealized scenario that is yet to exist, the premise is the following: Generators, Consumers, and the ISO get instantaneous information about the overall state of the market, and suitably respond to this information and adjust themselves, as much as is feasible, in order to optimize a suitable cost function. Such an idealized setting should then enable an efficient integration of the DERs, and impose the most suitable DR algorithm and stay at the desired equilibrium. Under any perturbations due to intermittency in the power generation, or adjustments in generation or demand, or congestion, the goal would be return to this desired equilibrium. That is, the underlying dynamic model should behave in a stable manner, with its equilibrium coinciding with the optimal operating point and have properties of asymptotic stability.

The communication infrastructure, a necessary tool for conveying price information to a real-time market and hence a stable operation of the grid, introduces certain challenges. The main challenge is the introduction of constant or time-varying delays due to the presence of Smart Meters, Smart Devices, and related information processing and communication lags. These delays in turn can endanger market operation and stability of the electricity grid.

The effect of delays in electricity markets has begun to be explored in [12–16]. Reference [13], to our knowledge, is one of the earliest papers to discuss a dynamic market model. In [12], an upper bound on the market clearing time and price signal delay is computed beyond which a time-discretized version of the single-supplier and single-consumer power market model used in [13] becomes unstable. The results in [12] indicate that the impact of the power market can be significant and should be anticipated by proper design of balancing mechanisms and market regulations. In [14], the authors continue the direction of [12], by investigating the limitations imposed by delays on large-scale ancillary service market for real-time balancing originating from a hierarchical tree-based communication topology. In our previous work in [15], inspired by the model in [13], a dynamic model was proposed for the wholesale market, whose stability was evaluated in the presence of a delay due to the presence of smart meter and other communicating devices. In [15, 16], dynamic market models that include transparent connections to Local Marginal Prices are developed, outlining a clear relationship between stability and delays. In this article, we carry out a similar analysis to [15,16], and adopt a discrete-time framework to derive the underlying dynamics, thereby providing a direct connection to the actual market practices that exist today. Using this framework, we then introduce delays due to the communication infrastructure and smart metering.

The main tool that we use to develop the proposed dynamic model stems from the optimal power flow (OPF) method, ubiquitous in operation planning in power systems. The OPF, originally introduced by Carpentier in the 1960s [17], is a highly useful methodology for power dispatch in a grid. Over the years, researchers have developed various algorithms for deriving solutions of the OPF method. References [17–20] present a comprehensive survey of the research in this area in the 1970s and 1980s, a time that witnessed the developed of several constrained optimization techniques such as Lagrange multiplier methods, penalty function methods, and sequential quadratic programming, coupled with gradient methods and Newton methods. In recent years, algorithms based on the primal–dual interior point method have gained popularity [21–23]. The use of game theory and Nash equilibrium is an alternate approach to determine how one can arrive at the market equilibrium. This approach allows a certain transparency in the process of reaching equilibrium, and the dynamic model that we propose is akin to this approach.

In the context of the wholesale electricity market, current practice is to exchange information between the Generating Company (GenCo), Consumers Company (ConCo), and Independent System Operator (ISO) only once. The methodology we propose is a significant departure from this exchange. Our thesis is that due to the huge volatility and uncertainty of the dynamic drivers such as wind and solar energy sources, and load in the market, such a single iteration will not suffice, and stability cannot be ensured. A continued iteration and exchange of information, similar to game theory, is needed in order to mitigate volatility in real-time price and ensure a stable market design. In the literature, game-theoretic models have been employed to investigate strategic behavior in restructured or deregulated electricity markets.

The Nash equilibrium concept has been mainly applied to Cournot models of electric markets with network constraints, as seen in [2–8]. In particular, [3] presents a dc transmission network model, but it does not include nonlinear losses. Games with incomplete information, where participants do not have full knowledge of other participants parameters, are shown in [3]. Finally, [9] presents a Nash bargaining game for transmission analysis, where power exchanges in a two-area system are analyzed. Our use of game theory in the current context is in determining the type of information that has to be exchanged in order to arrive at the equilibrium. This information is cast in the form of a dynamic model and analyzed.

The goal of this article is to derive the underlying dynamic model of an ideal real-time energy market, and analyze its stability properties in the presence of nonideal communication channels characterized by time-delays. While other papers on bilateral contracts, power exchanges, and Poolco markets [13, 24–28] have captured some of the relevant issues in a real-time market, relations between temporal mechanisms in a market and stability are yet to be quantified. This article makes an attempt in this direction, and uses a discrete-time framework to develop the model starting from an OPF formulation and using a primal–dual interior point method to derive the underlying dynamics. The stability of the resulting dynamical model which consists of three main participants, Generating Company (GenCo), Consumers Company (ConCo) and Independent System Operator (ISO) is investigated. Finally the upper limit on the delay of the price signal between these participants that can be tolerated by a stable market is established. In particular, it is shown in this article that the delay between the GenCo and ISO is quite crucial and should therefore have a high priority in the hierarchical tree-based communication topology.

This article has been organized as follows: In Sect. 2, we present the necessary preliminaries. In Sect. 3, the underlying dynamic model is derived and the stability implication of a dynamic market with time delay is presented in Sect. 4. In Sect. 5, numerical studies are presented. A summary is presented in Sect. 6.

2 Preliminaries

In this section, we provide some preliminaries related to the wholesale energy market structure and the underlying dynamic model. These include the fundamental theorem of convex optimization, and the primal-dual interior point method, which are presented in Sects. 2.1 and 2.2, respectively. Given the close relationship between convex optimization and Game theory, we present in Sect. 2.3 related definitions and theorems related to game theory and Nash equilibrium. In particular, the link between the convex optimization problem and uniqueness of the Pure Strategy Nash Equilibrium is presented including sufficient conditions for the latter. These in turn are directly used in establishing the equilibrium of the wholesale market in Sect. 3.

Definition 1. A set $K \subseteq \mathbb{R}$ is convex if for any two points $x, y \in K$,

$$\alpha x + (1 - \alpha)y \in K, \ \forall x, y \in K \text{ and } \alpha \in [0, 1]. \tag{1}$$

Definition 2. Given a convex set $K \subseteq \mathbb{R}$ and a function $f(x) : K \to \mathbb{R}$; f is said to be a convex function on K if, $\forall x, y \in K$ and $\alpha \in (0, 1)$,

$$f(\alpha x + (1 - \alpha)y) \leq \alpha f(x) + (1 - \alpha)f(y), \tag{2}$$

Furthermore, a function $f(x)$ is concave over a convex set if and only if the function $-f(x)$ is a convex function over the set.

Definition 3. Given a scalar-valued function $f(x) : \mathbb{R}^n \to \mathbb{R}$ we use the notation $\nabla f(x)$ to denote the gradient vector of $f(x)$ at point x, i.e.,

$$\nabla f(x) = \left[\frac{\partial f(x)}{\partial x_1}, \ldots, \frac{\partial f(x)}{\partial x_n} \right]^T. \tag{3}$$

Definition 4. Given a scalar-valued function $f(x) : \prod_{i=1}^{I} \mathbb{R}^{m_i} \to \mathbb{R}$ we use the notation $\nabla_i f(x)$ to denote the gradient vector of $f(x)$ with respect to x_i at point x, i.e.,

$$\nabla_i f(x) = \left[\frac{\partial f(x)}{\partial x_i^1}, \ldots, \frac{\partial f(x)}{\partial x_i^{m_i}} \right]^T. \tag{4}$$

2.1   Dual Decomposition

Consider a generic optimization problem (in the minimization form)[1]:

$$\text{Minimize} \quad f(x)$$

such that

$$g_n(x) = 0, \quad \forall n = 1, \ldots, N$$

$$\sum_{n=1}^{N} R_{mn} h_n(x) \leq c_m, \quad \forall m = 1, \ldots L \tag{5}$$

where $f(x)$ is called the objective function or cost function, R is a matrix of constants and c_m are constants. We assume that $f(x) : \mathbb{R}^n \rightarrow \mathbb{R}$ is a convex function to be minimized over the variable x, the functions $g_n(x)$ as equality constraints are affine, and the functions $h_n(x)$ as inequality constraints are convex. With these assumptions the optimization Problem 5 is termed a convex optimization problem. In order to derive the dual optimization problem, the Lagrangian function is defined as

$$L(x, \lambda, \mu) = f(x) + \sum_{n=1}^{N} \lambda_n g_n(x) + \sum_{m=1}^{L} \mu_m (R_{mn} h_n(x) - c_m) \tag{6}$$

where λ_n and $\mu_m \geq 0$ are (dual) Lagrangian multipliers for the equality and inequality constraints, and x is the primal variable. Denoting

$$D(\lambda, \mu) = \inf_x L(x, \lambda, \mu) \tag{7}$$

the dual optimization problem is formulated as

$$\text{Maximize} \quad D(\lambda, \mu)$$

such that

$$\mu \geq 0, \quad \forall m = 1, \ldots, L. \tag{8}$$

Under the condition that the original problem (5) is strictly feasible, then there is no duality gap (i.e., the original (5) and the dual problems (8) have the same optimum). In this case, the dual problem (8) can be solved instead of the original problem (5). In addition, the constraint set for the optimization problem is convex, which allows us to use the method of Lagrange multipliers and the Karush Kuhn Tucker (KKT) theorem, which we state below [29, 30].

---

[1] While these preliminaries are also addressed in [16], we introduce them here too for the sake of completeness.

Theorem 1. *Consider the optimization formulated in* (5), *where* $f(x)$ *is a convex function,* $g_n(x)$ *are affine functions, and* $h_n(x)$ *are convex functions. Let* x^* *be a feasible point, i.e. a point that satisfies all the constraints. Suppose there exist constants* λ_n *and* $\mu_m \geq 0$ *such that*

$$\nabla f(x^*) + \sum_{n=1}^{N} \lambda_n \nabla g_n(x^*) - \sum_{m=1}^{L} \mu_m (R_{mn} \nabla h_n(x^*) - c_m) = 0 \quad \forall n = 1 \ldots N$$

$$\mu_m (R_{mn} h_n(x^*) - c_m) = 0 \quad , \forall m = 1, \ldots, L \tag{9}$$

then x^* *is a global maximum. If* $f(x)$ *is strictly concave, then* x^* *is also the unique global maximum.*

Proof. See [30]. □

2.2 Subgradient Algorithm

Often it is simpler to determine the solution of the above optimization problem in an iterative manner. For this purpose, a gradient approach is often employed, and is briefly described below. Since the ultimate goal of constraint optimization problem in (5) is the minimization of a Lagrangian denoted as $L(x, \lambda, \mu)$ in (6), we progressively change x, λ, and μ so that minima-Lagrange multiplier pairs λ and μ satisfy the conditions of Theorem 1 denoted in (9). We do this by using Primal–Dual interior point method [30] which is given by

$$x(t_{k+1}) = x(t_k) - hk_x \nabla_x L(x_k, \lambda_k, \mu_k),$$

$$\lambda(t_{k+1}) = \lambda(t_k) + hk_\lambda \nabla_\lambda L(x_k, \lambda_k, \mu_k),$$

$$\mu(t_{k+1}) = \mu(t_k) + hk_\mu \left[\nabla_\mu L(x_k, \lambda_k, \mu_k) \right]_\mu^+, \tag{10}$$

where k_x, k_λ, and k_μ are step sizes which control the amount of change in the direction of the gradient, and $t_{k+1} = t_k + h$, with $\{t_k\}$ denoting the time instances at which information $x(t_k)$, $\lambda(t_k)$, and $\mu(t_k)$ are transmitted, and

$$\left[M(x, y) \right]_y^+ = \begin{cases} M(x, y) & \text{if } y > 0, \\ max(0, M(x, y)) & \text{if } y = 0. \end{cases} \tag{11}$$

The choice of $\left[M(x, y) \right]_y^+$ in (11) ensures that μ_m's in (10) are always non-negative.

Remark 1. Concepts related to duality arguments and subgradient algorithms have been used for congestion control in the Internet (see, for example, [31–33]). Here, we use the same concepts, but in the context of wholesale energy market and its stability in the presence of delays due to the overall cyber-communication infrastructure.

2.3 Game Theory and NASH Equilibrium

An alternative game-theoretic approach can be used to describe the above market equilibrium. A game is a formal representation of a situation in which a number of individuals interact in a setting of strategic interdependence [1]. To describe a game, there are four things to consider: (1) the players, (2) the rules of the game, (3) the outcomes, and (4) the payoffs and the preferences (utility functions) of the players. A player plays a game through actions. An action is a choice or election that a player takes, according to his (or her) own strategy. Since a game sets a framework of strategic interdependence, a participant should be able to have enough information about its own and other players' past actions. This is called the information set. A strategy is a rule that tells the player which action(s) it should take, according to its own information set at any particular stage of a game. Finally, a payoff function expresses the utility that a player obtains given a strategy profile for all players. More formally a strategic form game is defined as follows.

Definition 5. A strategic forms game is a triplet $\langle \mathscr{I}, (S_i)_{i \in \mathscr{I}}, (u_i)_{i \in \mathscr{I}} \rangle$ such that $\mathscr{I}$ is a finite set of players , i.e. $\mathscr{I} = \{1, \ldots, I\}$, S_i is the set of available actions for player i, $s_i \in S_i$ is an action for player i, and finally $u_i : S \to R$ is the payoff function of player i where $S = \prod_i S_i$ is the set of all action profiles.

In addition, we use the notation $s_{-i} = [s_j]_{j \neq i}$ as a vector of actions for all players except i, $S_{-i} = \prod_{j \neq i} S_j$ as the set of all action profiles for all players except i, and finally $(s_i, s_{-i}) \in S$ denoted a strategy profile of the game. Informally, a set of strategies is a Nash equilibrium if no player can not do better by unilaterally changing his or her strategy.

Definition 6. A Pure Strategy Nash Equilibrium of a strategic game $\langle \mathscr{I}, (S_i)_{i \in \mathscr{I}}, (u_i)_{i \in \mathscr{I}} \rangle$ is a strategy profile $s^* \in S$ such that for all $i \in \mathscr{I}$

$$u_i(s_i^*, s_{-i}^*) \geq u_i(s_i, s_{-i}^*) \ \forall s_i \in S_i. \tag{12}$$

Definition 7. Let us assume that for player $i \in \mathscr{I}$, the strategy set S_i is given by

$$S_i = \{x_i \in \mathbb{R}^m | h_i(x_i) \geq 0\} \tag{13}$$

where $h_i : \mathbb{R}^m \to \mathbb{R}$ is a concave function and the set of strategy profiles $S = \prod_{i=1}^{I} S_i \subset \prod_{i=1}^{I} \mathbb{R}_i^m$ is a convex set. The payoff functions $(u_1, \ldots, u_I)$ are said to be diagonally strictly concave for $x \in S$ if for every $x^*, \hat{x} \in S$, we have

$$(\hat{x} - x^*)^T \nabla u(x^*) + (x^* - \hat{x})^T \nabla u(\hat{x}) > 0. \tag{14}$$

Theorem 2. *Consider a strategic form game* $\langle \mathscr{I}, (S_i)_{i \in \mathscr{I}}, (u_i)_{i \in \mathscr{I}} \rangle$. *For all* $i \in \mathscr{I}$, *assume that the strategy sets* S_i *are given by* $S_i = \{x_i \in \mathbb{R}^m | h_i(x_i) \geq 0\}$, *where* h_i *is a concave function, and there exists some* $\tilde{x}_i \in \mathbb{R}^{m_i}$ *such that* $h_i(\tilde{x}_i) > 0$. *Assume* *also that the payoff functions $(u_1, \ldots, u_I)$ are diagonally strictly concave for $x \in S$. Then the game has a unique pure strategy Nash equilibrium identical to the optimal solution $x^* \in \prod_{i=1}^{I} \mathbb{R}_i^m$ of the following optimization problem*

$$\max_{y_i \in \mathbb{R}_i^m} u_i(y_i, x_{-i}^*) \tag{15}$$

$$s.t. \ h_i(x_i) \geq 0. \tag{16}$$

Proof. See [2]. □

We complete this section by giving a sufficient conditions that insure the payoff functions $(u_1, \ldots, u_I)$ are diagonally strictly concave. The condition is given in terms of the $m \times m$ matrix $U(x)$, which is the Jacobian of the gradient vector of $u(x)$ defined in Eq. 3.

Theorem 3. *For all $i \in \mathcal{I}$, assume that the strategy sets S_i are given by (13), where $h_i(x_i)$ is a concave function. Assume that the symmetric matrix $(U(x) + U^T(x))$ is negative definite for all $x \in S$, i.e. for all $x \in S$, we have*

$$y^T(U(x) + U^T(x))y < 0 \ \forall y \neq 0 \tag{17}$$

where the matrix $U(x)$ is the Jacobian of the gradient vector of $u(x)$, i.e. the jth column of $U(x)$ is $\partial u(x)/\partial x_j$, $j = 1, \ldots, I$, then the payoff functions $(u_1, \ldots, u_I)$ are diagonally strictly concave for $x \in S$.

Proof. See [2].

In the current context, the optimization problem in (5) can be viewed as a game played by x, λ, and μ for each agent distributively so that together they can arrive at a Nash equilibrium that satisfies the condition (9) in Theorem 1. The difference equation in (10) can therefore be viewed as a strategy for arriving at the Nash equilibrium, which happens to coincide with the equilibrium of (10).

3 Wholesale Energy Market Structure

The electricity market that we consider in this article is wholesale and is assumed to function as follows: First, each generating company (GenCo) submits the bidding stacks of each of its units to the pool. Similarly, each consumer (ConCo) submits the bidding stacks of each of its demands to the pool. Then the ISO clears the market using an appropriate market-clearing procedure resulting in prices and production and consumption schedules. In what follows, we model each of the components (GenCo, ConCo, and ISO) together with their constraints and the optimization goal.

3.1 Generating Company

It is assumed that the generating company consists of N_G generating units, and that the production of each generating unit $i \in G_f = \{1, 2, \ldots, N_G\}$ is denoted as P_{Gi}. The associated operating cost is denoted as $C_{G_i}(P_{Gi})$

$$C_{G_i}(P_{Gi}) = b_{Gi}P_{Gi} + \frac{c_{Gi}}{2}P_{Gi}^2, \tag{18}$$

where b_{Gi}, and c_{Gi} are generators cost coefficients. The goal of the company is to maximize its overall profit, π_{Gi}, and is stated as

$$\max_{P_{Gi}} \pi_{Gi} = \max_{P_{Gi}} \left[\rho_{n(i)}P_{Gi} - C_{G_i}(P_{Gi}) \right]$$

$$\text{s.t. } P_{Gi}^{min} \leq P_{Gi} \leq P_{Gi}^{max} \tag{19}$$

where $\rho_{n(i)}$ denotes the LMP of unit i at node n in the network, P_{Gi}^{min}, and P_{Gi}^{max} are lower and upper bounds for the production of GenCo i. To focus on the market equilibrium, minimum and maximum generation constraints are neglected in this study. By applying KKT condition, and the assumption that GenCo cannot manipulate the price, e.g., $\frac{\partial \rho_{n(i)}}{\partial P_{Gi}} = 0$, the optimal generation quantities can be obtained as

$$\left(-\rho_{n(i)}^* + \frac{d(C_{G_i}(P_{Gi}))}{dP_{Gi}}\Big|_{P_{Gi}^*} \right) = 0 \quad \forall i \in G_f. \tag{20}$$

3.2 Consumer Modeling

A consumer company (ConCo) is assumed to consist of N_D units, and the demand of each unit $j \in D_q = \{1, 2, \ldots, N_D\}$, is denoted as P_{Dj}. The associated utility function is denoted as $U(P_{Dj})$, which represents the value of using electricity for the consumer and is defined as

$$U(P_{Dj}) = b_{Dj}P_{Dj} + \frac{c_{Dj}}{2}P_{Dj}^2, \tag{21}$$

where b_{Dj} and c_{Dj} are consumers utility coefficients. The goal of the ConCo is to maximize the total profit, π_{Dj}, while consuming electricity. This profit, for a unit j connected to node n, is determined as the difference between the utility $U(P_{Dj})$ and the corresponding LMP, $\rho_n(j)$. Assuming that the corresponding power consumed is denoted as P_{Dj}, the maximization problem can be posed as

$$\max_{P_{Dj}} \pi_{Dj} = \max_{P_{Dj}} \left[U(P_{Dj}) - \rho_{n(j)}P_{Dj} \right]$$

$$\text{s.t. } P_{Dj}^{min} \leq P_{Dj} \leq P_{Dj}^{max} \tag{22}$$

where $\rho_{n(j)}$ denotes the LMP of ConCo j at node n in the network, P_{Dj}^{min}, and P_{Dj}^{max} are lower and upper bounds for the consumption of ConCo j. The decision variables of this problem are P_{Dj}, the amounts of power to be consumed by each demand j. The same as GenCo, the minimum and maximum consumption constraints are neglected in this study. As GenCo modeling, the optimal P_{Dj} can be obtained by KKT as

$$\left(\rho_{n(j)}^* - \frac{d(U_{Dj}(P_{Dj}))}{dP_{Dj}} |_{P_{Dj}^*} \right) = 0 \quad \forall j \in D_q. \tag{23}$$

3.3   ISO Market-Clearing Model

The market-clearing procedure consists of optimizing a cost function, subject to various network constraints. The most dominant network constraints are due to line capacity limits [34] and network losses [25]. The power flow through any line is often limited due to technical constraints and is said to be congested when it approaches its maximum limit [25]. This constraint is explicitly included in our model below. The second constraint is due to losses, most of which are due to the heat loss in the power lines. For ease of exposition, such ohmic losses are not modeled in this article.

The cost function that is typically used is referred to as Social Welfare. Denoted as S_w, Social Welfare is defined as

$$S_W = \sum_{j \in D_q} U_{Dj}(P_{Dj}) - \sum_{i \in G_f} C_{G_i}(P_{Gi}), \tag{24}$$

where the first and second terms denote the revenue due to surpluses stemming from bids from GenCo and ConCo, respectively. $U_{Dj}(P_{Dj})$ and $C_{G_i}(P_{Gi})$ correspond to utility of consumers and cost of generators company and is defined in (18) and (21). In summary, the market-clearing procedure is given by

$$\text{Maximize } S_W = \text{Minimize } -S_W \tag{25}$$

subject to

$$-\sum_{i \in \theta} P_{Gi} + \sum_{j \in \vartheta} P_{Dj} + \sum_{m \in \Omega} B_{nm}[\delta_n - \delta_m] = 0 \quad \rho_n, \quad \forall n \in N, \tag{26}$$

$$B_{nm}[\delta_n - \delta_m] \le P_{nm}^{max} \quad \gamma_{nm}, \quad \forall n \in N; \forall m \in \Omega, \tag{27}$$

where $\theta(\vartheta)$ is denoted the set of indices of generating units (demand) at node n, Ω is the set of indices of nodes connected to node n, δ_n is a voltage angle of bus n, B_{nm} is the susceptance of line $n - m$, and P_{nm}^{max} is the transmission capacity limit of line

$n - m$. The constraints (26) and (27) are due to power balance and capacity limits, respectively. It can be seen that the associated Lagrange multipliers, ρ_n and γ_{nm}, are indicated in each constraint. The underlying optimization problem of the ISO can therefore be defined as the optimization of (25) subject to constraints (26) and (27). The corresponding Lagrangian of the market clearing optimization problem is given by

$$
L(x, \rho_n, \gamma_{nm}) = \sum_{i \in G_f} C_{G_i}(P_{Gi}) - \sum_{j \in D_q} U_{Dj}(P_{Dj})
$$

$$
+ \sum_{n=1}^{N} \rho_n \left[-\sum_{i \in \theta} P_{Gi} + \sum_{j \in \vartheta} P_{Dj} + \sum_{m \in \Omega} B_{nm} [\delta_n - \delta_m] \right]
$$

$$
+ \sum_{n=1}^{N} \sum_{m \in \Omega} \gamma_{nm} \left[B_{nm} [\delta_n - \delta_m] - P_{nm}^{max} \right], \tag{28}
$$

where $x = [P_G \quad P_D \quad \delta]^T$ is the primial optimization variables. The resulting solution can be determined, using KKT conditions [29], as P_{Gi}^*, the amounts of power to be generated by each generating unit i, P_{Dj}^*, the amounts of power to be consumed by each consumer j, the locational marginal prices, ρ_n^*, and congestion price γ_{nm}^* that satisfies the following conditions:

$$
\left(\frac{d(C_{G_i}(P_{Gi}))}{dP_{Gi}} \Big|_{P_{Gi}^*} - \rho_{n(i)}^* \right) = 0 \quad \forall i \in G_f, \tag{29}
$$

$$
\left(\rho_{n(j)}^* - \frac{d(U_{Dj}(P_{Dj}))}{dP_{Dj}} \Big|_{P_{Dj}^*} \right) = 0 \quad \forall j \in D_q, \tag{30}
$$

$$
\left(\sum_{m \in \Omega} B_{nm} \left[\rho_n^* - \rho_m^* + \gamma_{nm}^* - \gamma_{mn}^* \right] \right) = 0 \quad \delta_n > 0 \ \forall n \in N, \tag{31}
$$

$$
\left(-\sum_{i \in \theta} \sum_{b=1}^{N_{Gi}} P_{Gi}^* + \sum_{j \in \vartheta} \sum_{k=1}^{N_{Dj}} P_{Dj}^* + \sum_{m \in \Omega} B_{nm} \left[\delta_n^* - \delta_m^* \right] \right) = 0 \quad \forall n \in N, \tag{32}
$$

$$
\gamma_{nm}^* \left(B_{nm} \left[\delta_n^* - \delta_m^* \right] - P_{nm}^{max} \right) = 0 \quad \forall n \in N; \ \forall m \in \Omega. \tag{33}
$$

We now connect these decision variables to a Nash equilibrium approach, and summarize the relation in Lemma 1.

Lemma 1. *The equilibrium of the market model in (29)–(33) is the Nash equilibrium for GenCo and ConCo, which collectively optimizes the overall benefit of GenCo and ConCo denoted in (19) and (22) while satisfying constraints of ISO in (26) and (27).*

Proof. From Theorems 4, and 3, it follows that the payoff functions denoted in (19), (22), and (24) are diagonally strictly concave and the corresponding constraints are concave functions. This implies that the market has a unique Pure Strategy Nash Equilibrium that is identical to the solution of the KKT denoted in (29)–(33). Furthermore, we note that (29) is identical to (20). This implies that at the extremum, P_{Gi}^*, P_{Dj}^*, δ_n^*, ρ_n^*, and γ_{nm}^*, the profit of Genco in (19) is maximized. That is,

$$\pi_{Gi}(P_{Gi}^*, P_{Dj}^*, \delta_n^*, \rho_n^*, \gamma_{nm}^*) \geq \pi_{Gi}(P_{Gi}, P_{Dj}^*, \delta_n^*, \rho_n^*, \gamma_{nm}^*)$$

Similarly, since (30) is identical to (23), it follows that the the the total profit of Conco is maximized. That is,

$$\pi_{Dj}(P_{Dj}^*, P_{Gi}^*, \delta_n^*, \rho_n^*, \gamma_{nm}^*) \geq \pi_{Dj}(P_{Dj}, P_{Gi}^*, \delta_n^*, \rho_n^*, \gamma_{nm}^*).$$

This implies that at the extremum, the best response of the players is given by P_{Gi}^*, P_{Dj}^*, δ_n^*, ρ_n^*, and γ_{nm}^*. Therefore, this extremum coincides with the Nash equilibrium.

3.4 Dynamical Model of Wholesale Electricity Market

In the actual power market, the market participants are not able to know the competitors' decision and profit functions. They are unable to reach the equilibrium condition at once. In fact, each participant is rational and can only decide the production strategy according to his expected profit at each iteration. For each market participant, the evaluation of his own profit is more accurate than the predication of the competitors' output. Therefore, market participants play a game with a dynamic adjustment to reach their Nash equilibrium. Now, we consider a reasonable dynamic model of the market participant's interactions. It is assumed that if the game is not at equilibrium, each participant will attempt to change his own strategy so as to obtain the maximum rate of change of his own payoff function with respect to a change in his own strategy.

Using (10) and (18), we can derive a difference equation for the ith GenCo $\forall i \in G_f$ as

$$P_{Gi_{k+1}} = P_{Gi_k} - hk_{P_{Gi}} \frac{\partial L(x, \rho_n, \gamma_{nm})}{\partial P_{Gi}} \quad \forall i \in G_f \tag{34}$$

with the goal of driving its solution P_{Gi} to the equilibrium P_{Gi}^* which solves (29). Similarly, using (10) and (21), a difference equation can be derived for the jth ConCo $\forall j \in D_q$ as

$$P_{Dj_{k+1}} = P_{Dj_k} - hk_{P_{Dj}} \frac{\partial L(x, \rho_n, \gamma_{nm})}{\partial P_{Dj}} \quad \forall j \in D_q \tag{35}$$

where $h = t_{k+1} - t_k$ where $h = T_M/N$, T_M denotes the market clearing time and N is a scaling factor greater than one, and $k_{P_{Gi}}$ and $k_{P_{Dj}}$ are constants correspond to the ramp limits. Finally, difference equations for the LMPs, congestion price and phase angles can be determined as

$$\delta_{n_{k+1}} = \delta_{n_k} - hk_{\delta_n} \frac{\partial L(x, \rho_n, \gamma_{nm})}{\partial \delta_n} \quad \forall n \in N \tag{36}$$

$$\rho_{n_{k+1}} = \rho_{n_k} + hk_{\rho_n} \frac{\partial L(x, \rho_n, \gamma_{nm})}{\partial \rho_n} \quad \forall n \in N \tag{37}$$

$$\gamma_{nm_{k+1}} = \gamma_{nm_k} + hk_{\gamma_n} \left[\frac{\partial L(x, \rho_n, \gamma_{nm})}{\partial \gamma_{nm}} \right]_\gamma^+ \quad \forall n \in N, \forall m \in \Omega \tag{38}$$

Equations (34)–(38) represent a dynamic model of the overall wholesale energy market.

Two important points should be made regarding the above model. The solution of this model P_{Gi_k}, P_{Dj_k}, δ_{n_k}, ρ_{n_k}, and γ_{nm_k} converges to the equilibrium in (29)–(33), as $k \to \infty$ if the overall system of equations is stable. At all other transient times, the trajectories P_{Gi_k}, P_{Dj_k}, δ_{n_k}, ρ_{n_k}, and γ_{nm_k} represent the specific path that these variables take, when perturbed, as they converge towards the optimal solution. In other words, P_{Gi_k}, P_{Dj_k}, δ_{n_k}, ρ_{n_k}, and γ_{nm_k} is distinct from the optimal solution $(P_{Gi}^*, P_{Dj}^*, \delta_n^*, \rho_n^*, \gamma_{nm}^*)$ and coincides with it as k tends to infinity if the market is stable.

A game theoretical interpretation of this model can be given as follows; $\{P_{Gi_k}\}_{i \in G_f}$, $\{P_{Dj_k}\}_{j \in D_q}$, and $\{\delta_{n_k}, \rho_{n_k}, \gamma_{nm_k}\}_{n \in N, m \in \Omega}$ can be viewed as strategies of the ith GenCo, j^{th} ConCo, and ISO for bus n, respectively. If these specific strategies are employed by these players in the electricity market, they will collectively arrive at the Pure Strategy Nash Equilibrium if the market dynamics denoted by Eqs. (34)–(38) is stable.

In order to analyze stability of the wholesale market, we assume that transmission lines are strong enough that the power flow is well within the capacity limit. Using the cost functions in (18) and (21) and the assumption of strong transmission lines, the model in (34)–(38) can be simplified as

$$P_{Gi_{k+1}} = P_{Gi_k} + hk_{P_{Gi}}(\rho_{n(i)_k} - c_{Gi}P_{Gi_k} - b_{Gi}) \quad \forall i \in G_f \tag{39}$$

$$P_{Dj_{k+1}} = P_{Dj_k} + hk_{P_{Dj}} \left(c_{Dj} P_{Dj_k} + b_{Dj} - \rho_{n(j)_k} \right) \quad \forall j \in D_q \quad (40)$$

$$\delta_{n_{k+1}} = \delta_{n_k} - hk_{\delta_n} \left(\sum_{m \in \Omega} B_{nm} [\rho_n - \rho_m] \right) \quad \forall n \in N \quad (41)$$

$$\rho_{n_{k+1}} = \rho_{n_k} + hk_{\rho_n} \left(-\sum_{i \in \theta} P_{Gi} + \sum_{j \in \vartheta} P_{Dj_k} + \sum_{m \in \Omega} B_{nm} [\delta_n - \delta_m] \right) \quad \forall n \in N$$
$$(42)$$

We now can simplify the overall wholesale market and state it compactly as

$$x(k + 1) = (I + A)x(k) + b, \quad (43)$$

where

$$x(k) = \begin{bmatrix} P_{Gi} & P_{Dj} & \delta_n & \rho_n \end{bmatrix}^{\mathrm{T}}_{(N_g+N_d+2N-1)\times 1}, \quad (44)$$

$$A = \begin{bmatrix} -hk_g c_g & 0 & 0 & hk_g A_g^T \\ 0 & hk_d c_d & 0 & -hk_d A_d^T \\ 0 & 0 & 0 & -hk_\delta Y_{bus}^T \\ -hk_\rho A_g & hk_\rho A_d & hk_\rho Y_{bus} & 0 \end{bmatrix}, \quad (45)$$

where B_{line} denotes the line admittance matrix (N_t by N_t diagonal matrix) with elements B_{nm} and let A denote the $N_t \times N$ bus incidence matrix, where

$$\begin{cases} A_{nm} = 1 & \text{if the } n\text{th transmission line is connected to} \\ & m\text{th bus and directed away from } m\text{th bus,} \\ A_{nm} = -1 & \text{if the } n\text{th transmission line is connected to} \\ & m\text{th bus and directed toward from } m\text{th bus,} \\ A_{nm} = 0 & \text{if the } n\text{th transmission line is not connected} \\ & \text{to } m\text{th bus.} \end{cases} \quad (46)$$

Let A_r denote the reduced bus incidence matrix ($N_t \times N - 1$), which is A with column corresponding to reference bus removed, and $Y_{bus} = A^T B_{line} A_r$ is a bus admittance matrix. A_g is generators incidence matrix, where $A_{g_{ij}} = 1$ if the ith generator is connected to jth bus and $A_{g_{ij}} = 0$ if the ith generator is not connected to jth bus, similarly for A_d which is load incident matrix, where $A_{d_{ij}} = 1$ if the ith

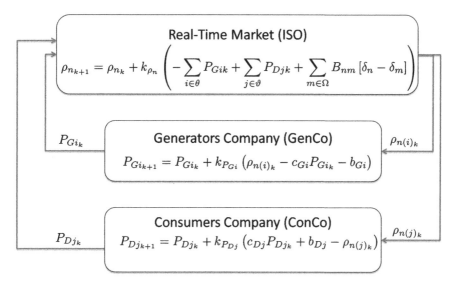

Fig. 1 Real-time market using demand feedback

consumer is connected to jth bus and $A_{d_{ij}} = 0$ if the ith consumer is not connected to jth bus. Finally,

$$b = \left[-b_g^T k_g \; b_d^T k_d \; 0 \right]^T. \tag{47}$$

The overall dynamics of Equations (39)–(42) are represented in the schematic shown in Fig. 1. These equations and Fig. 1 illustrate that the generators communicate the information about P_{Gi} to the ISO, the consumers communicate the demand P_{Dj} to the ISO, and the ISO communicates the price to the generators and consumers every $h = t_{k+1} - t_k$, where $h = T_M/N$, T_M denotes the market clearing time and N is a scaling factor greater than one.

3.5 Concept of Nash Equilibrium in the Wholesale Market

In the above game between three entities, GenCo, ConCo, and ISO, the market equilibrium denoted in (29)–(33) means that none of entities can get extra profit by unilaterally deviating from the equilibrium denoted in (29)–(33).

Lemma 2. *Let strong duality hold. Then (i) the equilibrium of (29)–(33) is identical to the Nash equilibrium of the game in (43); (ii) if all eigenvalues of matrix $(I + A)$ are inside unit circle, then the strategic distributed game in (43) will converge to the Nash equilibrium.*

Proof. (i) Since strong duality implies that there is no duality gap which in turn implies that primal–dual interior method for the convex problem of ISO in (25)–(27) has a unique solution, it follows that the unique equilibrium of (43) coincide with the optimal solution of the original ISO problem. Using Lemma 1, it follows that this equilibrium is Nash equilibrium.

(ii) When all eigenvalues of matrix $(I + A)$ are inside unit circle which implies that the difference system in (43) is asymptotically stable and the strategic distributed game in (43) will converge to the Nash equilibrium. □

The underlying wholesale market model has the following characteristics:

1. The market participants need not have global market information, such as the market demand and the competitors' cost. They decide their generation quantities by estimating their own marginal profit , as exemplified by c_{Gi} and b_{Gi}.
2. Convergence of the dynamic system to the equilibrium condition implies that each GenCo reaches its own maximum profit with no further improvement by virtue of its own generation strategies. That is, the market reaches the condition of Nash equilibrium.
3. At any given iteration, if the marginal profit of the GenCo is greater than zero, the GenCo will increase P_{G_i} to obtain a greater economic benefit; if the marginal profit of the GenCo is less than zero, the Genco will decrease P_{G_i}.

4 Impact of Time Delay on the Stability of Wholesale Market

In reality, information transfer between the different entities in the overall market occurs through a smart meter together with information transfer and information processing through nonideal communication channels. That is, this information transfer may not take place instantaneously but with a time-delay. Representing the delay from ISO to generators as d_G, from consumers to ISO as d_{P_D}, and ISO to consumers as d_D, a modified dynamic model, shown in Fig. 2, is derived compactly as

$$x(k + 1) = (I + A_0)x(k) + A_G x(k - d_G) + A_D x(k - d_D)$$
$$+ A_{pd} x(k - d_{pd}) + b, \qquad (48)$$

$$A_G = \begin{bmatrix} 0\,0\,0 & hk_g A_g^T \\ 0\,0\,0 & 0 \\ 0\,0\,0 & 0 \\ 0\,0\,0 & 0 \end{bmatrix}, \quad A_D = \begin{bmatrix} 0\,0\,0 & 0 \\ 0\,0\,0 & -hk_d A_d^T \\ 0\,0\,0 & 0 \\ 0\,0\,0 & 0 \end{bmatrix}, \quad A_{pd} = \begin{bmatrix} 0 & 0 & 0\,0 \\ 0 & 0 & 0\,0 \\ 0 & 0 & 0\,0 \\ 0 & hk_p A_d & 0\,0 \end{bmatrix},$$

and using (45) we have $A_0 = A - A_G - A_D - A_{pd}$. In what follows, we analyze (48) with the following simplification: At any time, only one of the delays d_G, d_{P_D}, and

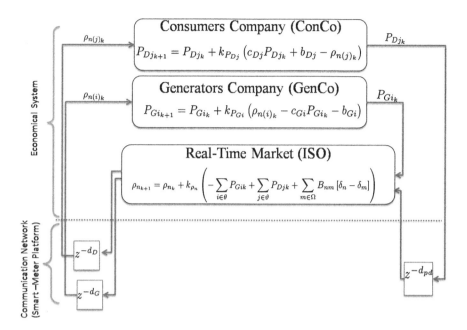

Fig. 2 Cyber-economical system; Notion of feedback in real-time wholesale electricity market due to communication networks time delay

d_D is nonzero. This simplification allows us to evaluate the effect of these delays on the overall stability of the market.

Theorem 4. *The system*

$$z(n+1) = Az(n) + A_d z(n-d)$$

$$d \leq \bar{d}, \quad d, \bar{d} \in Z^+$$

$$z(n) = \psi(n), \quad \forall n \in \{-2\bar{d}, -2\bar{d}+1, \ldots, -1, 0\} \tag{49}$$

is asymptotically stable for any delay $0 \leq d \leq \bar{d}$ if there exist a matrix W and positive definite matrices P, Q, and V solving the following LMI:

$$\begin{bmatrix} \phi_{11} & \phi_{12} & \phi_{13} & \phi_{14} \\ \phi_{12}^T & \phi_{22} & A_d A_d^T V & 0 \\ \phi_{13}^T & V A_d A_d^T & -V & 0 \\ \phi_{14}^T & 0 & 0 & -V \end{bmatrix} < 0, \tag{50}$$

where

$$\phi_{11} = (A + A_d)^T P(A + A_d) - P + A_d^T P A_d + Q$$
$$\qquad + (A + A_d)^T W^T A_d + A_d W(A + A_d),$$
$$\phi_{12} = -(A + A_d)^T W^T A_d - A_d^T P A_d,$$
$$\phi_{13} = (A - I)^T A_d^T V,$$
$$\phi_{14} = \bar{d}(A + A_d)^T (W^T + P),$$
$$\phi_{22} = -Q + A_d^T P A_d.$$

Proof. See [35].   □

Remark 2. The LMI condition (50) is rather complex, and depends on several system parameters. A necessary condition for (50) to be satisfied is the stability of the matrix $A + A_d$ [35]. A detailed articulation of sufficient conditions for (50) to hold is beyond the scope of this article.

Remark 3. We note that the same theorem can be applied to the case when all three delays are present and are equal to each other.

Theorem 4 can be applied to determine the stability of the solutions of (48) around $x = x_{eq}$, by choosing $z = x - x_{eq}$, where x_{eq} is the equilibrium of (48). In the next section, we address the implication of a dynamic model with time delay for the real-time market.

4.1 Implications of a Dynamic Market with Delay

The electricity wholesale market is a coupling of two constrained and highly complex dynamical systems; one physical and one economic. The physical system is a complex network consisting of power flowing through transmission lines, characterized by distributed generation units, Kirchhoff's laws, and operational and security constraints. Loads and generation are each subject to uncertainty, whereas these uncertainties cause the price volatility. Since the main participants of the markets are sensitive to the prices, in a "real-time" market, they respond to the price volatility. On the other hand, the economic system typically consists of coupled markets such as day-ahead market and real-time auctions. These auctions are also dynamic since prices vary because of the "rational agents" driving the economic side of the wholesale market [36]. The final complexity arises from the fact these two systems are coupled, through feedback. It is therefore important to analyze the behavior of the overall problem that includes the dynamics of these individual systems and their feedback interactions. The dynamic model that we have proposed in the previous section is one such encapsulation of the overall problem. In addition

Table 1 Generators cost and demand utilities coefficients

	GenCo1	GenCo2	ConCo1	Conco2
c_G [$\$/MWh$]	50	52	–	–
c_D [$\$/MWh$]	–	–	−40	−41
$k_{P_{G_i}}$ [$MW/\$$]	0.012	0.018	–	–
b_G [$\$/MWh$]	97.2	98.8	–	–
b_D [$\$/MWh$]	–	–	70	73
$k_{P_{D_j}}$ [$MW/\$$]	–	–	0.015	0.015

to the fact that a dynamic model allows us to capture the interactions between the market and the underlying physics in a more accurate manner, it also sheds light on the stability effects of these interactions.

To provide a numerical evaluation of how large a delay is allowable for the wholesale market to be stable, we briefly examine the typical ramp-up limit in a power system. In particular, we examine the capacity reserves in a power grid, which pertain to real-time market for compensating load fluctuation. We note that the underlying discrete-time model in (39)–(42) included constants $k_{P_{G_i}}$, $k_{P_{D_j}}$, k_{δ_n}, and k_{ρ_n} whose units are summarized in Table 1. Assuming a fixed rate of price, $c^{[h/\$]}$, we can relate the constants $k_{P_{G_i}}^{[MW/\$]}$ to ramp-up limits $R_{Gi}^{[MW/h]}$ as

$$k_{P_{G_i}}^{[MW/\$]} = c^{[h/\$]} R_{Gi}^{[MW/h]}.$$

This implies that entities with larger $R_{Gi}^{[MW/h]}$ can adjust their output faster to the changes of the price, which in turn corresponds to a larger $k_{P_{G_i}}^{[MW/\$]}$ for the same changing rate of price. The following observations can be made regarding capacity reserves [37]:

1. Fast reacting reserves are provided by Automatic Generation Control (AGC) based on real-time frequency deviation. Such reserves cater to about 0.1–0.3% of load with full deploying capability in 30 s, which corresponds to a ramp-up limit $R_{g_1} = 120^{[MW/h]}$.
2. Regulation services that ISO issues commands to every 5–8 s cater to 1–2% of load, with full deployment capability in less than 5 min, which corresponds to a ramp-up limit $R_{g_2} = 12^{[MW/h]}$.
3. Spinning or operating reserves that the ISO may call upon to provide energy within 15 min, and provides an additional 3–6% of peak load or the size of the largest contingency, and corresponds to the ramp-up limit $R_{g_3} = 4^{[MW/h]}$.
4. Slower reserves exits, with a ramp-up limit $R_{g_4} = 0.5 - 2^{[MW/h]}$.

The above numbers reveal that time-scales ranging from 30 s to 2 h are present in a power grid, with ramp-up limits ranging from $120^{[MW/h]}$ to $2^{[MW/h]}$. Therefore, any time-delays and their effects have to be placed in proper perspective.

In the next section, we will evaluate the dynamics of the model we proposed in the presence of various delays in the context of a numerical study.

5   Impact of Time Delay and Communication Topology on the Stability of Wholesale Market

The model in (48) was shown to include three delays, d_G, d_D, and d_{pd}. In what follows, we assume that only one of these delays is present at any given time, resulting in the following three scenarios:

1. The price signals undergo transmission and processing delays to generators $d_G \neq 0$ without delays for consumers ($d_D = 0$) and without delay of consumption information ($d_{pd} = 0$)
2. $d_D \neq 0, d_G = 0$, and $d_{pd} = 0$
3. $d_{pd} \neq 0, d_G = 0$, and $d_D = 0$

We evaluate the upper bound for time delay in each one of these scenarios that ensures the stability of (48). An IEEE-4 bus case is used for all simulation studies, whose interconnections and also the relevant parameters of the generators and consumers are chosen as in [16]. Prices coefficients b_{Gi} and c_{Gi} corresponding to cost functions of generators as well as coefficients b_{Dj} and c_{Dj} of the utility functions of consumers are shown in Table 1.

5.1   Simulation Results

An analysis of (48) showed that the stability is dependent on time delay, with each of the three scenarios described above leading to a different upper bound and are summarized in Table 2, and illustrated in Fig. 3.

As can be seen in this figure, delaying the price signal, in scenario 1, sent to the producer, can be catastrophic and delaying the consumption information to the ISO, considered in scenario 3, is equally critical. In comparison, delaying the price signal sent to the consumer considered in scenario 2 is less of a problem. Indeed, in scenario 3, delaying the consumption to the ISO is also critical. This analysis shed a light on the communication topology design of smart grid. Since GenCo have the smallest delay margin, they should have the highest priority in communication. The relative values of the delay margin in the three scenarios impose a certain structure in the communication topology and is described in Sect. 5.2.

Table 2 Range of delay for scenarios 1–3

	d_G	d_D	d_{pd}
Scenario 1	(0, 3)	0	0
Scenario 2	0	(0,18)	0
Scenario 3	0	0	(0,5)

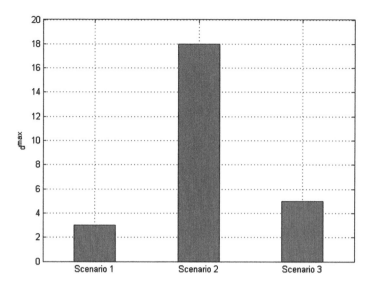

Fig. 3 Delay margin of the wholesale market in different scenarios

5.2 *A Hierarchical Tree-Based Communication Topology*

A hierarchical tree-based communication topology is assumed for coordinating large numbers of DERs and DR units [14]. Tree-structured communication networks, as can be seen in Fig. 4, are characterized by a central root node (at the top of the hierarchy) that is connected to the nodes in the last but one of the highest levels of the hierarchy. The nodes on the second level of the tree are connected to nodes on the third level, and so forth. Let the levels of the tree be denoted by $n \in N := \{1, 2, \ldots, N\}$, where the top root is the only node present in the 0-th level. Each node of the communication network corresponds to either a price-sensitive generator or load (see [38]), whereas the market-based feedback control law, i.e., the price-forming mechanism described by (48), is implemented in the root node. Assuming that the transmission of a message between two connected nodes in the n-th and $(n - 1)$-th level takes exactly T_c seconds, the time needed for price signals generated by the root node to reach a level-n node will be nT_c. In such a topology, based on the results of Fig. 3, it follows that the GenCos must be placed in the higher level of hierarchy so as to minimize nT_c.

6 Summary

The introduction and availability of cyber-communication technologies in a Smart Grid have introduced a paradigm shift in the analysis of power systems. This in turn necessitates a framework that includes a dynamic analysis of the underlying

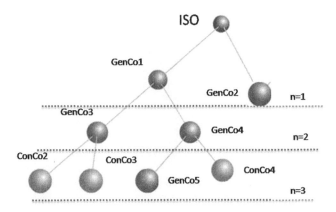

Fig. 4 Hierarchical tree-based communication topology for smart-meter platform

components of DERs, DR units, and cyber-communication components. In this article, we begin with Optimal Power Flow formulation and derive a dynamic model of the real-time market using Primal–Dual interior point method. In particular, a gradient-based algorithm is used to derive the dynamic evolution of the primal variables and dual variables to reach the optimum solution of the real-time market, using a discrete-time framework. The connection between the market equilibrium and Nash equilibrium is discussed.

The stability of the resulting dynamical model of the real-time market is investigated in the presence of delays introduced by the cyber-communication components. In particular, stability of the market is investigated for a range of time delays between the price signal to the GenCos as well as to ConCos and in the delays between ConCos and the ISO. An LMI approach is used to carry out the underlying analysis. The maximum tolerable time delay in each one of scenarios places an implicit limitation on the communication topology design.

Numerical results are included that validate the theoretical results using IEEE four-bus system. The simulation results show that the overall wholesale market stability has the least amount of robustness to the delays in communicating the price signal to the GenCos than delays elsewhere in the overall system. Dependence of the wholesale market stability on the time-delay serves as a design guideline in the underlying communication topology. In a hierarchical tree-based topology, where the time delay is parameterized by the level of hierarchy, the results of the article suggest to place GenCos in the highest level of hierarchy.

Acknowledgments This work was supported by the Technische Universität München—Institute for Advanced Study, funded by the German Excellence Initiative and by Deutsche Forschungsgemeinschaft (DFG) through the TUM International Graduate School of Science and Engineering (IGSSE).

References

1. Torre S, Contreras J, Conejo A (2004) Finding multiperiod nash equilibria in pool-based electricity markets. IEEE Trans Power Syst 19:643–651
2. Rosen J (1965) Existence and uniqueness of equilibrium points for concave N-person games. Econom 33:520–534
3. Berry CA, Hobbs BF, Meroney WA, ÓNeill RP, Stewart WR (1999) Understanding how market power can arise in network competition: A game theoretic approach. Utilities Policy 8:139–158
4. Cardell JB, Hitt CC, Hogan WW (1997) Market power and strategic interaction in electricity networks. Resource Energy, Econom 19:109–137
5. Hobbs BF, Metzler CB, Pang J-S (2000) Strategic gaming analysis for electric power systems: An MPEC approach. IEEE Trans Power Syst 15:637–645
6. Nasser T-O (1998) Transmission congestion contracts and strategic behavior in generation. Electricity J 11:32–39
7. Wei J-Y, Smeers Y (1999) Spatial oligopolistic electricity models with Cournot generators and regulated transmission prices. Oper Res 47:102–112
8. Cunningham LB, Baldick R, Baughman ML (2002) An empirical study of applied game theory: Transmission constrained Cournot behavior. IEEE Trans Power Syst 17:166–172
9. Ferrero RW, Rivera JF, Shahidehpour SM (1998) Application of games with incomplete information for pricing electricity in deregulated power pools. IEEE Trans Power Syst 13:184–189
10. Bai X, Shahidehpour SM, Ramesh VC, Yu E (1997) Transmission analysis by Nash game method. IEEE Trans Power Syst 12:1046–1052
11. H. level Advisory Group on ICT for Smart Electricity Distribution Networks (2009) Ict for a low carbon economy smart electricity distribution networks. European Commission, Technical Report
12. Nutaro J, Protopoescu V (2010) The impact of market clearing time and price signal delay on the sability of electric power market. IEEE Trans Power Syst 24:1337–1345
13. Alvarado F, Meng J, Mota W, DeMarco C (2000) Dynamic coupling between power markets and power systems. IEEE Trans Power Syst 4:2201–2205
14. Hermans R, Jokic A, van den Bosch P, Frunt J, Kamphuis I, Warmer C (2010) Limitations in the design of ancillary service markets imposed by communication network delays. In: 7th international conference on the european energy market (EEM), Madrid, Spain
15. Kiani A, Annaswamy AM (2010) The effect of a smart meter on congestion and stability in a power market. In: IEEE conference on decision and control, Atlanta, USA
16. Kiani A, Annaswamy A (2011) Wholesale energy market in a smart grid: Dynamic modeling, and stability. In: IEEE conference on decision and control (CDC), Florida, USA
17. Carpentier J (1985) Optimal power flow, uses, methods and development, planning and operation of electrical energy system. In: Proceedings of IFAC symposium, Rio de Janeiro, Brazil
18. Alsac O, Stott B (1974) Optimal load flow with steady-state security. IEEE Trans Power Syst 93:745–751
19. Hakim L, Kubokawa J, Yuan Y, Mitani T, Zoka Y, Yorino N, Niwa Y, Shimomura K, Takeuchi A (2009) A study on the effect of generation shedding to total transfer capability by means of transient stability constrained optimal power flow. IEEE Trans Power Syst 24:347–355
20. Wang H, Murillo-Schnchez CE, Zimmerman RD, Thomas RJ (2007) On computational issues of market-based optimal power flow. IEEE Trans Power Syst 22:1185–1193
21. Mehrotra S (1992) On the implementation of a primal-dual interior point method. SIAM J Optim 2(4):575–601
22. Wei H, Sasaki H, Kubakawa J, Yokoyama R (1998) An interior point nonlinear programming for an interior point nonlinear programming for optimal power flow problems with a novel data structure. IEEE Trans Power Syst 13:870–877

23. Wu Y, Debs A, Marsten R (1994) A direct nonlinear predictor corrector primal-dual interior point algorithm for optimal power flows. IEEE Trans Power Syst 9:876–883
24. Kiani A, Annaswamy AM (2010) Perturbation analysis of market equilibrium in the presence of renewable energy resources and demand response. In: IEEE innovative smart grid technologies (ISGT), Gothenburg, Sweden
25. Shahidehpour M, Yamin H, Marke ZL (2002) Operations in electric power systems: forecasting, scheduling and risk management. Wiley, New York
26. Roozbehani M, Dahleh M, Mitter S (2010) On the stability of wholesale electricity markets under real-time pricing. In: IEEE conference on decision and control, Atlanta, USA
27. MacCormack J, Hollis A, Zareipour H, Rosehart W (2010) The large-scale integration of wind generation: Impacts on price, reliability and dispatchable conventional suppliers. Energy Policy 38:3837–3846
28. dong Tang Y, Wu J, Zou Y (2001) The research on the stability of power market. Autom Electr Power Syst 25:11–16
29. Bertsekas DP (2001) Dynamic programming and optimal control. Athena Scientific, Belmont
30. Bertsekas DP (1999) Nonlinear programming. Athena Scientific, Belmont
31. Keith Briggs FK, Smith M (2010) Explicit congestion control: charging, fairness and admission management. Cambridge University Press, Cambridge
32. Shakkottai S, Johari R (2010) Demand-aware content distribution on the internet. IEEE ACM Trans Network 18:476–489
33. Papachristodoulou A, Jadbabaie A (2010) Delay robustness of nonlinear internet congestion control schemes. IEEE Trans Autom Contr 55:1421–1428
34. Shahidehpour M, Alomoush M (2001) Restructured electrical power systems, operation trading and volatility. Marcel Dekker, New York
35. Lee B, Lee J (1999) Delay-dependent stability criteria for discerete-time delay systems. In: American control conference, San Diego, USA
36. Meyn S, Negrete M, Wang G, Kowli A, Shafieepoorfard E (2010) The value of volatile resources in electricity markets. In: IEEE conference on decision and control, Atlanta, USA
37. Caramanis M, Foster J (2009) Management of electric vehicle charging to mitigate renewable energy intermittency and distribution network congestion. In: IEEE conference on decision and control, Shanghai, China
38. Kok K, Warmer C, Kamphuis R (2010) Intelligence in electricity networks for embedding renewables and distribution generation. In: Negenborn RR, Lukszo Z, Hellendoorn J (eds) Intelligent infrastructures, vol 42. Springer, Berlin, pp 179–210

Coordinating Regulation and Demand Response in Electric Power Grids: Direct and Price-Based Tracking Using Multirate Economic Model Predictive Control*

Haitham Hindi, Daniel Greene, and Caitlin Laventall

Abstract We propose a framework for reducing demand-supply imbalances in the grid, by jointly controlling both the supply-side electric power regulation together with the demand-side energy consumption by residential and commercial consumers demand response. We focus on performance improvements that arise from the complementary dynamics: regulation allows for frequent control updates but suffers from slower dynamics; demand response has faster dynamics but does not allow as frequent control updates. We propose a multirate model predictive control (MPC) approach for coordinating the two services, and we refer to this coordinator as an aggregator. Multirate MPC captures the varying dynamics and update rates, and nonlinearities due to saturation and ramp rate limits, and a total variation constraint limits the switching of the demand response signal. Our approach can operate with both direct reference or indirect market-price based imbalance signal. Numerical examples are presented to show the efficacy of this joint control approach.

*Based on "Coordinating regulation and demand response in electric power grids using multirate model predictive control," by H. Hindi, D. Greene, C. Laventall, which appeared in the IEEE Innovative Smart Grid Technologies Conference ISGT 2011, © 2011 IEEE.

H. Hindi (✉) • D. Greene
(Xerox) PARC, Palo Alto, CA, USA
e-mail: hhindi; greene@parc.com

C. Laventall
AOL.com, Palo Alto, CA, USA
e-mail: klaventall@aol.com

A. Chakrabortty and M.D. Ilić (eds.), *Control and Optimization Methods for Electric Smart Grids*, Power Electronics and Power Systems 3, DOI 10.1007/978-1-4614-1605-0_5, © Springer Science+Business Media, LLC 2012

1 Introduction

In this chapter, we explore the idea of reducing demand-supply imbalances in the grid, by jointly controlling both the supply-side electric power regulation together with the demand-side energy consumption by residential and commercial consumers [15, 16, 20]. Currently, the main instrument for regulating the supply demand imbalance is a set of supply-side generation reserves, known as *ancillary services*, that operate on various scales of time and frequency. The control of energy consumption by residential and commercial customers is known as *demand response* and has become a key idea in the Smart Grid vision.

Here, we focus specifically on the potential performance improvements that arise from the complementary nature of two sets of dynamics: *regulation* and *fast demand response*. Regulation is an ancillary service that operates on the second-to-minute timescale, and is traded in units of Mega Watts [19, 22–24]; while fast demand response refers to the kind of rapid demand cutbacks that can be achieved "within the flick of a switch", e.g. from turning off a home appliance, such as an airconditioner or dryer or light [15, 16, 20]. The complementary dynamics we wish to explore in this chapter are the following: regulation allows for frequent control updates but suffers from slower dynamics of large generation equipment; demand response has faster dynamics but does not allow as frequent control updates, due to potential wear and tear on appliances.

We propose *multirate model predictive control* (MPC) as an effective computational framework for coordinating regulation and demand response and for exploring the space of different design scenarios [3, 4, 7, 10, 18, 21, 25]. The multirate MPC approach captures the varying dynamics and update rates, as well as the nonlinearities due to saturation and ramp rate limits, and we use a total variation constraint to limit the switching of the demand response signal. The multirate MPC approach results in a quadratic program (QP) that must be solved at each time step [2, 4, 5, 10, 13] or a more complex optimization problem, e.g. when nonconvex costs are considered. We call this multirate MPC-based coordinator or controller an *aggregator*, because it combines and coordinates the two services into an effective joint ancillary service.

In addition, we show that our approach has the flexibility to be implemented in the two most likely deployment scenarios. In the first, a direct demand-supply imbalance reference tracking signal is available. In the second, an indirect market price-based tracking signal is available [1, 6, 9, 19, 22–25]. This market-based tracking approach is related to recent so-called economic MPC [8].

There are some practical applications, which would not have the resources to solve a QP at each time step. Therefore, we also present a much simpler heuristic controller, which delivers reasonably good performance in some operating regimes.

Numerical examples are presented to show the efficacy of this joint control approach. Specifically, it is shown that, under certain conditions, fast demand response can significantly enhance the quality of traditional supply-side regulation, to achieve better overall performance, in terms of minimizing demand-supply imbalance.

This chapter is a slightly expanded version of [12]; new material includes the derivation of the price-based MPC, which was only sketched briefly in the original paper, along with a numerical example to demonstrate its efficacy.

Notation: $||x||_p$ denotes the p-norm of x for $p > 1$. The *saturation* function with level $\alpha > 0$ and the *Kronecker delta* function are defined, respectively, as

$$\text{sat}_\alpha(x) = \begin{cases} -\alpha & \text{if } x < -\alpha, \\ x & \text{if } |x| \leq \alpha, \\ \alpha & \text{if } x > \alpha, \end{cases} \qquad \delta(x) = \begin{cases} 1 & \text{if } x = 0, \\ 0 & \text{otherwise.} \end{cases}$$

For compactness, we use $k \in I_{n_1}^{n_2}$ to denote $k \in \{n_1, n_1 + 1, \ldots, n_2\}$.

2 Problem Statement

The objective of this chapter is to design a controller, termed the aggregator, which simultaneously manages demand response and a regulation service. More specifically, we would like to design control signals which will enable our plant to track a time varying reference signal (i.e., the energy imbalance). The plant of our model (see Fig. 1) is composed of two subsystems, one for the demand response and one for the regulation service. Evolution of each of these subsystems is subject to a number of constraints, see Sect. 3. Denote x_t^{rf} as the state of the reference (imbalance) signal, x_t^{dr} as the state of the demand response, x_t^{rg} as the state of the regulation service, and

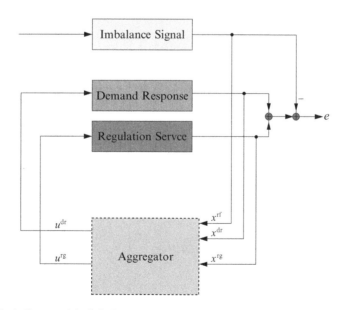

Fig. 1 Block diagram of the linked system

$e_t = x_t^{rg} + x_t^{dr} - x_t^{rf}$ as the tracking error. The control signals u_t^{rg} and u_t^{dr} are produced by the aggregator, which is a time-varying state feedback mapping $f_{agg} : \mathbf{R}^3 \rightarrow \mathbf{R}^2$. We model the closed loop nonlinear time-varying system as

$$
\begin{aligned}
x_{t+1}^{rf} &= x_t^{rf} + w_t^{rf}, \\
x_{t+1}^{rg} &= f_{rg}\left(x_t^{rg}, u_t^{rg}, t\right), \\
x_{t+1}^{dr} &= f_{dr}\left(x_t^{dr}, u_t^{dr}, w_t^{dr}, t\right), \\
e_t &= x_t^{rg} + x_t^{dr} - x_t^{rf}, \\
\left(u_t^{rg}, u_t^{dr}\right) &= f_{agg}\left(x_t^{rf}, x_t^{rg}, x_t^{dr}, t\right),
\end{aligned}
\tag{1}
$$

where f_{rg} and f_{dr} are decoupled, nonlinear functions, f_{agg} is the time varying state feedback control function, which will be computed using MPC or heuristic method. Details of these functions will be covered in Sect. 4.1. Precise modeling of the dynamics of imbalance reference signal neither is the focus of our chapter nor will affect our conclusions. This is because the multirate MPC framework is model based and does not place any restrictions on the reference dynamics. For simplicity, in this work we assume it evolves as a zero mean random walk, driven by a white noise w_t^{rf}. Historic data of this imbalance signal, and its associated price, are publicly available on numerous ISO and utility websites. We also include a noise term w_t^{dr} in the demand response dynamics, due to the expected higher uncertainty in the response of homes and small businesses.

3   Qualitative Description of Models and Specs

Numerous model constraints (e.g., capacity limitations, communication delays.) impede the performance of our system. In the section below, all model constraints we have incorporated into our system are outlined. A qualitative description of the constraints is discussed, followed by their mathematical representation. Once again, we note that the multirate MPC framework is capable of modeling a wide range of dynamical models, including: continuous, discrete, hybrid, and stochastic. Since our main goal here is to focus on the multirate dynamical aspects of regulation and demand response, we will use the standard abstractions of regulation and demand response dynamics as first-order processes with various capacity and ramp-rate limits [1, 6, 9, 19, 22–25].

3.1   Demand Response Constraints

- *Limited communication*: Because communication protocols have yet to be established for the demand response program, we assume limited communication between the aggregator and demand response. Part of this chapter's objective is to show that even when the performance of demand response is limited by infrequent control updates, it is still able to reduce the workload of a regulation service and contribute to grid stability.

- *Customer disutility*: One of the primary limitations of demand response is how much customers are willing to cut back before they become inconvenienced. From the aggregator's perspective, it must limit the total resource consumption by the demand response.
- *Mechanical wear-and-tear of appliances*: Excessive wear on appliances, specifically those with a duty cycle, should be avoided. This means the aggregator cannot send cutback signals to the demand response (homes and businesses) too frequently. We model this by limiting the total variation of the output by the demand response.
- *Uncertainty of response*: An essential component of the demand response program is the ability for a customer to override the aggregator's signal at any time. As the population of demand response participants increases, so does the variance of this uncertainty—hence the w_t^{dr} in (1).

3.2 Regulation Service Constraints

- *Maximum reserve capacity limitations*: Typically, a power plant has allocated a limited amount of its total capacity for regulation service. This means the peak value of regulation power output is limited.
- *Ramp rate*: The large inertia of plant generators limits how quickly they can ramp up or ramp down in response to the aggregator's input. This is modeled as a constraint on the peak value of the control input.

4 Quantitative Description of Models and Specs

4.1 System Model with Incorporated Constraints

Out next task is to put the constraints described in the previous section into a system theoretic framework. Each constraint adds nonlinear behavior to the dynamics. However, as we shall see in Sect. 5, our plant and controller can be reformulated as a linear system with linear constraints. We first define the following parameters.

- α_{max}^{rg}, α_{max}^{dr}: Maximum of the regulation service and demand response, respectively. Saturation point of maximum reserve capacity.
- α_{rmp}^{rg}: Ramp rate constant for the regulation. Saturation point of the input u^{rg}. For completeness, we will also introduce α_{rmp}^{dr}, but it will be set assumed infinite in this chapter.
- T_{rg}, T_{dr}: Input control update rates of the regulation service (demand response, respectively).

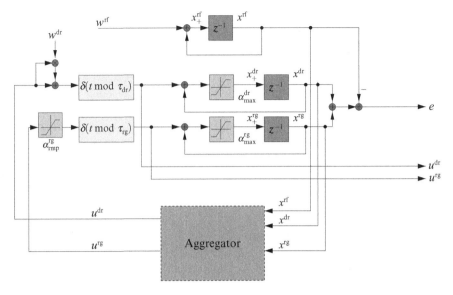

Fig. 2 Block diagram of the linked system with incorporated model constraints

With these parameters, we can describe the state space model (1) in more detail:

$$x_{t+1}^{\mathrm{rf}} = x_t^{\mathrm{rf}} + w_t^{\mathrm{rf}},$$

$$x_{t+1}^{\mathrm{rg}} = \mathrm{sat}_{\alpha_{\max}^{\mathrm{rg}}}\left(x_t^{\mathrm{rg}} + \delta(t \bmod T_{\mathrm{rg}})\, \mathrm{sat}_{\alpha_{\mathrm{rmp}}^{\mathrm{rg}}}(u_t^{\mathrm{rg}})\right),$$

$$x_{t+1}^{\mathrm{dr}} = \mathrm{sat}_{\alpha_{\max}^{\mathrm{dr}}}\left(x_t^{\mathrm{dr}} + \delta(t \bmod T_{\mathrm{dr}})\left(u_t^{\mathrm{dr}} + \sqrt{u_t^{\mathrm{dr}}}w_t^{\mathrm{dr}}\right)\right). \tag{2}$$

See Fig. 2 for a block diagram representation of this system. Note the periodic time-varying characteristic of the control signal: $\delta(t \bmod T_{\mathrm{xx}})u_t^{\mathrm{xx}} \neq 0$ only if t is a multiple of T_{xx}. Thus when $T_{\mathrm{rg}} \neq T_{\mathrm{dr}}$ we effectively have control output which directs each sub-plant at different rates. For the purposes of our model, we assume $T_{\mathrm{rg}} = cT_{\mathrm{dr}}$ with $c < 1$. Thus, demand response receives information from the aggregator less frequently than the regulation service.

Also, note that we have modeled the uncertainty in the demand response as a noise w_t^{dr}, which enters in a *multiplicative* rather than additive way, scaled by the square-root of the input. This is in anticipation that ultimately, the demand response is likely to be aggregated from a large number of participants, e.g. homes and small businesses, thus the variance would scale as the sum of random variables.

We will not pursue the demand response uncertainty modeling any further in this chapter. Observe that in the absence of this multiplicative noise, the structure of the regulation and the demand response dynamics is identical, albeit with different values for their respective parameters, of course. In addition, the regulation will have a finite ramp-rate limit $\alpha_{\mathrm{rmp}}^{\mathrm{rg}} \leq \infty$, while the demand response will be assumed to have effectively infinite ramp rate $\alpha_{\mathrm{rmp}}^{\mathrm{dr}} = \infty$.

4.2 Performance Measures

The following performance metrics measure the costs associated with each model constraint:

- $||x_t^{\text{rf}} - x_t^{\text{rg}} - x_t^{\text{dr}}||_2^2$: Euclidean norm of the tracking error. This is the primary measure we would like to keep small. A soft constraint.
- $||x_t^{\text{rg}}||_1$, $||x_t^{\text{dr}}||_1$: Total resource consumption by the regulation and demand response. This is the dominant cost deriving from regulation cost and consumer disutility. A soft constraint.
- $||x_t^{\text{rg}}||_\infty, ||x_t^{\text{dr}}||_\infty$: Maximum peak of the regulator (demand response, respectively). Restricts state trajectories since they cannot operate beyond full capacity. A hard constraint.
- $||x_t^{\text{dr}}||_{\text{TV}}$: Total variation of demand response. A secondary cost related to mechanical wear-and-tear of the load. A hard constraint.
- $||u_t^{\text{rg}}||_2$, $||u_t^{\text{dr}}||_2$: Weighted input cost for regulation (demand response, respectively). A soft constraint.
- $||u_t^{\text{rg}}||_\infty, ||u_t^{\text{dr}}||_\infty$: Ramp rate constraints. Limits the speed either service can ramp-up or ramp-down. A hard constraint.

Note that we represent the ramp rate constraints as slew-rate limits on the inputs; they could just as well be represented as direct rate limits on the state variables. Also, our choice of using the 1-norm and Euclidean norms to represent those costs are primarily for illustrative and computational convenience. Many other choices are possible for capturing the costs of regulation and demand response, depending on the generation resource being used; while many other choices are possible as metrics of tracking error. Although these other choices might change some of the computational properties of the optimization problem (e.g., convexity), they will still fall within the general framework here of optimizing an objective function with terms representing power costs of regulation and demand response, and terms representing an imbalance or tracking error metric (or price signal, as shown below).

Furthermore, the optimization framework presented here can also be used within a more general framework that considers other costs, prices, and constraints. These could include: fuel cost, startup/shutdown costs, regular operating costs; consumer and appliance utility and discomfort, minimum up/down time constraints; prices for other services, such as spinning and nonspinning reserve.

5 Reference Tracking Multirate MPC

We now design a MPC scheme where at each time step the aggregator solves a planning problem, which incorporates explicit knowledge of the plant model and feedback information into its formulation. For simplicity of notation, we consider a single regulation resource and a single demand response resource. However, the framework extends trivially to any number of either, by simply adding the obvious costs and constraints corresponding to each new agent.

Given the performance measures outlined in the previous section and a finite horizon N, the aggregator will solve the following planning problem at each time step $t = 0, 1, \ldots$:

$$
\underset{u^{\text{rg}}, u^{\text{dr}}, \hat{x}^{\text{rg}}, \hat{x}^{\text{dr}}}{\text{minimize}} \left\{
\begin{array}{l}
\displaystyle\sum_{k=0}^{N} ||\hat{x}_k^{\text{rf}} - \hat{x}_k^{\text{rg}} - \hat{x}_k^{\text{dr}}||_2^2 + \\[2mm]
\displaystyle\sum_{k=0}^{N} \rho_0 ||\hat{x}_k^{\text{rg}}||_1 + \rho_1 ||\hat{x}_k^{\text{dr}}||_1 + \\[2mm]
\displaystyle\sum_{k=0}^{N-1} \rho_2 ||\hat{u}_k^{\text{rg}}||_2^2 + \rho_3 ||\hat{u}_k^{\text{dr}}||_2^2
\end{array}
\right.
$$

subject to
$$\hat{x}_{k+1}^{\text{rg}} = \hat{x}_k^{\text{rg}} + \delta((t+k) \bmod T_{\text{rg}})\hat{u}_k^{\text{rg}}, \ k \in I_0^{N-1}$$

$$\hat{x}_{k+1}^{\text{dr}} = \hat{x}_k^{\text{dr}} + \delta((t+k) \bmod T_{\text{dr}})\hat{u}_k^{\text{dr}}, \ k \in I_0^{N-1}$$

$$\hat{x}_0^{\text{rg}} = x_t^{\text{rg}}$$

$$\hat{x}_0^{\text{dr}} = x_t^{\text{dr}}$$

$$||\hat{x}_k^{\text{rg}}||_\infty \leq \alpha_{\text{max}}^{\text{rg}}, \ k \in I_0^N$$

$$||\hat{x}_k^{\text{dr}}||_\infty \leq \alpha_{\text{max}}^{\text{dr}}, \ k \in I_0^N$$

$$||\hat{u}_k^{\text{rg}}||_\infty \leq \alpha_{\text{rmp}}^{\text{rg}}, \ k \in I_0^{N-1}$$

$$||\hat{u}_k^{\text{dr}}||_\infty \leq \alpha_{\text{rmp}}^{\text{dr}}, \ k \in I_0^{N-1}$$

$$\sum_{k=0}^{N-1} |\hat{x}_{k+1}^{\text{dr}} - \hat{x}_k^{\text{dr}}| \leq \beta_{\text{TV}}, \tag{3}$$

where $\hat{x}_1^{\text{rg}}, \ldots, \hat{x}_N^{\text{rg}}, \hat{x}_1^{\text{dr}}, \ldots, \hat{x}_N^{\text{dr}}, \hat{u}_0^{\text{rg}}, \ldots, \hat{u}_{N-1}^{\text{rg}}, \hat{u}_0^{\text{dr}}, \ldots, \hat{u}_{N-1}^{\text{dr}}$ are our variables and $x_t^{\text{rf}}, T_{\text{rg}}, T_{\text{dr}}, \alpha_{\text{max}}^{\text{rg}}, \alpha_{\text{max}}^{\text{dr}}, \alpha_{\text{rmp}}^{\text{rg}}, \alpha_{\text{rmp}}^{\text{dr}}$, and the initial states $x_t^{\text{rg}}, x_t^{\text{dr}}$ are given data (see Sect. 3); the constants $\rho_0, \rho_1, \rho_2, \rho_3$ allow us to weight the different terms in the cost function.

The reference imbalance signal $\hat{x}_k^{\text{rf}}, k = 0, \ldots, N$, with $\hat{x}_0^{\text{rf}} = x_t^{\text{rf}}$, is assumed to be given. In practice, it could come from an internal or external forecast, previously agreed upon contracts, day-ahead / hour-ahead markets, or other mechanisms. For the purposes of this chapter, we will use a naive certainty equivalent estimate. We will not view the reference signal as a state of the system, but instead as a zero mean random walk that we are trying to track. At any time t we only know the current value x_t^{rf}, but not the future values. So for each planning step we will track $\mathbf{E}[x_k^{\text{rf}}|x_t^{\text{rf}}] = x_t^{\text{rf}}$ for $k = t, t+1, \ldots$. In other words, at each time step t, the planning problem will track a *constant* $\hat{x}_k^{rf} \equiv x_t^{rf}, k = 0, \ldots, N$, that constant being our best estimate of the average value of the future values of x^{rf}, which is the current value x_t^{rf}, since it is a zero mean random walk.

The first summation in the objective function of the optimization problem above penalizes tracking error, while the second and third summations are meant to capture the input and output costs of the regulation and the fast demand response. Note that this MPC planning problem can be cast as a convex QP with linear constraints, which can be solved very efficiently, and to global optimality.

Thus at each time step t, given $(x_t^{rf}, x_t^{rg}, x_t^{dr})$, the aggregator solves the planning problem (3) and selects control signals using

$$f_{agg}\left(x_t^{rf}, x_t^{rg}, x_t^{dr}, t\right) = \left(u_t^{rg,mpc}, u_t^{dr,mpc}\right)$$

$$= \left(\hat{u}_0^{rg}, \hat{u}_0^{dr}\right).$$

The closed loop system with the MPC inputs will evolve as in (2). And because the MPC respects all the system saturation and ramp rate limits, in the absence of noise, (2) reduces to

$$x_{t+1}^{rf} = x_t^{rf} + w_t$$

$$x_{t+1}^{rg} = x_t^{rg} + \delta\left(t \bmod T_{rg}\right) u_t^{rg,mpc}$$

$$x_{t+1}^{dr} = x_t^{dr} + \delta\left(t \bmod T_{dr}\right) u_t^{dr,mpc}$$

$$e_t = x_t^{rf} - x_t^{rg} - x_t^{dr}.$$

Note again the time-varying characteristic of the control signal: $\delta(t \bmod T_{xx})$ $u_t^{xx,mpc} \neq 0$ only if t is a multiple of T_{xx}.

6 Market Price Based Multirate MPC

In this section, we model the scenario where the aggregator is operating off an indirect imbalance signal, a market price signal λ_t, rather than a direct reference signal x_t^{rf}. Again, for simplicity of notation, we consider a single regulation resource and a single demand response resource.

This indirect market price formulation can be rigorously derived from the direct reference tracking formulation above: one applies the standard economics method, of appending the market clearing constraint (demand=supply) to the objective in the first formulation, then duality is used to obtain a decomposition into the usual producer and consumer subproblems. The objective in our second formulation below is equivalent to producer subproblem, namely that of profit maximization; the constraints remain unchanged.

Toward this end, let us re-write our direct reference tracking problem (3) in an equivalent way, with an extra variable y:

$$
\underset{u^{\text{rg}}, u^{\text{dr}}, \hat{x}^{\text{rg}}, \hat{x}^{\text{dr}}, y}{\text{minimize}}
\begin{cases}
\displaystyle\sum_{k=0}^{N} ||\hat{x}_k^{\text{rf}} - y_k||_2^2 + \\[2mm]
\displaystyle\sum_{k=0}^{N} \rho_0 ||\hat{x}_k^{\text{rg}}||_1 + \rho_1 ||\hat{x}_k^{\text{dr}}||_1 + \\[2mm]
\displaystyle\sum_{k=0}^{N-1} \rho_2 ||\hat{u}_k^{\text{rg}}||_2^2 + \rho_3 ||\hat{u}_k^{\text{dr}}||_2^2
\end{cases}
$$

$$\text{subject to } y_k = \hat{x}_k^{\text{rg}} + \hat{x}_k^{\text{dr}}, \quad k \in I_0^N$$

$$\text{constraints of (3).} \tag{4}$$

Clearly, this formulation is equivalent to (3), as the new equality could be used to eliminate y from (4) and recover (3). The new variable, y, represents the amount of load actually fulfilled, as compared to x^{rf}, which is to be interpreted as the amount of load power "desired." The expression $||\hat{x}_k^{\text{rf}} - y_k||_2^2$ can be interpreted as demand disutility, or a penalty on unmet demand. The first line in the objective of (4) can be interpreted as the consumer benefit function, which models demand. The second and third lines of the objective can be interpreted as the production cost, as they measure the amount of inputs used and the amount of power produced. Hence, they model supply. The new constraint is simply enforcing that actual fulfilled demand y_k must equal supply $\hat{x}_k^{\text{rg}} + \hat{x}_k^{\text{dr}}$.

Forming the partial Lagrangian with the demand=supply constraint we obtain

$$
\begin{aligned}
L(u, x, y, \lambda) &= \sum_{k=0}^{N} ||\hat{x}_k^{\text{rf}} - y_k||_2^2 + \lambda_k(y_k - \hat{x}_k^{\text{rg}} - \hat{x}_k^{\text{dr}}) \\
&\quad + \sum_{k=0}^{N} \rho_0 ||\hat{x}_k^{\text{rg}}||_1 + \rho_1 ||\hat{x}_k^{\text{dr}}||_1 \\
&\quad + \sum_{k=0}^{N-1} \rho_2 ||\hat{u}_k^{\text{rg}}||_2^2 + \rho_3 ||\hat{u}_k^{\text{dr}}||_2^2 \\
&= \left\{ \sum_{k=0}^{N} ||\hat{x}_k^{\text{rf}} - y_k||_2^2 + \lambda_k\, y_k \right\} \\
&\quad + \left\{ \sum_{k=0}^{N} \rho_0 ||\hat{x}_k^{\text{rg}}||_1 + \rho_1 ||\hat{x}_k^{\text{dr}}||_1 - \lambda_k \left(\hat{x}_k^{\text{rg}} + \hat{x}_k^{\text{dr}} \right) \right. \\
&\quad \left. + \sum_{k=0}^{N-1} \rho_2 ||\hat{u}_k^{\text{rg}}||_2^2 + \rho_3 ||\hat{u}_k^{\text{dr}}||_2^2 \right\}.
\end{aligned}
\tag{5}
$$

The first line in the objective of (5) can be interpreted as the consumer surplus function, which trades off deviation penalty with payment $\lambda_k\, y_k$. The second and third lines of the objective can be interpreted as the (negative of) producer profit, as they measure the amount of inputs and power produced minus revenue $\lambda_k(\hat{x}_k^{\mathrm{rg}}+\hat{x}_k^{\mathrm{dr}})$. Note that, for a fixed λ, the consumer surplus is just a function of y, while the production cost is just a function of $\hat{x}$ and u. Therefore, the partial Lagrangian is seperable.

The dual function can now be defined as

$$
\begin{aligned}
q(\lambda) = \underset{y}{\mathrm{minimize}} &\left\{ \sum_{k=0}^{N} ||\hat{x}_k^{\mathrm{rf}} - y_k||_2^2 + \lambda_k\, y_k \right\} \\
+ \underset{u,\hat{x},(3)}{\mathrm{minimize}} &\left\{ \sum_{k=0}^{N} \rho_0||\hat{x}_k^{\mathrm{rg}}||_1 + \rho_1||\hat{x}_k^{\mathrm{dr}}||_1 - \lambda_k \left(\hat{x}_k^{\mathrm{rg}} + \hat{x}_k^{\mathrm{dr}}\right) \right. \\
&\left. + \sum_{k=0}^{N-1} \rho_2||\hat{u}_k^{\mathrm{rg}}||_2^2 + \rho_3||\hat{u}_k^{\mathrm{dr}}||_2^2 \right\}.
\end{aligned}
\tag{6}
$$

Thus for a given price λ, computing the dual function amounts to solving two separate optimization problems: a consumer surplus maximization and a producer profit maximization.

The role of the market maker, aggregator, or ISO, would be to compute the optimal price $\lambda^\star$, which solves the dual optimization problem

$$
\underset{\lambda}{\mathrm{maximize}} \quad q(\lambda).
\tag{7}
$$

Then, under suitable conditions (e.g. convexity of costs, and polytopic constraints) strong duality will hold, so if the consumers and producers solve their individual optimizations using the optimal prices $\lambda^\star$, they will also solve (4), with the demand=supply constraint intact, and hence the equivalent (3). Therefore, $\lambda^\star$ can be used for price-based tracking.

Hence, in the price-based market scenario, the aggregator would solve the following QP planning problem at each time step $t = 0, 1, \dots$:

$$
\underset{u^{\mathrm{rg}},u^{\mathrm{dr}},\hat{x}^{\mathrm{rg}},\hat{x}^{\mathrm{dr}}}{\mathrm{minimize}}
\left\{
\begin{aligned}
&\sum_{k=0}^{N} -\hat{\lambda}_k \cdot \left(\hat{x}_k^{\mathrm{rg}} + \hat{x}_k^{\mathrm{dr}}\right) + \\
&\sum_{k=0}^{N} \rho_0||\hat{x}_k^{\mathrm{rg}}||_1 + \rho_1||\hat{x}_k^{\mathrm{dr}}||_1 + \\
&\sum_{k=0}^{N-1} \rho_2||\hat{u}_k^{\mathrm{rg}}||_2^2 + \rho_3||\hat{u}_k^{\mathrm{dr}}||_2^2
\end{aligned}
\right.
$$

$$
\text{subject to } \hat{x}_{k+1}^{\mathrm{rg}} = \hat{x}_k^{\mathrm{rg}} + \delta((t+k) \bmod T_{\mathrm{rg}})\hat{u}_k^{\mathrm{rg}}, \; k \in I_0^{N-1}
$$

$$
\hat{x}_{k+1}^{\mathrm{dr}} = \hat{x}_k^{\mathrm{dr}} + \delta((t+k) \bmod T_{\mathrm{dr}})\hat{u}_k^{\mathrm{dr}}, \; k \in I_0^{N-1}
$$

$$
\hat{x}_0^{\mathrm{rg}} = x_t^{\mathrm{rg}}
$$

$$
\hat{x}_0^{\mathrm{dr}} = x_t^{\mathrm{dr}}
$$

$$||\hat{x}_k^{\mathrm{rg}}||_\infty \le \alpha_{\max}^{\mathrm{rg}}, \ k \in I_0^N$$

$$||\hat{x}_k^{\mathrm{dr}}||_\infty \le \alpha_{\max}^{\mathrm{dr}}, \ k \in I_0^N$$

$$||\hat{u}_k^{\mathrm{rg}}||_\infty \le \alpha_{\mathrm{rmp}}^{\mathrm{rg}}, \ k \in I_0^{N-1}$$

$$||\hat{u}_k^{\mathrm{dr}}||_\infty \le \alpha_{\mathrm{rmp}}^{\mathrm{dr}}, \ k \in I_0^{N-1}$$

$$\sum_{k=0}^{N-1} |\hat{x}_{k+1}^{\mathrm{dr}} - \hat{x}_k^{\mathrm{dr}}| \le \beta_{\mathrm{TV}}, \tag{8}$$

where $\hat{\lambda}_k, k = 0, \ldots, N$ is the market price signal, with $\hat{\lambda}_0 = \lambda_t$; it is a surrogate for the market's best estimate of $\lambda^\star$. The other constants and variables are as defined earlier. The price signal $\hat{\lambda}_t$ could come from market clearing price, internal or external price forecasts, previously agreed upon contracts, day-ahead / hour-ahead markets, or other market mechanisms.

In this formulation, the objective function is minimizing the difference between the cost of regulation and demand response, captured in the second two summations, and the revenue, captured in the first summation. Minimizing this difference between cost and revenue is equivalent to maximizing profit, which is defined as revenue minus cost.

Thus at each time step t, given $(\lambda_t, x_t^{\mathrm{rg}}, x_t^{\mathrm{dr}})$, the aggregator solves the planning problem (8) and selects control signals using

$$f_{\mathrm{agg}}\left(\lambda_t, x_t^{\mathrm{rg}}, x_t^{\mathrm{dr}}, t\right) = \left(u_t^{\mathrm{rg,mpc}}, u_t^{\mathrm{dr,mpc}}\right)$$

$$= \left(\hat{u}_0^{\mathrm{rg}}, \hat{u}_0^{\mathrm{dr}}\right).$$

As in the direct reference tracking case above, the resulting closed loop system with the MPC inputs will evolve according to (2).

A couple of comments regarding the practical implementation of this price-based controller are in order. We note that the seperability technique described here is applicable to any number of generators (producers) and loads (consumers). Under convexity and strong duality, the same optimal price vector will clear the entire market with multiple consumers and producers. On the other hand, it is well known that real-world power system market optimizations contain nonconvexities, e.g. due to generation constraints such as startup/shutdown, minimum up/down time, and general unit commitment issues. Regarding this issue we have two comments. First, it can be shown that as the number of participants becomes large, the effect of these nonconvexities diminishes, to the point where the duality gap can be negligible. Second, even if this is not the case, one can view our method as being instantiated after the unit commitment is done, i.e., during the economic dispatch phase, where the integer variables associated with the nonconvexities have been predetermined. Finally, the problem of estimating or forecasting the surrogate optimal market price schedule $\hat{\lambda}$ would need be adequately addressed, before a method such as the one we propose here could be deployed in practice.

7 Heuristic Control

In scenarios where it is not possible to solve a QP at each time step, we developed a heuristic controller which can still deliver good performance, in terms of linking demand response with regulation. This was achieved by designing the following control signals

$$
\begin{aligned}
u_t^{\text{rg,heu}} &= x_t^{\text{rf}} - x_t^{\text{dr}} - x_t^{\text{rg}}, \\
u_t^{\text{dr,heu}} &= x_t^{\text{rf}} - x_t^{\text{dr}} - \text{sat}_{\alpha_{\text{res}}}\left(x_t^{\text{rg}}\right),
\end{aligned} \tag{9}
$$

where α_{res} is a parameter which adjusts how much of the imbalance is taken on by the regulation service. The closed loop system will then evolve following (2).

Roughly speaking, in the absence of saturation, this controller feeds back the tracking error between the reference and the sum of the regulation and demand response. Thus, the closed loop system essentially acts like a multirate integral controller for rejecting the reference disturbance. The more frequently acting regulation service handles small variations below α_{res}, while the less frequent but potentially larger demand response handles larger variations that could create ramping problems for the regulation service. Beyond the level α_{res}, the regulation control input is saturated explicitly, so that more of the control effort must come from the demand response.

Section 8 provides simulation and performance evaluation of this system. We will see that in certain operating regimes, this controller performs surprisingly well, considering its simplicity.

8 Numerical Examples

Simulations of our model using both the heuristic and MPC control schemes are generated below. Using the MPC framework, a Pareto optimal trade-off curve between tracking error and total resource consumption by the demand response is also generated. Note that our examples are all implemented with dynamics on the timescale of seconds to emphasize the fast demand response aspects of this study. However, our multirate MPC framework does not depend on this in any way. Energy markets are continually evolving, so it is important to maintain generality.

8.1 Multirate MPC Versus Heuristic Controller

Figure 3 shows a simulation of (2) with the multirate MPC with the following parameters:

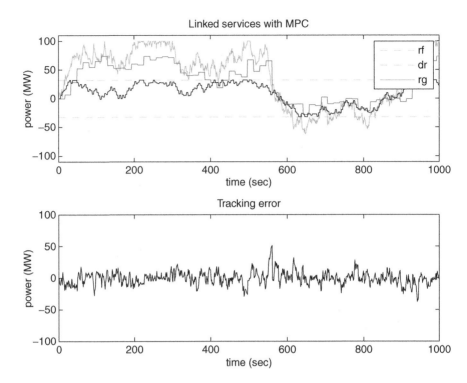

Fig. 3 State trajectories with the MPC controller for $T = 1,000$ s. The reference has the fastest update rate, followed by regulation, followed by demand response, with the slowest update rate

Parameter	Value
α_{max}^{rg}	32 mw
α_{max}^{dr}	100 mw
α_{rmp}^{rg}	6 mw
α_{rmp}^{dr}	∞
α_{res}	N/A
β_{TV}	∞
T_{rg}	4 s
T_{dr}	16 s

Note the reference signal is shown as having the fastest update rate (1 s). Next is the regulation, with slightly "blockier" looking response (4 s). Slowest in terms of update rate is the demand response, which has the "blockiest" response (16 s).

In order to compare the performance of the heuristic model with the MPC model, we matched as many parameters as possible. The MPC model enables us to control more constraints than the heuristic model (e.g., total variation, total

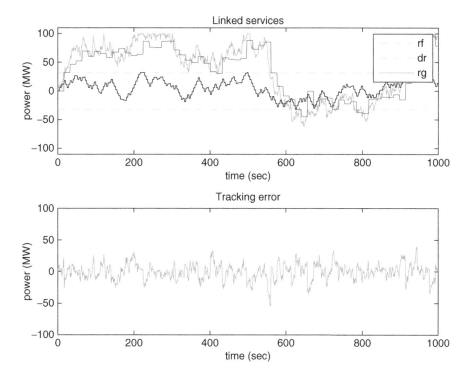

Fig. 4 State trajectories with the heuristic controller for $T = 1,000$ s

resource consumption.), so parameters not available to the heuristic model were left unbounded. Figure 4 shows a simulation of (9) and (2) with the following parameters:

parameter	value
α_{max}^{rg}	32 mw
α_{max}^{dr}	100 mw
α_{rmp}^{rg}	6 mw
α_{rmp}^{dr}	N/A
α_{res}	12 mw
β_{TV}	N/A
T_{rg}	4 s
T_{dr}	16 s

State trajectories are plotted in Fig. 4 over $T = 1,000$ s.

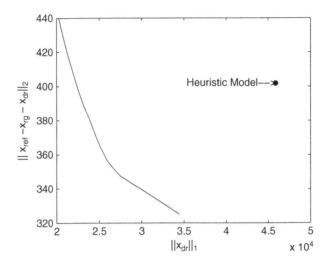

Fig. 5 Trade-off curve between the tracking error and total resource consumption by the demand response for $\rho_1 \in [1 \times 10^{-8}, 10]$, and $\rho_0 = \rho_2 = \rho_3 = 0$

8.2 Pareto Optimal Performance Curve

The following chart compares the performance of the heuristic model versus MPC with $\rho_1 = 1 \times 10^{-8}$, 2.5 and 9.5 and $\rho_0 = \rho_2 = \rho_3 = 0$. All other measures were adjusted to be approximately equal. Note that for $\rho_1 = 9.5$ both controllers have comparable tracking error cost, but the MPC controller reduces total resource consumption by more than 50%.

	Heuristic	$\rho_1 = 10^{-8}$	$\rho_1 = 2.5$	$\rho_1 = 9.5$
Tracking error	401.25	325.21	365.70	439.01
$\|x^{dr}\|_1$	45,870	34,415	25,020	20,322

A Pareto-optimal curve modeling the trade-off in cost between tracking error and resource consumption by the demand response is generated in Fig. 5. This curve defines the limits of performance of our system and can be used as a benchmark for measuring the performance of other controllers. This curve was generated via the scalarized multi-criterion optimization problem defined in Sect. 5 with $\rho_1 \in [10^{-8}, 10]$ and $\rho_0 = \rho_2 = \rho_3 = 10^{-8}$. More specifically, a full MPC simulation was run for 20 samples of $\rho_1 \in [10^{-8}, 10]$. For each sample, the metrics $\|x^{rf} - x^{rg} - x^{dr}\|_2^2$ and $\|x^{dr}\|_1$ were computed. A graph of these values generates the curve.

One might wonder if the performance of the heuristic controller could be improved, somewhat, by further tuning, or using a different heuristic controller. The Pareto curve gives us the answer, by showing us potentially how much more there would be to gain from such tuning or from another controller—quite a bit in this case.

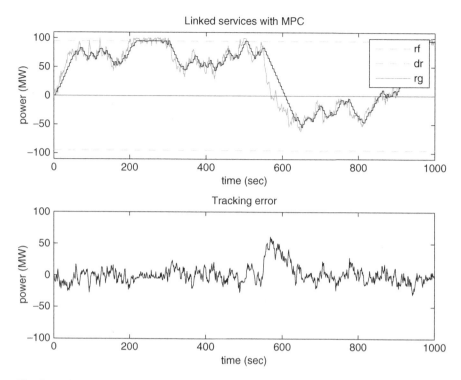

Fig. 6 MPC controller with regulation only; no demand response

8.3   Regulation Versus Demand Response

Figures 6 and 7 compare the performance of pure regulation versus combined regulation plus demand response. The results highlight how, in spite of the slower update rate, the speed with which demand response can react significantly reduces the error in the region around $t = 600$, where the imbalance reference transitions quickly from positive to negative. This is because the demand response has no limitation on its ramp rate: $\alpha_{rmp}^{dr} = \infty$, i.e., demand response can react almost instantly.

8.4   Price-Based Tracking via Economic MPC

Figure 8 illustrated tracking via the economic MPC method. The bottom plot shows the price signal, which is used as an input to the economic MPC of (8). Note that the price can go negative, e.g. around time 560 s. This happens when supply (DR+regulation) exceeds demand (load), in which case ancillary services are paid to spend more energy. Similar requirements exist already in today's ancillary markets, known as "reg-up"/"reg-down". The middle plot shows the multirate

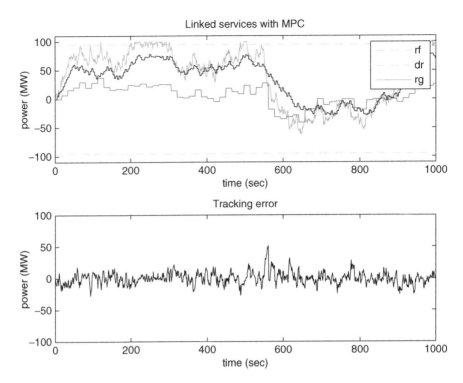

Fig. 7 MPC controller with equal regulation and demand response

control, with the larger less frequent stems being demand response, while the more frequent smaller stems are the regulation. The top plot demonstrates that effective tracking is possible using this price as a reference signal.

9 Conclusion

In this chapter, we have explored the question of whether it is possible to reduce demand-supply imbalances in the grid, by jointly controlling both the supply-side electric power regulation together with the demand-side energy consumption by residential and commercial consumers. Specifically, we focused on the potential performance improvements that arise from the complementary nature of the dynamics of the two: regulation allows for frequent control updates but suffers from slower dynamics; demand response has faster dynamics but does not allow as frequent control updates.

We proposed a multirate MPC approach. This captures the varying dynamics and update rates, as well as the nonlinearities due to saturation and ramp rate limits, and we use a total variation constraint to limit the switching of the demand response

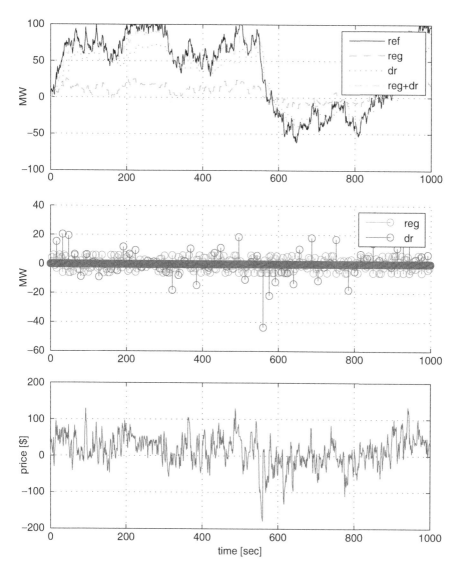

Fig. 8 Economic MPC: tracking using the price as the reference input. (*Top*) Imbalance signal along with regulation, DR, and regulation+DR. (*Mid*) Multirate control: regulation and DR. (*Bottom*) Price signal

signal. The multirate MPC approach results in a QP that must be solved at each time step. We also presented a much simpler heuristic controller which delivers reasonably good performance. In addition, we showed that our approach has the flexibility to be implemented in the two most likely deployment scenarios: where a direct demand-supply imbalance reference tracking signal is available; or where an indirect market price based imbalance signal is available.

Numerical examples were presented to show the efficacy of this joint control approach. Specifically, it was shown that there are indeed conditions under which fast demand response can significantly enhance the quality of traditional supply-side regulation, to achieve better overall performance.

In closing, we also point out that the multirate MPC framework presented here is not limited to modeling only regulation and demand response. It can just as well model any combination of loads and power sources, with differing dynamics, control update rates, and sample rates.

Acknowledgments We are grateful to Dr. Anna Osepayshvili for very helpful discussions about power system economics and markets. Our computations were done in Matlab, using the CPLEX [14] and Gurobi [11] solvers. We used the Yalmip [17] and Tomlab [13] solver interfaces to code up our optimization problems. We are grateful to Johan Löfberg for his help with our Yalmip implementations, and Per Rutquist for his help with our Tomlab implementations.

References

1. Arroyo J, Conejo A (2000) Optimal response of a thermal unit to an electricity spot market. IEEE Trans Power Syst 15(3):1098–1104
2. Bertsekas D (1999) Nonlinear programming. Athena Scientific, Belmont
3. Bittanti S, Colaneri P (2008) Periodic systems: filtering and control. Springer, Hiedelberg
4. Borrelli F, Bemporad A, Morari M (2010) Model predictive control. Unpublished book manuscript
5. Boyd S, Vandenberghe L (2004) Convex optimization. Cambridge Press, Cambridge
6. Carrion M, Arroyo J (2006) A computationally efficient mixed-integer linear formulation for the thermal unit commitment problem. IEEE Trans Power Syst 21(3):1371–1378
7. Constantinescu E, Zavala V, Rocklin M, Lee S, Anitescu M (2009) Unit commitment with wind power generation: Integrating wind forecast uncertainty and stochastic programming. Argonne National Laboratory, Technical Report
8. Diehl M, Amrit R, Rawlings J (2011) A Lyapunov function for economic optimizing model predictive control. IEEE Trans Autom Contr 56(3):703–707
9. Galiana FD, Motto AL, Conejo AJ, Huneault M (2001) Decentralized nodal-price self-dispatch and unit commitment. In: Hobbs B, Rothkopf M, O'Neill R, Chao H (eds) The next generation of electric power unit commitment models, vol 15. Springer, Heidelberg
10. Gondhalekar R, Jones C (2009) A unified framework for multirate and multiplexed model predictive control via periodic systems. In: IEEE conference on decision and control, Shanghai, China
11. Gurobi Optimization, Houston, TX, USA, http://gurobi.com
12. Hindi H, Greene D, Laventall C (2011) Coordinating regulation and demand response in electric power grids using multirate model predictive control. In: IEEE conference on innovative smart grid technologies ISGT, Anaheim, California, USA
13. Holmstrom K, Goran A, Edvall M (2005) TOMLAB/CPLEX Optimization Software for Matlab, TOMLAB Optimization, http://tomopt.com, 2005
14. IBM ILOG CPLEX Optimizer, IBM, CA, USA, http://www.ibm.com/CPLEX
15. Kirby B, Hirst E (1999) Load as a resource in providing ancillary services. Oak Ridge National Labortory, Technical Report
16. Kirby B, Milligan M (2008) An examination of capacity and ramping impacts of wind energy on power systems. Electricity J 21(7):30–42. [Online]. Available: http://www.sciencedirect.com/science/article/B6VSS-4TG8KCD-1/2/d4b9b37f9a250605a790c46c8d07c63f

17. Löfberg J (2004) YALMIP: A Toolbox for Modeling and Optimization in MATLAB. Proceedings of the 2004 IEEE International Symposium on Computer Aided Control System Design, CACSD, Taipei, Taiwan
18. Mayne DQ, Rawlings JB, Rao CV, Scokaert POM (2000) Constrained model predictive control: stability and optimality. Automatica 36(6):789–814. [Online]. Available: http://dx.doi.org/10.1016/S0005-1098(99)00214-9
19. Papavasiliou A, Hindi H, Greene D (2010) Market based control mechanisms for electric power demand response. In: 49th IEEE conference on decision and control (CDC 2010) December 15–17 2010, Atlanta, GA
20. Piette M, Kiliccote S, Ghatiker G (2008) Linking continuous energy management and open automated demand response. Lawrence Berkeley National Laboratory, Paper LBNL-1361E
21. Primbs J, Sung C (2009) Stochastic receding horizon control of constrained linear systems with state and control multiplicative noise. IEEE Trans Autom Contr 54(2):221–230
22. Schweppe F, Caramanis M, Tabors R, Bohn R (1988) Spot pricing of electricity. Kluwer Academic Press, Boston
23. Shahidehpour M, Yamin H, Li Z (2002) Market operations in electric power systems. Wiley, New York
24. Wood A, Wollenberg B (1996) Power generation, operation, and control. Wiley, New York
25. Xie L, Ilic M (2009) Model predictive economic/environmental dispatch of power systems with intermittent resources. In: IEEE PES general meeting, Calgary, Alberta, Canada

Smart Vehicles in the Smart Grid: Challenges, Trends, and Application to the Design of Charging Stations

I. Safak Bayram, George Michailidis, Michael Devetsikiotis,
Fabrizio Granelli, and Subhashish Bhattacharya

Abstract Future "smart electric vehicles", expected to evolve from emerging electric and plug-in hybrid electric vehicles (PHEV) are becoming increasingly attractive. However, the current electric grid is not considered capable of handling the power demand increase required by a large number of charging stations, especially during peak loads. Furthermore, the envisioned critical infrastructure for such vehicles must include the capability for information exchange involving energy availability, distances, congestion levels and possibly, spot prices or priority incentives. In this chapter we discuss current trends and challenges in this fascinating and rapidly developing area of research. Our emphasis is on topics related to control, demand-response, infrastructure provisioning and the communications framework necessary to accomplish all of these "smart" features. As a particular application and a form of case study, we zero in on the design and development of charging stations. We describe a candidate PHEV charging station architecture, and a quantitative stochastic model that allows the analysis of its performance, using queuing theory and economics. The architecture we envision has the capability to store excess power obtained from the grid. Our goal is to promote a general architecture able to sustain grid stability, while providing a required level of quality of service; and to further the development of a general methodology to analyze the performance of such stations with respect to traffic characteristics, energy storage size, pricing and cost parameters.

I.S. Bayram (✉) • M. Devetsikiotis • S. Bhattacharya
NC State University, Raleigh, NC, USA
e-mail: isbayram@ncsu.edu; sbhatta4@ncsu.edu; mdevets@ncsu.edu

G. Michailidis
University of Michigan, Ann Arbor, MI, USA
e-mail: gmichail@umich.edu

F. Granelli
University of Trento, Trento, Italy
e-mail: granelli@disi.unitn.it

A. Chakrabortty and M.D. Ilić (eds.), *Control and Optimization Methods for Electric Smart Grids*, Power Electronics and Power Systems 3, DOI 10.1007/978-1-4614-1605-0_6, © Springer Science+Business Media, LLC 2012

1 Smart Vehicles in the Era of the Smart Grid

The development of a smart grid offers the possibility to increase the connectivity, automation and coordination between power suppliers, consumers and the corresponding transportation and distribution networks, with the purpose of increasing efficiency in power utilization. The traditional paradigm of the power grid is fast changing with the more decentralized power generation even at the level of small businesses and residential units that are becoming capable of covering some of their energy needs and also sell back surplus power back to the utility companies. Further, there is a significant shift to more volatile power generating resources, such as wind and solar, that in turn pose new challenges for real time management of power flows, metering technologies, and coordination between the various agents involved.

On the demand side, there is a strong push by governments and car manufacturers toward Electrical Vehicles (EV) and Plug-in Hybrid Electrical Vehicles (PHEV), in order to reduce dependence on fossil fuels together with emissions. However, as a number of studies indicate [1–4], the penetration of EV/PHEVs will strain the available capacity of the power grid and especially its capability to handle peak loads. For example, [1] concludes that by considering daily average driving requirements, existing US electric power infrastructure can support up to 73% of the energy requirement of light-duty vehicle fleets if nighttime charging is employed. Similarly, [4] shows that the existing Vermont electric grid could support up to 100,000 PHEVs recharging during off-peak hours. On the other hand, the EV/PHEVs themselves can act as power storage units that can be tapped if necessary to accommodate demand peaks, or temporary loss of generating capacity.

We believe that the introduction of distributed and volatile modes of power generation and the presence of PHEVs require enhanced communication technologies and a move towards "smarter vehicles." Already, in the case of EV/PHEVs there are numerous efforts on developing vehicle-to-vehicle (V2V) and vehicle-to-infrastructure (V2I) communication technologies, so that they will be able to interact among themselves and with the existing power infrastructure, so that they can accomplish better utilization of available power and faster recharge times, which in turn would speed up their social acceptance.

The integration of the above concepts leads to the so-called Vehicle-to-grid (V2G) scenario: a system in which EV/PHEVs communicate with the power grid to sell demand response services by either delivering electricity into the grid or by throttling their charging rate. The overall concept is to enable vehicles to become major actors in the smart grid scenario, allowing V2G vehicles to provide power to help balance loads by "valley filling" (charging at night when demand is low) and "peak shaving" (sending power back to the grid when demand is high). This could provide utilities new ways to provide regulation services (by maintaining stability in voltage and frequency) and provide spinning reserves (by meeting sudden demands for power).

This approach could provide a most efficient utilization of intermittent renewable energy sources such as wind power, with vehicles able to store excess energy produced during windy periods and to provide it back to the grid during high load periods. This scenario offers several challenges and interesting open problems crossing the fields of power management and communications. This chapter provides a brief overview of some of these issues and challenges and also examines the role of charging stations and their storage capacity in handling the increased power demands of EV/PHEVs. The main challenges can be classified into the following areas:

1.1 Demand Response Support

Demand response support allows generators and loads to interact in an automated fashion in real time, coordinating demand to flatten spikes. The main purpose is to avoid the deployment of reserve generators by shaving the fraction of demand generating such spikes, and to allow users to cut their energy bills by controlling the use of low priority appliances and devices during off peak hours when charging rates are low. Currently, power grid systems have varying degrees of communication within control systems, including generator plants, transmission lines, substations and major energy users. In general, information flows one way, from the users and the loads back to the utilities. The total amount of power demand by the users can have a very wide probability distribution, which requires spare generating capacity in standby mode to respond to the rapidly changing power usage. "This one-way flow of information is expensive; the last 10% of generating capacity may be required as little as 1% of the time, and brownouts and outages can be costly to consumers" [5]. The V2G paradigm fits this area by enabling vehicles to become active actors in the smart grid arena. On the one hand, electric or hybrid vehicles could be considered as "moving batteries" that the grid could exploit when needed. This point is consistent with the observation that 90–95% of the time vehicles are parked. On the other hand, plug-in battery charging or switching stations would be required to enable efficient penetration of EV/PHEVs in the mobility market – to define an eco-friendly alternative mobility solution.

1.2 Communication Technologies

Integrated and efficient communication technologies are required to support the finer control and information gathering processes at the basis of the smart grid concept. A key technology in this framework is represented by power-line communications, which enable us to exploit the energy distribution system as a signaling infrastructure. Fiber-optics can also be used as a reliable, very high speed medium for smart grid communications.

However, in the case of EV/PHEVs, the need for wireless communications is clearly a must. Along these lines, recent advancements in Vehicular Ad-Hoc

Networking (VANET) in forms of V2V and V2I communications represent a promising technology to support the V2G interaction. Nevertheless, further research is required along this line to address smart-grid-specific issues such as security and reliability. Moreover, wireless communications would enable the fixed infrastructure connected with the smart grid to inform customers or vehicles themselves about incentives or price fluctuations in order to control the demand response pattern, and maintain stability of the power request over time.

1.3 Pricing

Pricing is a key control variable, since policies for efficient usage of power can be designed around price incentives. In many countries, the use of electric utilities has led to installing double tariff electricity meters in many homes, to encourage people to use their electric power during nighttime or weekends, when the overall demand from big power users (e.g., industry) is very low. During off-peak time the price is reduced significantly as demand decreases, enabling utilization of power for heating storage radiators but also for domestic appliances. In the framework of the smart grid, the above idea could be further explored and detailed on a finer timescale, enabling potential price changes in short timescales, based on a real-time estimation of the overall demand–including the potential contribution of customers from renewable sources. Personal preferences of customers, for example to use only green energy, can be incorporated in this novel paradigm of the power grid.

2 Smart Vehicles: Current Trends and Developments

As argued in the introductory section, PHEVs will play an important role in the emerging smart grid, both as power users, but also as storage units. According to the Electric Power Research Institute [1] and the International Energy Agency [6], with current policies and regulations, 50% of the new car sale will be green energy vehicles in the near future. However, a model where charging occurs only during at night at the user's residence will not be practical; hence, the emergence of public charging stations coupled with the development of fast charging technologies.

The introduction of public charging stations would lead to a charge depleting range as low as 13 miles, thus considerably enhancing their usability at all times and across wider regions. However, their adoption would be heavily influenced by the time it takes to charge them away from the owner's residence and not during the night. Regarding standards for charging systems, SAE J1772 [7] defined a conductive charging system architecture for electric vehicles. SAE recommends three types of charging methods, namely AC level 1, AC level 2 and DC charging. Table 1 summarizes these requirements. In [3], researchers present the impacts of PHEV charging under different charging scenarios.

Table 1 Charge method electrical ratings in North America [7]

Charge method	Nominal supply voltage (V)	Maximum current (Amps continuous)	Branch circuit breaker rating (A)
AC level	120 vac, 1-Phase	12	15 (minimum)
AC level	208–240 vac, 1-phase	32	40
DC charging	600 vdc maximum	400 (maximum)	As required

Table 2 Impacts of PHEV charging [3]

Charging scenario	Increase in%	
	Total load	Peak demand
Uncontrolled	2.7	2.5
Continuous	4.8	4.6

To compare with public charging stations, we are presenting the results of uncontrolled charging and continuous time charging scenarios (see Table 2). However, results presented in [1–4] are based on the assumptions that PHEV are charged using residential outlets and PHEV-20 was assumed to be the base case vehicle. We believe that in order to speed up the usage of PHEVs, the development and deployment of fast charging stations is a necessary requirement. In order to compete with gas stations, DC charging seems to be the strongest candidate in fast charging stations [8], since charging duration is shorter than AC Level 1 or AC Level 2.

Some of the concerns about lack of public recharging stations are fast fading away, at least for the pilot cities. [9, 10] have already installed hundreds of fast charging stations together with battery switching stations in less than a year. It is now possible to travel for hundreds of miles using only electric power, given the current availability of public charging stations [11]. At present, there is a strong need for daytime charging stations, which uses level-2 charging infrastructure (typical charging duration 4–5 h). PHEV owners can charge their vehicles while at work, during shopping or even while parking on the street. Wide presence of such infrastructures will considerably extend the range of the vehicles.

There are two different daytime public recharging architectures: fast recharge systems and battery switching stations, which quickly replaces a discharged battery with a fully charged one. PHEV/EV owners may see battery swapping stations more alluring, the since total switching process takes only a few minutes. On the other hand, initial investment for battery switching stations is much higher than fast charging stations. The switching process requires advanced automatic robots (around $300 million for each switching station) and the stations require more physical space with construction costs being 2–3 times higher than fast charging stations [12]. Another handicap of switching stations is that they require compatibility and the same performance for all batteries. Nevertheless, they represent an emerging business model where customers can pay both for the swapped batteries and the capital cost of the battery on an incremental basis. Thus, it will help to

Table 3 Charging speed (approx.) in DC fast charge station [20]

Charging power (kW)	80% capacity charging time (min)		
	Compact EV	SUV/sedan	Heavy truck
50	15	22	46
75	11	15	32
100	8	12	22
125	6.5	9	19
150	5	8	16

reduce battery ownership costs for EV/PHEVs. In a similar manner, even for garage charging, innovative business models could be developed, which enable EV/PHEV owners to pay for their batteries over time via bundling that cost with recharging costs. With the current battery cost, decoupling of vehicle purchase cost and battery costs will be helpful to promote the usage of EV/PHEVs [13].

In both infrastructures, there is a need for demand monitoring related to the number of charged vehicles by the facility. In the fast charging station paradigm, this is a trivial task, since the amount of power requested and the number of slots occupied (typically in the dozens) need to be tracked, whereas in a switching station the inventory of batteries needs to be properly managed. Further, according to [14], car manufacturers in the U.S. prefer fast charging stations instead of battery swapping stations.

Fast, reliable, and robust communications between vehicles and the charging/switching stations will be important in order for drivers to find the nearest available station. From the charging station's operator point of view, it is vital to be able to predict and control power demand. On the other hand, already rolled-out standards-compliant technologies (e.g., 3G, and GPS) can be easily adapted. Nevertheless, we believe there is a need for the development of novel networking and communication strategies, coupled with pricing and smart grid architectural designs, that will allow the fast and reliable operation of this large, complex, "cyber-physical" system.

Even though availability of today's electricity system in the U.S. is 99.97% [15], utilities have concerns about possible negative impacts of fast charging stations on the power grid. It is believed that fast charging stations can overload the grid and unpredictable peaks in the electricity demand may impact grid reliability [15, 16]. In [17], authors concluded that if EV/PHEVs use fast charging stations and 5% of the vehicles charge at the same time, there will be a 5 GW increase in total power demand by year 2018 in the VACAR region (Virginia–North Carolina–South Carolina) [18].

Several industry leaders have already introduced their DC fast charging station technologies [10, 19]. The duration of charging an electric vehicle is proportional to the available power. A recharged range of more than 70 miles can be achieved with less than 10 min of charging, for specific battery types [20]. Table 3 shows a typical charging duration of a PHEV by using DC fast charging. In addition, the IEEE Standards Association and SAE International plan to work together to develop standards on smart grid and electric vehicles [21].

We describe next an EV/PHEV fast charging station architecture, which aims to reduce the unpredictability of the power demand, while at the same time provide the vehicle owners with a certain level of *quality-of-service* by employing a local energy storage system. In this study, the focus is on the relation between size of the local energy storage unit and the quality of service the vehicle owner receives in the form of a *blocking probability*. The evaluation of the performance of the proposed charging station architecture is done with respect to the traffic characteristics, stochastic power demand, local energy storage size, pricing and cost parameters.

3   Application to Storage Modeling for EV/PHEV Charging Stations

The proposed architecture for the charging station is presented in Fig. 1. The core idea is that power demand originating from EV/PHEVs can be covered either by constant power supplied by the utility or with a local energy storage unit. Excess power from the grid is employed to charge the local energy storage, which in turn is used to support temporary high loads on the charging station. The local energy storage component in this architecture plays an important role in satisfying customers' demand. However, since it is a costly component, selecting the correct size becomes an important issue [22]. In general, energy storage is considered a critical component of the Smart Grid, since it can contribute to offset peak loads, reduce electricity transmission congestion and support demand response resources [23].

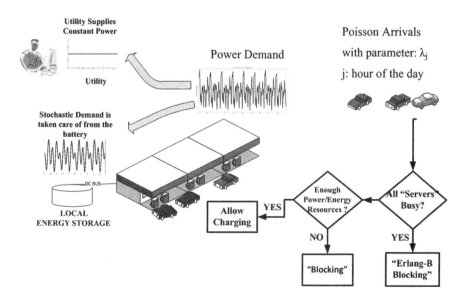

Fig. 1 Proposed architecture

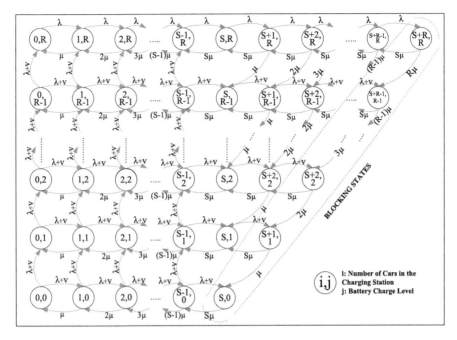

Fig. 2 Continuous time Markov chain

The problem of sizing the energy storage system in a charging station has received some attention in the literature. For example, [8] discusses an architecture for a DC charging station, where the size energy storage system is determined using Monte Carlo simulations. [24] models the charging station as an $M/M/\infty$ queueing system. The goal is to avoid overloading the power grid, by controlling the arrival rate of the vehicles through broadcasting information about available power.

In our approach, we also adopt a queueing model. Specifically, vehicles arrive to the charging station according to a Poisson process of rate λ. The station can simultaneously accommodate up to S vehicles provisioned by the power grid. Further, the station is equipped with a storage unit, which when charged can accommodate an additional R vehicles with a fully charged storage unit. The amount of time to charge a vehicle is exponentially distributed of rate μ, while charging the storage unit from the grid is also exponentially distributed of rate ν. Finally, it is assumed that a vehicle is charged from the grid as long as the number of vehicles does not exceed S, while additional vehicles are charged from the storage unit; otherwise, they are blocked.

The dynamics of this system are captured by a continuous time birth-death Markov chain with finite two-dimensional state space, with one dimension corresponding to the number of vehicles that can be charged by the station and the second one to the charge level of the storage unit. The birth and death rates are given in Fig. 2. Note that the set of states $(S + j, j)$, $0 \leq j \leq R$ represents

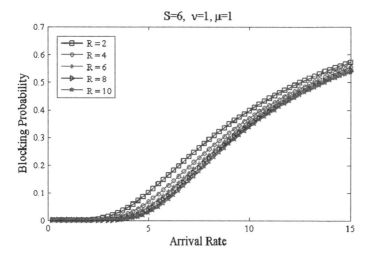

Fig. 3 Arrival rate vs. blocking probability S = 6, R = 1–9

the blocking ones, where the charging station rejects new vehicle arrivals. It can easily be seen that given its structure, the Markov chain is irreducible and positive recurrent, hence ergodic. Its unique stationary distribution π can be calculated by solving the equation $\pi Q = 0$ subject to the constraint $\pi e = 1$, where e denotes a vector comprising of ones, using standard numerical techniques [25]. More details about the model are provided in [26].

Some performance characteristics of the proposed system are illustrated next; a more detailed evaluation is provided in [26]. In Fig. 3, the blocking probability as a function of the arrival rate of the vehicles for $S = 6$ and varying R is shown for a system with $\mu = \nu = 1$. It can be seen that he greater the R the smaller the blocking probability, as expected, but the differences for $R \geq 6$ tend to be rather small (Fig. 4).

Next, we examine combinations of S and R that achieve a prespecified blocking probability and the resulting maximum allowable arrival rate, fixing again $\mu = 1, \nu = 1$ (see Fig. 5). It can be seen that systems with a large S are uniformly preferable. However, such charging stations configurations are not desirable in practice, since they put a lot of strain on the grid, as previously outlined.

The previous brief evaluation provides insight into the tradeoffs between the size of the storage energy unit, the blocking probability and the arrival rate of the vehicles. Next, we present a simple model that based on simple financial principles relates pricing parameters to the queuing model's parameters. This model can provide guidance for choosing the best operating range for the charging station, as well as possible control strategies (Fig. 6).

It assumes that the charging station's operator obtains revenue from each vehicle successfully charged. However, blocked vehicles constitute dissatisfied customers

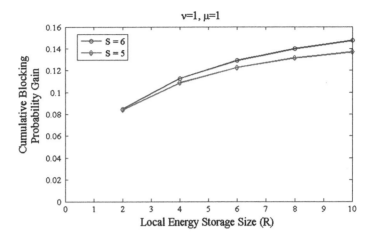

Fig. 4 Average gain in blocking probability vs. battery size

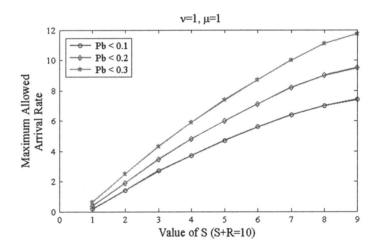

Fig. 5 Maximum supported arrival rate for a given blocking probability

that must be compensated. Let R_g and R_l be the revenue obtained per vehicle when charged from the grid and the storage unit, respectively. Further, let C_b denote the cost per blocked vehicle, C_0 the fixed installation cost of the storage unit and C_a an acquisition cost proportional to its capacity R.

Let $\rho^{(g)} = \{(i, j) : 0 \le i \le S, 0 \le j \le R\}$ denote the grid charging states and $\rho^{(l)} = \{(i, j) : S + j \le i \le S + R, 1 \le j \le R\}$ the storage unit charging ones. Also, we denote by $\rho^{(bl)}$ the blocking states, respectively. Finally, $\pi(s)$ denotes the stationary probability for generic state $s \equiv (i, j)$ and $i(s)$ the number of vehicles receiving service in states s. Then the proposed profit function can be written as

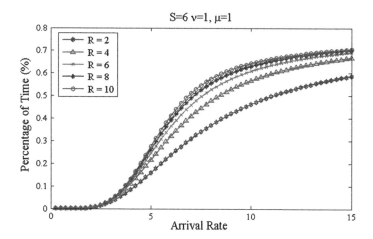

Fig. 6 Percentage of time PHEV charging made by local energy storage

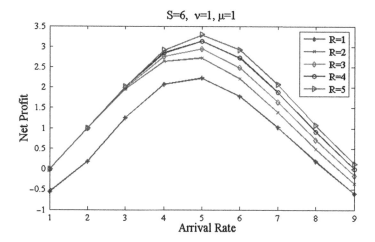

Fig. 7 Net profit of the charging station

$$P = \sum_{s \in \rho^{(g)}} R_g i(s)\pi(s) + \sum_{s \in \rho^{(l)}} R_l i(s)\pi(s)$$

$$-(C_0 + RC_a) - \sum_{s \in \rho^{(bl)}} C_b i(s)\pi(s). \qquad (1)$$

Under the following specification of the various model parameters ($R_g = R_l = 1$, $C_b = 1.25$, $C_0 = 2$ and $C_a = 0.001$), setting the charging capacity from the grid to $S = 6$, charging rates to $\mu = \nu = 1$ and varying the size of the storage unit, we plot the profit as a function of the arrival rate in Fig. 7 and set the grid capacity

$S = 6, \mu = \nu = 1$ and vary both the arrival rate from 1 to 9 and R from 5 to 9. The results are shown in Fig. 7. It can be seen that for very small arrival rates the station is not profitable due to the small number of vehicles served, while for large arrival rates the same thing happens due to the large number of rejected vehicles. It is worth noting that for all sizes R the profit is maximized at the same arrival rate, although for larger sizes R the profit is very similar for a larger range of arrival rates.

This model offers the possibility for selecting the size of the station (R, S) based on financial principles. It also allows one to evaluate different incentive schemes to vehicle owners to forgo charging their vehicle at present, thus lowering the arrival rate and hence the blocking probability, at an additional cost to the station's operator.

4   Concluding Remarks

In this chapter, we discussed a number of issues and challenges regarding smart vehicles in the era of the Smart Grid, primarily related to control, communications, and infrastructure demands. Regarding the latter, we argued that there is a need for charging stations and discussed an architecture that incorporates a storage unit, so as to smoothly accommodate peak loads, while providing an acceptable quality of service to the vehicles. A stochastic model was introduced that allowed us to assess performance of the charging station with respect to the traffic characteristics, energy storage size, pricing and cost parameters. Such insights are important in designing future charging stations.

References

1. Scott M, Meyer M, Elliot D, Warwick W (2007) Impacts of plug-in hybrid vehicles on electric utilities and reginal U.S. power grids. Pasific Northwest National Laboratory, Palo Alto, CA, Technical Report
2. Denholm P, Short W (2006) An evaluation of utility system impacts and benefits of optimally dispatched plug-in hybrid electric vehicles. National Renewable Energy Labotory, Technical Report
3. Parks K, Denholm P, Markel T (2007) Costs and emissions associated with plug-in hybrid electric vehicle charging in the xcel energy colorado service territory. National Renewable Energy Labotory, Technical Report
4. Letendre S, Watts R, Cross M (2008) Plug-in hybrid vehicles the vermont grid: a scoping analysis. University of Vermont Transportation Center, Technical Report
5. Smart grid. http://en.wikipedia.org/wiki/Smart_grid. Accessed June 2011
6. Technology roadmap: electric and plug-in hbyrid electric vehicles. International Energy Agency, Technical Report, June 2011
7. Standard J1772 (2001) Sae electric vehicle conductive charge coupler. November 2001
8. Bai S, Yu D, Lukic S (2010) Optimum design of an ev/phev charging station with dc bus and storage system. In: 2010 IEEE Energy conversion congress and exposition (ECCE), pp 1178–1184
9. http://www.betterplace.com/. Accessed June 2011
10. http:chademo.com/. Accessed March 2011

11. Gordon-Bloomfield N (2011) Electric car fan takes 2011 leaf on 480 mile road trip. http://www.allcarselectric.com/news/1062240_electric-car-fan-takes-2011-leaf-on-480-mile-road-trip. Accessed June 2011
12. Yongxiang L, Fuhui H, Ruilin X, Tao C, Xin X, Jie L (2011) Investigation on the construction mode of the charging station and battery-exchange station. In: 2011 Asia-Pacific power and energy engineering conference (APPEEC), March 2011, pp 1–2
13. Becker T (2009) Electric vehicles in the United States: a new model with forecasts to 2030
14. Joos G, de Freige M, Dubois M (2010) Design and simulation of a fast charging station for phev/ev batteries. In: 2010 IEEE electric power and energy conference (EPEC), August 2010, pp 1–5
15. The U.S. Department and of Energy (2008)The smart grid: an introduction
16. Woody T (2010) Plans for fast-charging stations raise concerns among california utilities [Online]. Available: http://green.blogs.nytimes.com/2010/01/28/plans-for-fast-charging-stations-raise-concerns-among-california-utilities/
17. Song J, Toliyat A, Turtle D, Kwasinski A (2010) A rapid charging station with an ultracapacitor energy storage system for plug-in electrical vehicles. In: 2010 international conference on electrical machines and systems (ICEMS), October 2010, pp 2003–2007
18. Hadley WH (2006) Impact of plug-in hybrid vehicles on the electric grid. Oak Ridge National Labs, Technical Report, October 2006
19. http://www.protoscar.com/. Accessed April 2011
20. http://www02.abb.com/global/dkabb/dkabb505.nsf/0/8cd0fca2ab2da0b2 c12577f400428ce7/$file/DC+Fast+Charge+Station+160910.pdf. Accessed March 2011
21. IEEE standards association and sae intrnational agree to collaborate on smart grid and vehicle-electrification standards. http://smartgrid.ieee.org/ieee-smart-grid-news/4323-ieee-standards-association-and-sae-international-agree-to-collaborate-on-smart-grid-and-vehicle-electrification-standards. Accessed May 2011
22. Wade N, Taylor P, Lang P, Jones P (2010) Evaluating the benefits of an electrical energy storage system in a future smart grid. Energy Policy 38(11):7180–7188. Energy efficiency policies and strategies with regular papers [Online]. Available: http://www.sciencedirect.com/science/article/B6V2W-50T94GD-2/2/47b370fece990212a23b0bb7314ba39d
23. Hoffman M, Sadovsky A, Kintner-Meyer M, DeSteese J (2010) Anaysis tools for sizing and placement of energy storage in grid applications: a literature review. Pasific Northwest National Laboratory, Technical Report, September 2010
24. Turitsyn K, Sinitsyn N, Backhaus S, Chertkov M (2010) Robust broadcast-communication control of electric vehicle charging. In: 2010 first IEEE international conference on smart grid communications (SmartGridComm), pp 203–207
25. Stewart WJ (2009) Probability, Markov chains, queues, and simulation: the mathematical basis for performance modeling. Princeton University Press, New Jersey
26. Bayram IS, Michailidis G, Devetsikiotis M, Bhattacharya S, Chakrabortty A, Granelli F (2011) Local energy storage sizing in plug-in hybrid electric vehicle charging stations under blocking probability constraints. In: Architectures and models for the smart grid (IEEE SmartGridComm) (IEEE SmartGridComm 2011 track-architectures and models), Brussels, Belgium, October 2011

Part II
Modeling and Analysis

Models for Impact Assessment of Wind-Based Power Generation on Frequency Control

Alejandro D. Domínguez-García

Abstract This chapter develops a modeling framework for studying the impact of variability and uncertainty in wind-based electricity generation on power system frequency. The focus is on time-scales involving governor response (primary frequency control) and automatic generation control (AGC) (secondary frequency control). The framework includes models of synchronous generators, wind-based electricity sources, the electrical network, and the AGC system. The framework can be used to study the impact of different renewable penetration scenarios on system frequency performance metrics. In order to illustrate the framework, a simplified model of the Western Electricity Coordinating Council (WECC) system is developed.

1 Introduction

Driven by initiatives like the DoE SmartGrid [1], electrical energy systems are undergoing a radical transformation in structure and functionality in a quest to increase efficiency and reliability. Such transformations are enabled by the introduction of new technologies such as advanced communication and control; new loads, such as plug-in hybrid electric vehicles (PHEV); advanced power electronics devices for power-flow control, such as flexible AC transmission systems (FACTS), and integration of new renewable-based electricity generation sources, e.g., wind and solar.

A.D. Domínguez-García (✉)
Department of Electrical Engineering, University of Illinois at Urbana Champaign,
341 Everitt Laboratory, MC-702, Urbana, IL 61801, USA
e-mail: aledan@ILLINOIS.EDU

A. Chakrabortty and M.D. Ilić (eds.), *Control and Optimization Methods for Electric Smart Grids*, Power Electronics and Power Systems 3,
DOI 10.1007/978-1-4614-1605-0_7, © Springer Science+Business Media, LLC 2012

Focusing on renewable-based electricity resources, it has long been acknowledged that deep penetration of these resources poses major challenges in power system operations [2–4]. For example, the high variability of wind speed not only makes wind-based electricity generation intermittent but also presents major difficulties in its forecast [5]. Thus, wind-based electricity generation is an additional source of uncertainty that impacts power system operation across different timescales, i.e., unit commitment, economic dispatch (optimal power flow), automatic generation control (AGC), and governor control. The impact of wind uncertainty has received a great deal of attention in both unit commitment and optimal power flow problems. The impact of wind variability and uncertainty on AGC and governor response for frequency control has received less attention (see [6–11] and the references therein); thus, the focus of this work.

The AGC system is responsible for maintaining the system frequency and keeping the power exchanges between areas at their schedules by controlling the power settings of the units participating in AGC to follow the load profile throughout the day and account for errors in load forecast. Traditional AGC models available in the literature abstract out the electrical network structure, only consider synchronous generators, and aggregate system loads by treating them as a disturbance [12–15]. Additionally, to the author's knowledge, there are no AGC system models available in the open literature that include wind-based electricity generation. One might be tempted to extend traditional AGC models by treating wind-based generation as a negative load and subtract it from the actual system load. We believe that it is necessary to take a step back and rethink how AGC systems are to be modeled in order to properly account for the unique features of wind-based generation and its possible impact on system frequency. First, wind variability ramp-down events are not as abrupt as the sudden loss of a conventional generating unit [7]. However, they are not as smooth as load ramp-down daily cycles, especially in the presence of weather fronts that result in high winds (entire wind power plants might shut down in just a few minutes). On the other hand, the time constants associated with wind-based electricity generation can be faster than the daily load cycle time constants [7]. In regards to the effect of the electrical network on AGC, we believe it is necessary to explicitly model it to understand its possible effects on mitigating wind variability. Given a certain amount of wind-based power present in a system, it is clear that the impact on system frequency response will be different depending on where wind-based power is injected in the network. For example, if all the power is injected in the same geographical area, it might be the case that due to network constraints, the conventional units in that area participating in AGC must be the ones compensating for wind-based variability. The network might also have some smoothing effects as it does affect the response of the overall closed-loop system dynamic behavior, which is something that traditional AGC models do not capture. In other words, if we disregard the network, the impact of wind-based variability is felt by the machines right away, whereas this does not happen in reality, i.e., the network acts as a filter.

This work builds on our previous work [16], where we were concerned with studying short-term impact of wind variability on power system dynamics without explicitly accounting for the AGC system. This chapter introduces an analytically tractable model for quantifying the impact of wind variability and uncertainty in power system frequency response. This model extends traditional AGC models by explicitly including models for wind-based electricity generation sources, and the electrical network. The model is formulated by using a nonlinear differential-algebraic equation (DAE) formalism. The differential part includes both the conventional synchronous generating units and the wind-based electricity sources, and the AGC system dynamics. The conventional units are described by a third-order model that includes the mechanical equations of motion and the governor dynamics. Since we are only interested in frequency response, it is appropriate to assume that the voltage regulator and other dynamics within the machine are fast enough compared to the phenomena of interest, justifying the use of this third-order model. The wind-based electricity resources are modeled as a piecewise linear first-order dynamical system describing the relation between power and wind speed. This piecewise linear model naturally captures the nonlinear nature of wind generator power characteristic. The power settings of synchronous generators participating in AGC (it is assumed that wind-based units do not participate in AGC) are modeled following [14, pp. 352–355]. The algebraic part describes the network electrical behavior using standard power flow equations, but separates the power injection at a node into the contributions of conventional synchronous generating units and wind-based electricity sources.

For the timeframes of interest for AGC (seconds to minutes), we assume that forecast errors are small. Then the nonlinear DAE model is linearized along a nominal system trajectory that results from wind and load forecasts. Then by using Kron reduction, it is possible to reduce the resulting linear DAE model to an ordinary differential equation (ODE) model. This ODE model describes how variations around wind and load forecast will affect the system dynamics and therefore the frequency. Following [17], it is assumed that, due to the fact that the system is never in steady-state, the system frequency is not homogenous across the system. Thus, frequency on every bus of the network is defined as the nominal system frequency plus the derivative of the voltage angle with respect to time. Therefore, this model can be used to study the impact of wind variability on frequency across the system, which might be relevant for deep penetration scenarios, e.g., 50% of the total power at any given time is served by wind-based resources. We believe that the model can be useful for understanding fundamental limits on the amount of wind that can be integrated in a power system without violating frequency performance metrics. Additionally, the model can be used to design advanced controls to augment AGC. For example, we envision that within the current hierarchy of frequency response controls comprised of governor control and AGC, it may be possible to introduce an intermediate control, acting faster than AGC but slower than governors, to coordinate groups of coherent generators to smooth the variability of wind-based generation in a particular area of the system where these generators are located.

The remainder of this chapter is organized as follows. Section 2 provides the mathematical models for each of the system components. Section 3 provides the formulation of the system nonlinear DAE model and its linearization along a nominal trajectory. Section 4 illustrates the application of the modeling framework to a simplified Western Electricity Coordinating Council (WECC) model. Concluding remarks are presented in Sect. 5.

2 Model Building Blocks

This section provides the mathematical models of the system components, which are conventional synchronous generating units, renewable-based electricity sources, the network, and the AGC system.

2.1 Conventional (Synchronous) Generating Units

For the timescales of interest, it is shown in [17] that a nine-state machine model, including models for damper-windings, mechanical equations of motion, exciter, voltage regulator, turbine, and the governor, can be reduced to a three-state model that only includes the mechanical equations and the governor dynamics.

For the ith synchronous machine, let δ_i (rad) denote the rotor electrical angular position (with respect to a synchronous reference rotating at ω_s (rad/s)), ω_i (rad/s) denote the rotor electrical angular velocity, and P_i (pu) denote the turbine power. Let V_i (pu) denote the machine terminal voltage magnitude, and θ_i (rad) denote the machine terminal voltage angle. Let P_i^{ref} (pu) denote the input to the generating unit control logic. If we neglect the motor changer controller dynamics, P_i^{ref} is just the unit power setting. Then, the machine dynamics can be described by

$$
\frac{d}{dt}\begin{bmatrix} \delta_i \\ \omega_i \\ P_i \end{bmatrix} = \begin{bmatrix} 0 & 1 & 0 \\ 0 & -\frac{D_i}{M_i} & \frac{1}{M_i} \\ 0 & -\frac{1}{\tau_i R_i \omega_s} & -\frac{1}{\tau_i} \end{bmatrix}\begin{bmatrix} \delta_i \\ \omega_i \\ P_i \end{bmatrix} + \begin{bmatrix} 0 \\ -\frac{E_i V_i}{M_i X_i}\sin(\delta_i - \theta_i) \\ 0 \end{bmatrix}
$$
$$
+ \begin{bmatrix} -1 \\ \frac{D_i}{M_i} \\ \frac{1}{\omega_s \tau_i R_i} \end{bmatrix}\omega_s + \begin{bmatrix} 0 \\ 0 \\ \frac{1}{\tau_i} \end{bmatrix}P_i^{ref}, \tag{1}
$$

where ω_s (rad/s) is the machine electrical synchronous speed, D_i (s/rad) is a damping coefficient, M_i (s²/rad) is the scaled machine inertia constant, E_i (pu) is the voltage behind reactance (or machine "internal voltage"), τ_i (s) is the governor time constant, and R_i (pu) is the slope of the machine speed-droop characteristic.

As explained in the next section, the generating unit i power setting P_i^{ref} is a function of the unit base-point generation P_i^{ed} (determined by economic dispatch in intervals of 5–15 min), and also by the AGC system generation allocation to the unit ith to account for deviations in frequency, area power exchange, and load (from the value used in the economic dispatch calculation).

2.2 Wind-Based Electricity Sources

For the timescales of interest, it was shown in [18, 19] that, under normal system operation (no faults in the network or sudden loss of synchronous generators), the most important interaction of wind farms with the system is through the interchange of power. Also in [18, 19], it was shown by using model-order reduction, that (1) a low-order dynamical model (up to three states) yields a simple yet accurate description of the relation between wind speed and power generated by a wind turbine, and (2) this model can be extended to also describe the power generated by a group of turbines.

In this work, we follow the ideas above and assume that the power generated by a wind farm can be described by a nonlinear dynamical system. For simplicity of subsequent developments and following [19], we assume that a first-order model suffices (the ideas can be easily extended to higher-order models), and that wind power plants are operated in power factor mode, with unity power factor. Although not explicitly shown, this model can also be easily extended to take into account wind power plant operation in voltage control mode. For the ith wind power plant, let P_i^w be the active power generation, Q_i^w the reactive power generation, and w_i be the representative wind speed of the site where the plant is located. Additionally, following the terminology used for single turbine power characteristics [20], denote by w_i^c, w_i^r, and w_i^f the wind plant representative cut-in, rated, and cut-out (furling) wind speeds.

Then plant active and reactive power are respectively given by

$$P_i^w = \phi_i(v_i, w_i) := \begin{cases} 0 & \text{for } w_i < w_i^c, \\ \beta_i(v_i) & \text{for } w_i^c \le w_i < w_i^r, \\ P_i^r & \text{for } w_i^r \le w < w_i^f, \\ 0 & \text{for } w \ge w_f, \end{cases} \tag{2}$$

$$Q_i^w = 0, \tag{3}$$

where P_i^r is a constant (assuming all turbines are operational, this is the plant nameplate power), $\beta_i : \mathbb{R} \mapsto \mathbb{R}^+$, and the evolution of $v_i \in \mathbb{R}$ in (2) described by

$$\dot{v}_i = \alpha_i(v_i, w_i), \tag{4}$$

where $w_i \in \mathbb{R}^+$ and $\alpha_i : \mathbb{R} \times \mathbb{R}^+ \mapsto \mathbb{R}$.

Although not explicitly described in (2) and (4), w_i can be modeled as a stochastic process. Since $\phi(\cdot)$ is a piecewise-defined function, the evolution of P_i^w can be formalized using stochastic hybrid modeling formalisms (see, e.g., [21, 22]).

2.3 Network

The electrical network is modeled by the standard power flow equations polar-coordinate formulation. At each node, it is assumed that bus power injections can be due to both synchronous generators and wind-based electricity sources. Loads are modeled as active and reactive power withdraws, and the formulation is general to account for the possibility of having loads at every bus of the network.

For the ith bus, let V_i denote the ith bus voltage magnitude, and let θ_i denote the bus voltage angle. Let P_i^s and Q_i^s denote active and reactive power injections from the ith synchronous generator, and P_i^w denote active power injections from the ith wind power plant. Finally, let P_i^d and Q_i^d be the active and reactive power demand. The power flows equations are

$$P_i^s + P_i^w - P_i^d = \sum_{k=1}^{n} V_i V_k \big(G_{ik} \cos(\theta_i - \theta_k) + B_{ik} \sin(\theta_i - \theta_k) \big), \qquad (5)$$

$$Q_i^s - Q_i^d = \sum_{k=1}^{n} V_i V_k \big(G_{ik} \sin(\theta_i - \theta_k) - B_{ik} \cos(\theta_i - \theta_k) \big), \qquad (6)$$

where G_{ik} and B_{ik} are the real and imaginary parts of the network admittance matrix (i, k) entry, respectively. It follows from the model in (1) that P_i^s and Q_i^s are given by

$$P_i^s = \frac{E_i V_i}{X_i} \sin(\delta_i - \theta_i), \qquad (7)$$

$$Q_i^s = \frac{E_i V_i}{X_i} \cos(\delta_i - \theta_i) - \frac{V_i^2}{X_i}, \qquad (8)$$

whereas the model for P_i^w was given in (2).

2.4 Automatic Generation Control System

For AGC modeling, it is assumed that there are m different balancing areas within the interconnected system, and within each area, only synchronous generators can participate in AGC. The ith generator participating in AGC will adjust its power setting P_i^{ref} according to the corresponding area control error (ACE) and its ACE participation factor. We follow the AGC model in [14, pp. 352–355], except for the assumption made when deriving (1), where the motor changer controller dynamics were neglected.

Denote by $\mathscr{A} = \{1, 2, \ldots, m\}$ the set that indexes all balancing areas. For every $k \in \mathscr{A}$, denote by $\mathscr{A}_k \subset \mathscr{A}$ the set of balancing areas that have tie lines with area k. Denote by P_{kj} the actual power interchange between areas k and j (positive for power leaving area k), and by f^k the actual frequency of area k. Then ACE for area k is given by

$$ACE_k = -\sum_{j \in \mathscr{A}_k} (P^{kj} - P^{kj}_{sch}) - b_k(f^k - f^{nom}), \tag{9}$$

where P^{kj}_{sch} is the scheduled power interchange between areas k and j, B_k is the bias factor for area k, and f^{nom} is the system nominal frequency.

Let $\mathscr{B}_{kj}$ be the set of buses in area k with tie lines to buses in area j, and $\mathscr{B}_{jk}$ be the set of buses in area j with tie lines to buses in area k. Denote by P_{lm} the power flow from bus l to bus m. Then the power exchange between areas k and j is

$$P^{kj} = \sum_{l \in \mathscr{B}_{kj}, m \in \mathscr{B}_{jk}} P_{lm} = \sum_{l \in \mathscr{B}_{kj}, m \in \mathscr{B}_{jk}} V_l V_m \big(G_{lm} \cos(\theta_l - \theta_m) + B_{lm} \sin(\theta_l - \theta_m)\big). \tag{10}$$

Let α_l^k denote the ACE participation factor for the lth generator in control area k, and denote by $\mathscr{G}_k$ the set that indexes all the generators in area k. Then the power setting of the lth generator in area k is given by

$$\frac{dz_k}{dt} = ACE_k - \sum_{i \in \mathscr{G}_k} \left(P_i^{ed} + \alpha_i^k \left(z_i - \sum_{j \in \mathscr{G}_k} P_j^{ed} \right) - P_i^s \right),$$

$$P_l^{ref} = P_l^{ed} + \alpha_l^k \left(z_k - \sum_{j \in \mathscr{G}_k} P_j^{ed} \right), \tag{11}$$

with $\sum_{i \in \mathscr{G}_k} \alpha_i^k = 1$, and where $P_i^s = \frac{E_i V_i}{X_i} \sin(\delta_i - \theta_i)$, i.e., power generated by unit i. Note that (11) is general enough to also include generators that do not participate in AGC by setting their participation factor to zero, which would result in $P_l^{ref} = P_l^{ed}$.

The area k frequency f_k in (9) can be related to the derivatives of the bus voltage angles as follows. Let f_i be the frequency on bus i, then

$$f_i = f^{nom} + \frac{1}{2\pi} \frac{d\theta_i}{dt}, \tag{12}$$

where θ_i is the ith bus voltage angle [17]. Then the frequency in area k can be obtained as a weighted linear combination of the frequencies in several buses, i.e.,

$$f^k = \sum_{i \in \mathscr{B}_k} \gamma_i f_i \tag{13}$$

with $\Sigma_{i \in \mathscr{B}_k} \gamma_i = 1$, where $\mathscr{B}_k$ is the set of buses in area k. In the simplest case, if the frequency of a single bus is used in (9), the summation in (13) reduces to a single term.

3   System Model

We introduce now a system-level nonlinear differential-algebraic model using the building blocks described in the previous section. We then linearize the differential-algebraic model along a nominal trajectory and use Kron reduction to reduce the differential-algebraic model linearization to a time-varying linear system model.

3.1   Nonlinear Differential-Algebraic Model

The system dynamic behavior can be described by a DAE. The differential part arises from both the individual synchronous machine dynamics as described in (1), the wind power plants as described in (2) and (4), and the AGC system as described in (9)–(13). The algebraic part results from the power flow equations as described in (5) and (6).

Define the vector of synchronous machine state variables as $x = [x_1', x_2', \ldots, x_m']'$, with $x_i = [\delta_i, \omega_i, T_i]'$; the vector of synchronous machine power settings as $u = [P_1^{ref}, P_2^{ref}, \ldots, P_n^{ref}]'$; the vector of algebraic variables as $y = [y_1', y_2', \ldots, y_n']'$, with $y_i = [\theta_i, V_i]'$; the vector of wind-power-plant generation as $p_w = [P_1^w, P_2^w, \ldots, P_n^w]'$; the vector of load demands as $p_d = [P_1^d, P_2^d, \ldots, P_n^d]'$; the vector of wind-power-plant representative wind speeds as $w = [w_1, w_2, \ldots, w_n]'$, and $v = [v_1, v_2, \ldots, v_n]'$ the vector of wind power plant internal variables defining plants power output evolution between cut-in and rated speed $\big($see (4)$\big)$. Denote by p^{sch} the vector of scheduled power interchange between areas; let $z = [z_1, z_2, \ldots, z_m]'$ be the vector of inputs to the generation allocation logic of the units participating in AGC; and let $p_{ed} = [P_1^{ed}, P_2^{ed}, \ldots, P_n^{ed}]'$ be the vector that results from the solution to the economic dispatch. Then the system dynamic behavior can be described by

$$\frac{dx}{dt} = f(x, y, u), \tag{14}$$

$$\frac{dz}{dt} = h(x, y, z, p_{sch}, p_{ed}), \tag{15}$$

$$u = k(z, p_{ed}), \tag{16}$$

$$\frac{dv}{dt} = \alpha(v, w), \tag{17}$$

$$p_w = \phi(v, w), \tag{18}$$

$$0 = g(x, y, p_d, p_w). \tag{19}$$

The function $f : \mathbb{R}^{3n} \times \mathbb{R}^{2n} \times \mathbb{R}^n \mapsto \mathbb{R}^n$ results from the synchronous generator dynamics as described in (1). The function $h : \mathbb{R}^n \times \mathbb{R}^n \times \mathbb{R}^m \times \mathbb{R}^n \times \mathbb{R}^n \mapsto \mathbb{R}^m$ results from the synchronous generators power settings as described in (9)–(13). For the jth with their power setting scheduled solely based on economic dispatch, it follows that $u_j = P_j^{ed}$. The functions $\alpha : \mathbb{R}^n \times (\mathbb{R}^+)^n \mapsto \mathbb{R}^n$ and $\phi : \mathbb{R}^n \times (\mathbb{R}^+)^n \mapsto \mathbb{R}^n$ result from wind power plant dynamics as described in (2) and (4). The function $g : \mathbb{R}^{3n} \times \mathbb{R}^{2n} \times \mathbb{R}^n \mapsto \mathbb{R}^n$ results from the network power flow equations.

The formulation in (14)–(19) is made general enough to allow the possibility of having wind power plants, loads, and synchronous generator connected to all the buses in the network. If this is not the case, the dimensions of x, p_w, and P_d should be modified accordingly as illustrated in an example in Sect. 4.

3.2 Linearized Model

In the nonlinear DAE model described in (14)–(19), there are two inputs that are subject to uncertainty, which are the load demand vector p_d and the wind-power-plant averaged wind speed vector w. These uncertain inputs can be characterized by their forecast p_d^* and w^*. For the timescales of interest, we assume that the forecast error is small and linearize the system in (14)–(19) along a nominal trajectory (x^*, y^*, u^*, p_w^*) that results from the forecasted p_d^* and w^*. Since the focus here is on the evolution of the system dynamics in the time elapsed between two economic dispatch calculation, and we assume that the economic dispatch solution p_{ed}^* (as well as the power interchange between areas p_{sch}^*) remain constant.

Let $x = x^* + \Delta x$, $y = y^* + \Delta y$, $u = u^* + \Delta u$, $p_w = p_w^* + \Delta p_w$, $w = w^* + \Delta w$, and $p_d = p_d^* + \Delta p_d$. Then small variations in system trajectories around (x^*, y^*, u^*, p^*) arising from small variations around p_d^* and w^* can be approximated by

$$\frac{d\Delta x}{dt} = A_1(t)\Delta x + A_2(t)\Delta y + B_1(t)\Delta u, \tag{20}$$

$$\frac{d\Delta z}{dt} = A_3(t)\Delta x + A_4(t)\Delta y + A_5(t)\Delta z, \tag{21}$$

$$\Delta u = B_2(t)\Delta z, \tag{22}$$

$$\frac{d\Delta p_w}{dt} = F(t)\Delta p_w + G(t)\Delta w, \tag{23}$$

$$0 = C_1(t)\Delta x + C_2(t)\Delta y + D_1(t)\Delta p_d + D_2(t)\Delta p_w, \tag{24}$$

where

$$A_1(t) = \frac{\partial f(x, y, u)}{\partial x}\Big|_{x^*, y^*, u^*}, \quad A_2(t) = \frac{\partial f(x, y, u)}{\partial y}\Big|_{x^*, y^*, u^*},$$

$$B_1(t) = \frac{\partial f(x, y, u)}{\partial u}\Big|_{x^*, y^*, u^*},$$

$$A_3(t) = \frac{\partial h(x, y, p_{sch}, p_{ed})}{\partial x}\Big|_{x^*, y^*, p_{sch}^*, p_{ed}^*},$$

$$A_4(t) = \frac{\partial h(x, y, p_{sch}, p_{ed})}{\partial y}\Big|_{x^*, y^*, p_{sch}^*, p_{ed}^*}$$

$$A_5(t) = \frac{\partial h(x, y, p_{sch}, p_{ed})}{\partial z}\Big|_{x^*, y^*, p_{sch}^*, p_{ed}^*}, \quad B_2(t) = \frac{\partial k(z, p_{ed})}{\partial z}\Big|_{z^*, p_{ed}^*},$$

$$C_1(t) = \frac{\partial g(x, y, p_d, p_w)}{\partial x}\Big|_{x^*, y^*, p_d^*, p_w^*}, \quad C_2(t) = \frac{\partial g(x, y, p_d, p_w)}{\partial y}\Big|_{x^*, y^*, p_d^*, p_w^*},$$

$$D_1(t) = \frac{\partial g(x, y, p_d, p_w)}{\partial p_d}\Big|_{x^*, y^*, p_d^*, p_w^*}, \quad D_2(t) = \frac{\partial g(x, y, p_d, p_w)}{\partial p_w}\Big|_{x^*, y^*, p_d^*, p_w^*},$$

$$F(t) = \text{diag}\big(F_1(t), F_2(t), \ldots, F_n(t)\big), \quad G(t) = \text{diag}\big(G_1(t), G_2(t), \ldots, G_n(t)\big),$$

with

$$F_i(t) = \begin{cases} \dfrac{\partial \alpha_i(v_i, w_i)}{\partial v_i}\Big|_{v_i^*, w_i^*}, & w_i^c \leq w_i < w_i^r, \\ 0 & \text{otherwise,} \end{cases} \tag{25}$$

and

$$G_i(t) = \begin{cases} \dfrac{\partial \beta_i(v_i)}{\partial v_i}\Big|_{v_i^*} \dfrac{\partial \alpha_i(v_i, w_i)}{\partial w_i}\Big|_{v_i^*, w_i^*}, & w_i^c \leq w_i < w_i^r, \\ 0 & \text{otherwise.} \end{cases} \tag{26}$$

In (24), as long as $C_2(t)$ is invertible, we can solve for Δy to obtain

$$\Delta y = -C_2^{-1}(t)\big(C_1(t)\Delta x + D_1(t)\Delta p_d + D_2(t)\Delta p_w\big). \tag{27}$$

A closer look at $C_2(t) = \frac{\partial g(x, y, p_d, p_w)}{\partial y}\Big|_{x^*, y^*, p_d^*, p_w^*}$ reveals that $C_2(t)$ is invertible if and only if the power flow Jacobian is invertible. We assume that for the nominal system trajectory (x^*, y^*, u^*, p^*), invertibility of the power flow equations Jacobian always holds. Then we can use (27) to substitute Δu in (20) and (21), and obtain an ODE model of the form

$$\frac{d}{dt}\begin{bmatrix} \Delta x \\ \Delta z \\ \Delta p_w \end{bmatrix} = \begin{bmatrix} A_{11}(t) & A_{12}(t) & A_{13}(t) \\ A_{21}(t) & A_{22}(t) & A_{23}(t) \\ A_{31}(t) & A_{32}(t) & A_{33}(t) \end{bmatrix} \begin{bmatrix} \Delta x \\ \Delta z \\ \Delta p_w \end{bmatrix} + \begin{bmatrix} B_{11}(t) & B_{12}(t) \\ B_{21}(t) & B_{22}(t) \\ B_{31}(t) & B_{32}(t) \end{bmatrix} \begin{bmatrix} \Delta w \\ \Delta p_d \end{bmatrix}, \tag{28}$$

where

$$A_{11}(t) = A_1(t) - A_2(t)C_2^{-1}(t)C_1(t), \qquad A_{12}(t) = B_1(t)B_2(t),$$

$$A_{13}(t) = -A_2(t)C_2^{-1}(t)D_1(t),$$

$$A_{21}(t) = A_3(t) - A_4(t)C_2^{-1}(t)C_1(t), \quad A_{22}(t) = A_5(t),$$

$$A_{23}(t) = -A_4(t)C_2^{-1}(t)D_2(t),$$

$$A_{31}(t) = 0, \quad A_{32}(t) = 0, \quad A_{33}(t) = F(t), \quad B_{11}(t) = 0,$$

$$B_{12}(t) = -A_2(t)C_2^{-1}(t)D_1(t),$$

$$B_{21}(t) = 0, \quad B_{22}(t) = -B_2(t)C_2^{-1}(t)D_1(t), \quad B_{31}(t) = G(t), \quad B_{32}(t) = 0.$$

4 Example

In this section, the modeling ideas presented in the previous section are illustrated
with the three-machine, six-bus system of Fig. 1, which is an adaption of the three-
machine, nine-bus model of the Western System Coordinating Council (WECC)
system [17]. As depicted in Fig. 1, we consider two balancing areas. We also

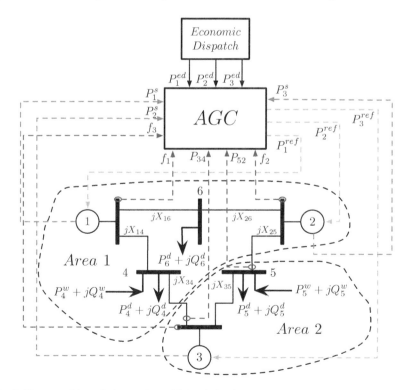

Fig. 1 Three-machine, six-bus system with two balancing areas

introduce wind power plants at buses 4 and 5. Also depicted in the figure are the frequency and power exchange feedback loops, as well as the generator power setting control loops.

4.1 Dynamics of Conventional (Synchronous) Generating Units

As described in (14), the dynamics of the synchronous generators in the system can be compactly written as $\frac{dx}{dt} = f(x, y, u)$. In this example, $x = [x_1, x_2, x_3]'$, with $x_i = [\delta_i, \omega_i, P_i]'$; $y = [y_1, y_2, y_3, y_4, y_5, y_6]'$, with $y_i = [\theta_i, V_i]'$; and $u = [P_1^{ref}, P_2^{ref}, P_3^{ref}]'$, which results in

$$\frac{d}{dt}\begin{bmatrix} x_1 \\ x_2 \\ x_3 \end{bmatrix} = \begin{bmatrix} \Phi_1 & 0 & 0 \\ 0 & \Phi_2 & 0 \\ 0 & 0 & \Phi_3 \end{bmatrix}\begin{bmatrix} x_1 \\ x_2 \\ x_3 \end{bmatrix} + \begin{bmatrix} \Gamma_1(x_1, y_1) \\ \Gamma_2(x_2, y_2) \\ \Gamma_3(x_3, y_3) \end{bmatrix} + \begin{bmatrix} \Upsilon_1 \\ \Upsilon_2 \\ \Upsilon_3 \end{bmatrix} + \begin{bmatrix} \Lambda_1 & 0 & 0 \\ 0 & \Lambda_2 & 0 \\ 0 & 0 & \Lambda_3 \end{bmatrix}\begin{bmatrix} P_1^{ref} \\ P_2^{ref} \\ P_3^{ref} \end{bmatrix},$$

$$(29)$$

where

$$\Phi_i = \begin{bmatrix} 0 & 1 & 0 \\ 0 & -\frac{D_i}{M_i} & \frac{1}{M_i} \\ 0 & -\frac{1}{\tau_i R_i \omega_s} & -\frac{1}{\tau_i} \end{bmatrix}, \quad \Gamma_i(x_i) = \begin{bmatrix} 0 \\ -\frac{E_i V_i}{M_i X_i}\sin(\delta_i - \theta_i) \\ 0 \end{bmatrix}, \quad \Upsilon_i = \begin{bmatrix} -\omega_s \\ \omega_s \frac{D_i}{M_i} \\ \frac{1}{\tau_i R_i} \end{bmatrix}, \quad \Lambda_i = \begin{bmatrix} 0 \\ 0 \\ \frac{1}{\tau_i} \end{bmatrix},$$

which is of the form in (14).

4.2 Dynamics of Wind Power Plants

As described in (17) and (18), the dynamics of the wind power plants can be compactly written as $\frac{dv}{dt} = \alpha(v, w)$, $p_w = \phi(v, w)$. In this example, there are only wind power plants at buses 4 and 5. It is assumed that the wind speed at the location of the wind power plant connected to bus 4 is such that the plant is operating at its rated power. Also, it is assumed that the wind power plant connected to bus 5 is operating between its cut in and rated speeds. Thus, $v = [v_4, v_5]'$, $p_w = [p_4^w, p_5^w]'$, and $w = [w_4, w_5]'$, which results in

$$\frac{d}{dt}\begin{bmatrix} v_4 \\ v_5 \end{bmatrix} = \begin{bmatrix} \alpha_4(v_4, w_4) \\ \alpha_5(v_5, w_5) \end{bmatrix},$$

$$\begin{bmatrix} P_4^w \\ P_5^w \end{bmatrix} = \begin{bmatrix} P_4^r \\ \beta_5(v_5, w_5) \end{bmatrix},$$

$$(30)$$

where P_4^r is the nameplate power of the wind power plant connected to bus 4.

4.3 Network

As described in (19), the network power flow equations can be compactly written as $0 = g(x, y, p_d, p_w)$. In this example, $p_d = [P_4^d, \ P_5^d, \ P_6^d]'$, and $p_w = [P_4^w, \ P_5^w]'$. Denote by $Y_{ik} = 1/X_{ik}$ the admittance of the transmission line between buses i and k. Then the power balance equations for bus 1 are

$$\frac{E_1 V_1}{X_1} \sin(\delta_1 - \theta_1) = Y_{14} V_1 V_4 \sin(\theta_1 - \theta_4) + Y_{16} V_1 V_6 \sin(\theta_1 - \theta_6),$$

$$\frac{E_1 V_1}{X_1} \cos(\delta_1 - \theta_1) - \frac{V_1^2}{X_1} = -Y_{14} V_1 V_4 \cos(\theta_1 - \theta_4) - Y_{16} V_1 V_6 \cos(\theta_1 - \theta_6)$$

$$+ (Y_{14} + Y_{16} + Y_1) V_1^2. \tag{31}$$

The power balance equations for bus 2 are given by

$$\frac{E_2 V_2}{X_2} \sin(\delta_2 - \theta_2) = Y_{25} V_2 V_5 \sin(\theta_2 - \theta_5) + Y_{26} V_2 V_6 \sin(\theta_2 - \theta_6),$$

$$\frac{E_2 V_2}{X_2} \cos(\delta_2 - \theta_2) - \frac{V_2^2}{X_2} = -Y_{25} V_2 V_5 \cos(\theta_2 - \theta_5) - Y_{26} V_1 V_6 \cos(\theta_2 - \theta_6)$$

$$+ Y_{25} + Y_{26} + Y_2) V_2^2. \tag{32}$$

The power balance equations for bus 3 are given by

$$\frac{E_3 V_3}{X_3} \sin(\delta_3 - \theta_3) = Y_{34} V_3 V_4 \sin(\theta_3 - \theta_4) + Y_{35} V_3 V_5 \sin(\theta_3 - \theta_5),$$

$$\frac{E_3 V_3}{X_3} \cos(\delta_3 - \theta_3) - \frac{V_3^2}{X_3} = -Y_{34} V_3 V_4 \cos(\theta_3 - \theta_4) - Y_{35} V_3 V_5 \cos(\theta_3 - \theta_5)$$

$$+ (Y_{34} + Y_{35} + Y_3) V_3^2. \tag{33}$$

The power balance equations for bus 4 are given by

$$P_4^w - P_4^d = Y_{14} V_4 V_1 \sin(\theta_4 - \theta_1) + Y_{34} V_4 V_3 \sin(\theta_4 - \theta_3),$$

$$Q_4^w - Q_4^d = -Y_{14} V_4 V_1 \cos(\theta_4 - \theta_1) - Y_{34} V_4 V_3 \cos(\theta_4 - \theta_3) + (Y_{14} + Y_{34}) V_3^2. \tag{34}$$

The power balance equations for bus 5 are given by

$$P_5^w - P_5^d = Y_{25} V_5 V_2 \sin(\theta_5 - \theta_2) + Y_{35} V_5 V_3 \sin(\theta_5 - \theta_3),$$

$$Q_5^w - Q_5^d = -Y_{25} V_5 V_2 \cos(\theta_5 - \theta_2) - Y_{35} V_5 V_3 \cos(\theta_5 - \theta_3) + (Y_{25} + Y_{35}) V_5^2. \tag{35}$$

The power balance equations for bus 6 are given by

$$- P_6^d = Y_{16} V_6 V_1 \sin(\theta_6 - \theta_1) + Y_{26} V_6 V_2 \sin(\theta_6 - \theta_2),$$

$$- Q_6^d = -Y_{16} V_6 V_1 \cos(\theta_6 - \theta_1) - Y_{26} V_6 V_2 \cos(\theta_6 - \theta_2) + (Y_{16} + Y_{26}) V_6^2. \quad (36)$$

4.4 Automatic Generation Control

The AGC block diagram for this example is depicted in Fig. 2. As described in (15) and (16), the power settings of generators participating in AGC can be compactly written as $\frac{du}{dt} = h(x, y, p_{sch}, p_{ed})$, $u = k(z, p_{ed})$. In this example, $u = [P_1^{ref}, P_2^{ref}, P_3^{ref}]'$. For area 2, we assume that the frequency measurement used in the ACE is the average of the frequencies on buses 1 and 3. Thus, the power setting of generators 1 and 2 in balancing area 1 are given by

$$ACE_1 = -(P^{12} - P_{sch}^{12}) - b_1(f^1 - f^{nom}),$$

$$P^{12} = P_{43} + P_{25} = Y_{34} V_4 V_3 \sin(\theta_4 - \theta_3) + Y_{25} V_2 V_5 \sin(\theta_2 - \theta_5),$$

$$\frac{dz_1}{dt} = ACE_1 - \left[P_1^{ed} + \alpha_1^1 (z_1 - (P_1^{ed} + P_2^{ed})) - P_1^s \right]$$

$$+ \left[P_2^{ed} + \alpha_2^1 (z_1 - (P_1^{ed} + P_2^{ed})) - P_2^s \right], \quad (37)$$

$$P_1^{ref} = P_1^{ed} + \alpha_1^1 (z_1 - (P_1^{ed} + P_2^{ed})),$$

$$P_2^{ref} = P_2^{ed} + \alpha_2^1 (z_1 - (P_1^{ed} + P_2^{ed})), \quad (38)$$

where $f^1 = \frac{f_1 + f_3}{2}$. The frequencies in buses 1 and 3 can be related to the voltage angles in buses 1 and 3 as follows:

$$f_1 = f^{nom} + \frac{1}{2\pi} \frac{d\theta_1}{dt},$$

$$f_2 = f^{nom} + \frac{1}{2\pi} \frac{d\theta_2}{dt}. \quad (39)$$

For area 2, we assume that the frequency used in the ACE is the frequency on bus 6. Thus, the power setting of generator 3 in balancing area 2 is given by

$$ACE_2 = -(P^{21} - P_{sch}^{21}) - b_2(f^2 - f^{nom}),$$

$$P^{12} = P_{34} + P_{52} = Y_{34} V_3 V_4 \sin(\theta_3 - \theta_4) + Y_{25}, V_5 V_2 \sin(\theta_5 - \theta_2),$$

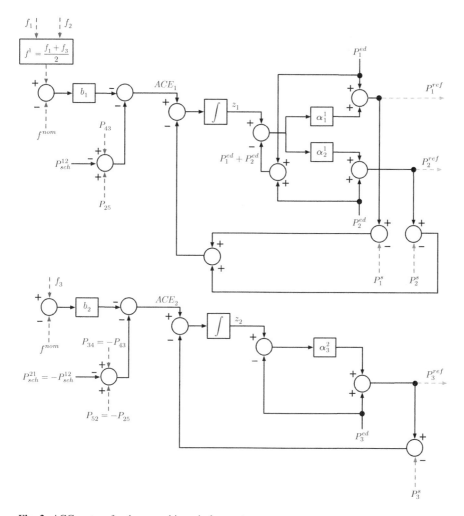

Fig. 2 AGC system for three-machine, six-bus system

$$\frac{dz_2}{dt} = ACE_2 - \left[P_3^{ed} + \alpha_3^2 (z_2 - P_3^{ed}) - P_3^s \right],$$

$$P_3^{ref} = P_3^{ed} + \alpha_3^2 (z_2 - P_3^{ed}) - P_3^s, \tag{40}$$

where $f^2 = f_3$, and $\alpha_3^2 = 1$. By rearranging (38) and (40), we obtain an expression of the form in (15) and (16).

5   Concluding Remarks

This chapter developed a dynamical model for capturing the impact of deep penetration of wind resources on system frequency response. This model includes the electrical network, which in AGC models is usually suppressed. Our hypothesis is that, for deep penetration scenarios, the network may also have some substantial smoothing effects as it does affect the response of the overall closed-loop system dynamic behavior. We believe that the model can be useful for understanding fundamental limits on the amount of wind that can be integrated in a power system without violating frequency performance metrics. Future work will include the comparison of the behavior of this model and classical models available in the literature to verify whether or not our hypothesis is indeed correct or on the contrary, the effect of the network to smooth out wind variability can be neglected.

Acknowledgments This work has been supported by the Global Climate and Energy Project. Any opinions, findings, and conclusions or recommendations expressed in this publication are those of the author and do not necessarily reflect the views of Stanford University, the Sponsors of the Global Climate and Energy Project, or others involved with the Global Climate and Energy Project.

References

1. U.S. Department of Energy. The smart grid: an introduction [Online]. Available: http://www.oe.energy.gov/smartgrid.htm
2. Hirst E (2002) Integrating wind output with bulk power operations and wholesale electricity markets. Wind Energ 5(1):19–36
3. Eriksen P, Ackermann T, Abildgaard H, Smith P, Winter W, Garcia JR (2005) System operation with high wind penetration. IEEE Power Energ Mag 3(6):65–74
4. DeMeo E, Jordan G, Kalich C, King J, Milligan M, Murley C, Oakleaf B, Schuerger M (2007) Accommodating wind's natural behavior. IEEE Power Energ Mag 5(6):59–67
5. Ahlstrom M, Jones L, Zavadil R, Grant W (2005) The future of wind forecasting and utility operations. IEEE Power Energ Mag 3(6):57–64
6. Eto J et al (2010) Use of frequency response metrics to assess the planning and operating requirements for reliable integration of variable renewable generation. Lawrence Berkeley National Laboratory, Technical Report LBNL-4142E, December 2010
7. Undrill J (2010) Power and frequency control as it relates to wind-powered generation. Lawrence Berkeley National Laboratory, Technical Report LBNL-4143E, December 2010
8. Martinez C, Xue S, Martinez M (2010) Review of the recent frequency performance of the eastern, western and ercot interconnections. Lawrence Berkeley National Laboratory, Technical Report LBNL-4144E, December 2010
9. Illian H (2010) Frequency control performance measurement and requirements. Lawrence Berkeley National Laboratory, Technical Report LBNL-4145E, December 2010
10. Mackin P et al (2010) Dynamic simulations studies of the frequency response of the three U.S. interconnections with increased wind generation. Lawrence Berkeley National Laboratory, Technical Report LBNL-4146E, December 2010
11. Coughlin K, Eto J (2010) Analysis of wind power and load data at multiple time scales. Lawrence Berkeley National Laboratory, Technical Report LBNL-4147E, December 2010

12. Debs A (ed) (1988) Modern power systems control and operation. Kluwer Academic Publishers, Boston, MA
13. Kundur P (1994) Power system stability and control. McGraw Hill Inc, New York
14. Wood A, Wollenberg B (eds) (1996) Power generation, operation, and control. Wiley, New York
15. Ilic M, Zaborzky J (eds) (2000) Dynamics and control of large electric power systems. Wiley, New York
16. Chen YC, Domímguez-García A (2011) Assessing the impact of wind variability on power system small-signal reachability. In: Proceedings of Hawaii international conference on system sciences, Kauai, HI, January 2011
17. Sauer P, Pai A (1998) Power system dynamics and stability. Prentice Hall, Upper Saddle River, NJ
18. Pulgar H (2010) Wind farm model for power system stability analysis. Ph.D. Dissertation, University of Illinois at Urbana-Champaign, Urbana, IL
19. Pulgar-Painemal HA, Sauer PW (2011) Towards a wind farm reduced-order model. Electric Power Systems Research 81(8):1688–1695
20. Ackermann T (2005) Wind power in power systems. Wiley, Chichester, UK
21. Hespanha J (2005) A model for stochastic hybrid systems with application to communication networks. Nonlinear Anal Spec Issue Hybrid Syst 62(8):1353–1383
22. Hespanha J (2007) Modeling and analysis of stochastic hybrid systems. IEE Proc Contr Theor Appl Spec Issue Hybrid Syst 153:520–535

Multi-Dimensional Modal Analysis in Large Power Systems from Ambient Data Based on Frequency-Domain Optimization

Xueping Pan[1] and Vaithianathan "Mani" Venkatasubramanian

Abstract This article proposes an algorithm denoted *Frequency-Domain Optimization* (FDO) for real-time modal estimation of power system oscillatory modes based on multiple synchronized Phasor Measurement Units (PMUs). The proposed method combines Fast Fourier Transform (FFT) with least-square optimization to estimate the mode parameters of electromechanical oscillations in power systems. Multiple signals are analyzed simultaneously to improve the accuracy of estimation, and the mode shape can also be determined from analyzing these multiple signals. Results from simulated and measured ambient PMU data show that this FDO method is able to estimate the system modal parameters results effectively.

1 Introduction

With vast applications of Wide-Area Monitoring System (WAMS) based on phasor measurement unit (PMU) technology, modal estimation using output-only responses has drawn great attention in power system since 1990s [1]. Compared to traditional model-based analysis methods, the main advantage of PMU-based modal analysis is that it allows the identification of realistic modal parameters in real-time for power

[1]Research in this article was carried out at Washington State University (WSU) while the author was a visiting scholar at WSU.

X. Pan (✉)
School of Energy and Electrical Engineering, Hohai University, Nanjing, China
e-mail: pxp_2002@hotmail.com

V. "Mani" Venkatasubramanian
School of Electrical Engineering and Computer Science, Washington State University, Pullman, WA 99164, USA
e-mail: mani@eecs.wsu.edu

A. Chakrabortty and M.D. Ilić (eds.), *Control and Optimization Methods for Electric Smart Grids*, Power Electronics and Power Systems 3,
DOI 10.1007/978-1-4614-1605-0_8, © Springer Science+Business Media, LLC 2012

systems. Accordingly, the identified modal parameters can be used directly in online stability monitoring and for feedback control.

Modal identification based on random system response data where the input is unmeasured ambient excitation is known as Operational Modal Analysis (OMA) [2]. OMA has been studied in many areas such as civil engineering, mechanical engineering, and aeronautics engineering. OMA is also much useful in power systems because the ambient data is easier to obtain compared to post-disturbance data.

The methods applied in OMA can be divided into two-categories: time-domain-based analysis [3–6] and frequency-domain-based analysis [7, 8]. The simplest method in frequency domain is the peak-picking method [9]. This method is valid only under the assumption that the modes are well separated and the damping levels are low. The main disadvantage is that it cannot estimate the damping ratio correctly. To overcome the disadvantages of the peak-picking method, Frequency-Domain Decomposition (FDD) method is presented based on Singular Value Decomposition (SVD) of the spectral matrix at every discrete frequency. The natural frequencies can be located at the peaks from the singular plot.

Maximum Likelihood (ML) identification is an optimization-based method that estimates the parameters of a model by minimizing an error norm [10]. Hermans et al. [11] discussed the use of ML estimator to identify parametric frequency-domain models. In [12], the curve fitting method is used to identify the modal parameters of the system.

In this article, an optimization-based method is applied to estimate parametric frequency-domain models from ambient data. Some study results for frequency-domain optimization (FDO) were reported in [12]. This article extends the method from [12] in the following ways: (1) Multiple signals are analyzed simultaneously in modal analysis to improve the accuracy of modal estimation as well as for estimating the mode shape, (2) simplified frequency-domain models are introduced in modal identification and (3) weighted objective function and the damped nonlinear Newton iteration method are applied in curve fitting. The purpose of this article is to introduce FDO method to power system mode and mode shape identification, and multiple signals are applied simultaneously to improve the accuracy of the estimation.

The method is currently being integrated into the Oscillation Monitoring System (OMS) [13] previously implemented at Tennessee Valley Authority (TVA) for real-time oscillation monitoring.

2 Frequency Domain Characteristics of Linear Systems

The state equation of a linear system is expressed as follows:

$$\begin{cases} \dot{\mathbf{x}}(t) = \mathbf{A}\mathbf{x}(t) + \mathbf{b}u(t) \\ \mathbf{y}(t) = \mathbf{C}\mathbf{x}(t) \end{cases}, \tag{1}$$

where $\mathbf{x}(t)$ is the n-order state vector, $\mathbf{y}(t)$ is the m-order output vector and $\mathbf{u}(t)$ is the k-order input vector. $\mathbf{A}$, $\mathbf{b}$, and $\mathbf{C}$ are system coefficient matrices.

Suppose ϕ is the right eigen-vector of $\mathbf{A}$, which satisfies $\mathbf{A}\phi = \lambda\phi$. Here, ϕ is a diagonal matrix.

Linear system (1) can be transformed into a decoupled system using $\mathbf{x} = \phi\mathbf{z}$.

$$\dot{\mathbf{z}}(t) = \lambda\mathbf{z}(t) + \phi^{-1}\mathbf{bu}(t) = \lambda\mathbf{z}(t) + \mathbf{u}'(t), \qquad (2)$$

where

$$\mathbf{u}'(t) = \phi^{-1}\mathbf{bu}(t) = \mathbf{b}'\mathbf{u}(t). \qquad (3)$$

The transfer function of the i th mode is given by

$$H_i(s) = \frac{z_i(s)}{u_i'(s)} = \frac{1}{s - \lambda_i}. \qquad (4)$$

The response of $\mathbf{x}(t)$

$$\mathbf{x}(t) = \phi_1 z_1(t) + \phi_2 z_2(t) + \cdots + \phi_n z_n(t). \qquad (5)$$

Its corresponding frequency domain expression:

$$\mathbf{x}(s) = \phi_1 z_1(s) + \phi_2 z_2(s) + \cdots + \phi_n z_n(s). \qquad (6)$$

Frequency domain expression of output signals

$$\mathbf{y}(s) = \mathbf{Cx}(s) = \mathbf{C}[\phi_1 z_1(s) + \phi_2 z_2(s) + \cdots + \phi_n z_n(s)] \qquad (7)$$

The kth variable in $\mathbf{y}(s)$

$$y_k(s) = C_k \phi_1 H_1(s) u_1'(s) + C_k \phi_2 H_2(s) u_2'(s) + \cdots + C_k \phi_1 H_k(s) u_k'(s), \quad (8)$$

where $\mathbf{C} = [\mathbf{C}_1, \mathbf{C}_2, \ldots, \mathbf{C}_m]^{\mathrm{T}}$.

2.1 Simplification of the Frequency-Domain Models

If the parameterized spectrum matrix is calculated as a whole from output signals shown in (8), it will increase the computational burden and the method is not suited for handling large amounts of data. Therefore, the simplification of the frequency-domain model is presented without overly sacrificing model accuracy.

2.1.1 Interactions from Different Modes

Let us consider a fourth-order linear system as an example, and we set

$$\mathbf{A} = \begin{bmatrix} 0 & 1 & 0 & 0 \\ 0 & 0 & 1 & 0 \\ 0 & 0 & 0 & 1 \\ a_{41} & a_{42} & a_{43} & a_{44} \end{bmatrix}, \tag{9}$$

where

$a_{41} = -\omega_1^2 \omega_2^2, a_{42} = -2\left(\xi_1 \omega_1 \omega_2^2 + \xi_2 \omega_2 \omega_1^2\right)$
$a_{43} = -\left(\omega_1^2 + 4\xi_1 \xi_2 \omega_1 \omega_2 + \omega_2^2\right), a_{44} = -2\left(\xi_1 \omega_1 + \xi_2 \omega_2\right),$

$$\mathbf{b} = \begin{bmatrix} 1 & 1 & 1 & 1 \end{bmatrix}^{\mathrm{T}}, \tag{10}$$

$$\mathbf{c} = \begin{bmatrix} 0 & 0 & 0 & 1 \end{bmatrix}, \tag{11}$$

$$\mathbf{u}(t) = [u(t)], \tag{12}$$

with the parameters
$\xi_1 = 0.2, \omega_1 = 1.4\pi \ (f = 0.7\,\text{Hz}), \xi_2 = 1.0$, and $\omega_1 = 2.4\pi \ (f = 1.2\,\text{Hz})$.
The transfer function between output signal $y(\omega)$ and input signal $u(\omega)$ is

$$H(\omega) = \frac{y(\omega)}{u(\omega)} = \frac{x_4(\omega)}{u(\omega)} = \sum_{j=1}^{4} \frac{R_i}{j\omega - \lambda_i}. \tag{13}$$

The two parts in (13) corresponding to two modes are

$$H(\omega_1) = \sum_{j=1}^{2} \frac{R_i}{j\omega - \lambda_i}, \tag{14}$$

$$H(\omega_2) = \sum_{j=3}^{4} \frac{R_i}{j\omega - \lambda_i}, \tag{15}$$

where $\lambda_{1,2} = 0.0880 \pm j4.3973, \lambda_{3,4} = 0.7540 \pm j7.5020, \omega_1, \omega_2$ are corresponding to the two oscillation angle frequency, and $R_i = \phi_{4i} b_i'$.

By plotting the complete response of $H(\omega)$ in Fig. 1, we can see clearly two peaks corresponding to the system oscillation frequency: one is at $f_1 = 0.7\,\text{Hz}$, corresponding to eigenvalue $\lambda_{1,2}$; the other is corresponding to $\lambda_{3,4}$. This indicates the natural frequencies of the power system can be estimated from frequency-domain models of output signals.

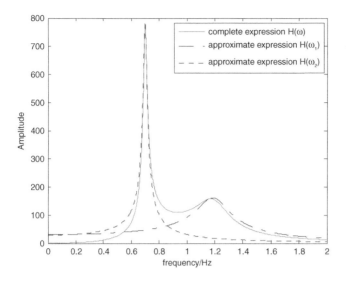

Fig. 1 Interactions between mode 1 and mode 2

Figure 1 also shows that at the natural frequency ω_1 (or ω_2), the transfer function $|H(\omega)|$ is dominated by the term $|H_1(\omega)|$ (or $|H_2(\omega)|$). This means that we can estimate the frequency response $y(s)$ at the natural frequency using $y(\omega_i) = H(\omega_i)u(\omega_i)$.

2.1.2 Effects of Neglecting One Term

Because $\lambda_{i,i+1} = -\sigma_i \pm j\omega_i$ are a pair of complex eigenvalues

$$|H(\omega_i)| = \frac{R_i}{j\omega - \lambda_i} + \frac{R_i^*}{j\omega - \lambda_{i,i+1}}, \tag{16}$$

and the two individual components

$$H_1(\omega_i) = \frac{R_i}{j\omega - (-\sigma_i + j\omega_i)}, \tag{17}$$

$$H_2(\omega_i) = \frac{R_i^*}{j\omega - (-\sigma_i - j\omega_i)}. \tag{18}$$

By plotting the magnitude of $H(\omega_2)$ and its two components $H_1(\omega_2)$ and $H_2(\omega_2)$, which is shown in Fig. 2, it is demonstrated that $H(\omega_2)$ is symmetrical, and magnitude of $H_1(\omega_2)$ can be used to approximate the magnitude of $H(\omega_2)$ at the plane of positive frequency.

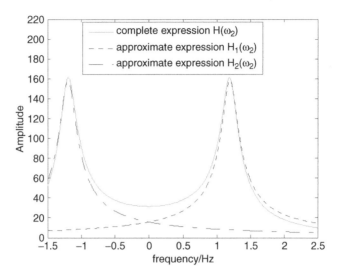

Fig. 2 Error from the approximate expressions

Therefore, we can obtain reliable estimates of modal components in power systems from the frequency-domain of output signals, provided the following conditions are met:

(1) Under quasi-steady-state conditions, power systems can be reasonably approximated as linear system when excited by random loads excitation, which assumed to be stationary Gaussian white noise.
(2) The modes of interest are well separated and they are lightly damped. Thus, the response at a natural frequency is dominated by the corresponding mode shape and the peaks in the frequency domain can be used to identify natural frequency.

3 Frequency-Domain Optimization-Based Modal Estimation

In the following, Fourier transform is applied to attain the frequency spectrum of the output signals, and the optimization method is used to estimate the mode parameters.

3.1 Fourier Transform

Using the above fourth-order linear system (as expressed from (9) to (12)) as an example, Fourier transform is applied to output signal $y(t)$ to attain its frequency spectrum, which is shown in Fig. 3.

We can see from Fig. 3 that the Fourier spectrum of $y(t)$ reaches local maxima near natural frequencies.

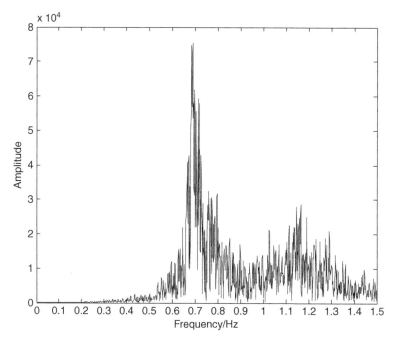

Fig. 3 Fourier spectrum of output signal $y(t)$

3.2 Oscillation Frequency and Damping Estimation

In the following, a least square optimization method is used for curve fitting within the neighborhood of each natural frequency ω_i. Simplified frequency-domain model, which is expressed in (17), is used to fit the curve.

The expression of estimated oscillation frequency and damping ratio are

$$f_i = \frac{\omega_i}{2\pi}, \quad \xi_i = \frac{\sigma_i}{\sqrt{\sigma_i^2 + \omega_i^2}}.$$

3.2.1 Least Square Optimization Method for Nonlinear Curve Fitting

To estimate oscillate frequency and damping ratio directly, (17) is changed as follows:

$$|H(\omega_i)| \approx \frac{R_i}{\sqrt{\xi_i^2 \omega_i^2 + (\omega - \omega_i)^2}} = \frac{k}{\sqrt{\xi_i^2 f_i^2 + (f - f_i)^2}}. \tag{19}$$

Here, we use the least square method to search for the parameter vector $\mathbf{x} = [k, \xi_i, f_i]$, the objective function is

$$L = \frac{1}{2} \sum_{j=1}^{n} \left(C_j - H(\omega) \right)^2 = \frac{1}{2} \sum_{j=1}^{n} \left(C_j - kh \right)^2, \tag{20}$$

where C_j is the jth coefficient of Fast Fourier Transform (FFT), n is the number of FFT within neighborhood of natural frequency f_i.

$$h = \frac{1}{\sqrt{\xi_i^2 f_i^2 + (f - f_i)^2}}. \tag{21}$$

Based on nonlinear Newton iteration method, Taylor expansion of L to its second order

$$L = L_0 + \mathbf{J}\Delta\mathbf{x} + \frac{1}{2}\Delta\mathbf{x}^T H \Delta\mathbf{x}, \tag{22}$$

where

$$\mathbf{J} = \left[J_1, J_2, J_3 \right]^T = \left[\frac{\partial L}{\partial k}, \frac{\partial L}{\partial \xi_i}, \frac{\partial L}{\partial f_i} \right]^T, \tag{23}$$

$$H = \begin{bmatrix} H_{11} & H_{12} & H_{13} \\ H_{21} & H_{22} & H_{23} \\ H_{31} & H_{32} & H_{33} \end{bmatrix} = \begin{bmatrix} \frac{\partial^2 L}{\partial k^2} & \frac{\partial^2 L}{\partial k \partial \xi_i} & \frac{\partial^2 L}{\partial k \partial f_i} \\ \frac{\partial^2 L}{\partial \xi_i \partial k} & \frac{\partial^2 L}{\partial \xi_i^2} & \frac{\partial^2 L}{\partial \xi_i \partial f_i} \\ \frac{\partial^2 L}{\partial f_i \partial k} & \frac{\partial^2 L}{\partial f_i \partial \xi_i} & \frac{\partial^2 L}{\partial f_i^2} \end{bmatrix}, \tag{24}$$

$$\Delta\mathbf{x} = \left[\Delta k, \Delta\xi_i, \Delta f_i \right]^T. \tag{25}$$

To find the minimal value of L, the derivation of L should satisfy

$$L' = \mathbf{J} + H\Delta\mathbf{x} = \mathbf{0}, \tag{26}$$

$$\Delta\mathbf{x} = -\mathbf{H}^{-1}\mathbf{J}. \tag{27}$$

Here the damped Newton iteration method is introduced when Hessen matrix $\mathbf{H}$ is singular. In this situation, an identity matrix is added with a suitable value of α to increase the value of diagonal elements to prevent $\mathbf{H}$ from being singular

$$\Delta\mathbf{x} = -(\mathbf{H} + \alpha\mathbf{I})^{-1}\mathbf{J}. \tag{28}$$

3.2.2 Least Square Optimization Method for Nonlinear Curve Fitting

Newton method is sensitive to the initial values. In this algorithm, we first estimate the initial value of oscillation frequency f_{i0} based on the peaks on the Fourier spectrum, and the initial value of damping value is set to be $\xi_{i0} = 0.05$.

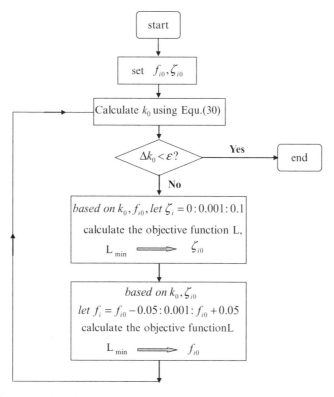

Fig. 4 Initial value calculation

Based on these two initial values, initial value of parameter k_0 can be calculated based on equation $\frac{\partial L}{\partial k} = 0$, which is

$$\frac{\partial L}{\partial k} = -\sum_{j=1}^{n} \left(C_j - k_0 h\right) h = 0. \tag{29}$$

Then

$$k_0 = \frac{\sum\limits_{j=1}^{n} C_j h}{\sum\limits_{j=1}^{n} h^2}. \tag{30}$$

The procedure to search for the initial values of $\mathbf{x_0} = \left[k_0, \ f_{i0}, \ \xi_{i0} \right]^{\mathrm{T}}$ is shown in Fig. 4.

3.2.3   Weight Coefficient Selection

When there are multiple modes in the signal, the modes may influence each other (such as in Fig. 3), especially at the region between the natural frequencies, this may lead to the big identified error in estimation.

To overcome this problem, the weighted objective function is applied; (20) is improved to the following format:

$$L = \frac{1}{2} \sum_{j=1}^{n} w_j \left(C_j - H\left(\omega\right)\right)^2 = \frac{1}{2} \sum_{j=1}^{n} w_j \left(C_j - kh\right)^2. \tag{31}$$

The weight coefficient will be selected based on experiences, here we choose

$$w = \begin{cases} a \text{ if } |f - f_i| < \varepsilon, \\ 1 \text{ if } |f - f_i| \geq \varepsilon, \end{cases} \tag{32}$$

where a and ε are tunable values that can be adjusted for different systems.

At the same time, (31) will be changed to the following:

$$k_0 = \frac{\sum_{j-1}^{n} w_j C_j h}{\sum_{j-1}^{n} w_j h^2}. \tag{33}$$

3.3   Identification of Mode Shape

The mode shape estimation corresponding to the natural frequency f_i is determined using a series of simultaneously recorded ambient data $y_k\left(t\right)$, $k = 1, 2, \ldots, n$. Based on (8) and (14),the mode shape of the k th output signal relative to the selected reference signal can be obtained as shown in (34).

$$\frac{y_k\left(\omega_i\right)}{y_{ref}\left(\omega_i\right)} \approx \frac{\mathbf{c_k}\boldsymbol{\phi_i} H_i\left(\omega_i\right) u'\left(\omega_i\right)}{\mathbf{c_{ref}}\boldsymbol{\phi_i} H_i\left(\omega_i\right) u'\left(\omega_i\right)} = \frac{\mathbf{c_k}\boldsymbol{\phi_i}}{\mathbf{c_{ref}}\boldsymbol{\phi_i}} = \frac{A_k}{A_{ref}} \angle \left(\varphi_{y_k} - \varphi_{y_{ref}}\right). \tag{34}$$

Equation (34) is routinely used in frequency domain based compensator designs in classical linear system theory [15]. Therefore, the mode shape of output signals can be attained by using the Fourier transform coefficient at the natural frequency. Better estimation of mode shape is obtained by averaging several data segments, which will be shown in the following simulation example.

4   Tests on a Linear System

4.1   Mode Estimation

The system is the same as the example in (9)–(12). The test is repeated for 100 times using different random inputs, and mode identification using proposed frequency

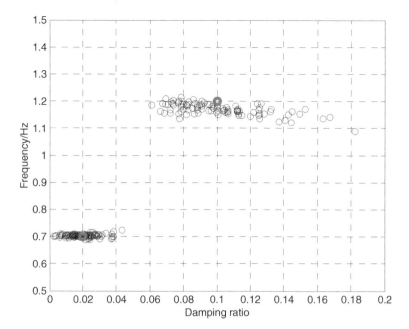

Fig. 5 Mode estimates by FDO for the LTI system

domain optimization (FDO) method is repeated in each case. The results are shown in the complex s-plane in Fig. 5 to show the values of frequency and damping identified.

For mode 1 with frequency 0.7 Hz and damping ratio at 0.02, the mean and standard variance for frequency from 100 identified samples are 0.7032 Hz and 2.9394×10^{-5} Hz; the mean and standard variance of damping ratio in 100 identified samples are 0.0196 and 7.3213×10^{-5}.

For mode 2 with frequency 1.2 Hz and damping ratio 0.10, the mean and standard variance of mode frequency from 100 identified samples are 1.1702 Hz and 4.3131×10^{-4} Hz; the mean and standard variance of damping ratio from 100 identified samples are 0.1009 and 6.1073×10^{-4}.

The estimated mode frequency and damping ratios are reasonable for both mode 1 and mode 2 in this example.

4.2 Mode Shape Estimation

Plots for the beginning ten second time-window of $x_1(t)$ $x_4(t)$ signals are shown in Fig. 6.

The averaged mode shape of 100 times simulations of dominate mode (mode 1) is shown in Fig. 7.

From Fig. 7, we can find that the identified mode shape is quite close to the analytical results, in which case the four signals should show $90°$ phase differences.

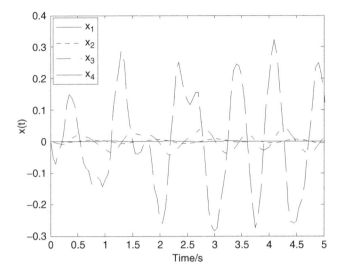

Fig. 6 Signals of $x_1(t) \sim x_4(t)$ excited by white noise

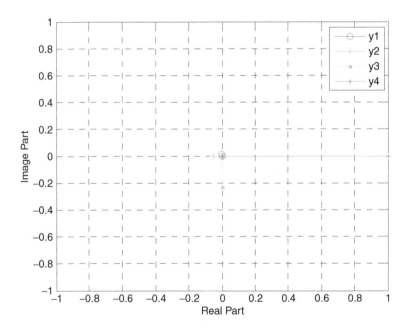

Fig. 7 Identified mode shape of mode 1

5 Two-Area Test System Results

The two-area Kundur test system from [14] is shown in Fig. 8, the network parameters and generator's parameters are the same as in reference [14]. The generators are modeled with two-axis models and exciters equipped with PSSs. Total base loading level is 2734 MW and 200 MVar, and they are modeled as constant impedance loads.

The eigenvalues of the system at the stable equilibrium point computed by the Small-signal Stability Analysis Tool (SSAT) software from Powertech Labs [14] are shown in Table 1.

The system is simulated in time-domain using Transient Stability Analysis Tool (TSAT) [14] with 1% of the load at L7 and L9 represented as Gauss white noise with variance 1 p.u. and 99% of original load as base load. The Fourier spectrum of each generator power angle referred to system center of inertia (COI) from one simulation is shown in Fig. 9.

From Fig. 9, we can find that mode 1 shows up in every power angle signal, while mode 2 is immersed in the random output and cannot be identified from the above output signals, and mode 3 can only be weakly seen in δ_4 because the damping ratios of mode 2 and mode 3 are much higher.

In the following estimation, mode 1 is identified from signals δ_1 through δ_4 while local mode 3 is identified from signal δ_4. The simulation is performed independently for 100 times and mode identification with FDO is applied for each simulation run. The mode identified results are shown in the complex s-plane, which is in Fig. 10.

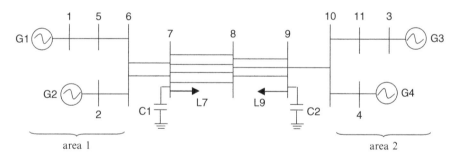

Fig. 8 Two-area Kundur test system [15]

Table 1 Eigenvalues

Mode	f (Hz)	ζ (%)	Mode shape
1	0.7062	2.38	$G_{1,2}$ vs. $G_{3,4}$
2	1.1598	10.87	G_1 vs. G_2
3	1.1948	10.38	G_3 vs. G_4

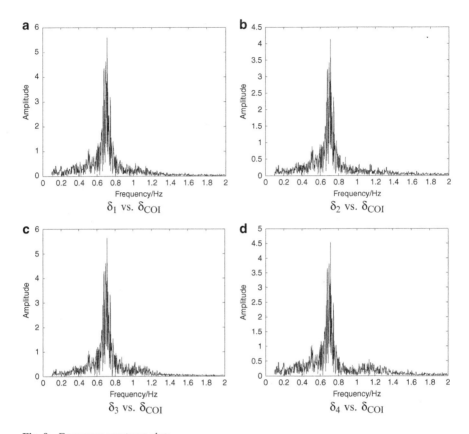

Fig. 9 Frequency spectrum plots

For model, the mean and standard variance of 100 estimated frequencies is 0.6986 Hz and 2.6549×10^{-5}; the mean and standard variance of 100 estimated damping ratio is 0.0235 and 8.1499×10^{-5}. For mode 3, the mean and standard variance of 100 estimated frequencies is 1.1270 Hz and 1.6108×10^{-4}; The mean and standard variance of 100 estimated damping ratio is 0.1009 and 2.5485×10^{-4}.

Mode shape results from 100 simulations and its average result for mode 1 are shown in Fig. 11.

Mode shape of local modes (G_1 vs. G_2) and (G_3 vs. G_4) cannot be estimated correctly because: (1) they cannot be clearly shown in the frequency spectrum because of their high damping ratio; and (2) the frequencies of two local modes are too close to be distinguished. However, if we select different reference signals, the local mode can still be identified, which is shown in Fig. 12.

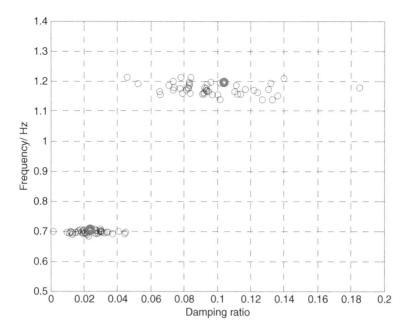

Fig. 10 Identified mode distribution from 100 tests

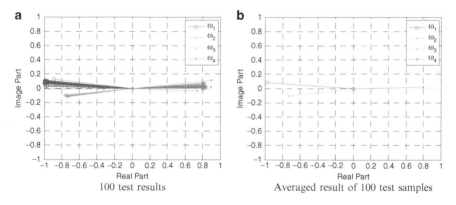

Fig. 11 Mode shape plots of mode 1 (G_{12} vs. G_{34})

6 Case Study

On August 10, 1996, a major blackout occurred in western American power grid. Here, we apply FDO method to identify the modal parameters from ambient data measured by PMUs after Keel_Allston line tripping and before the tripping of

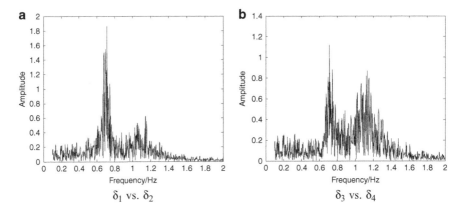

Fig. 12 Frequency spectrum plots

Ross-Lexington line. A total of nine power signals are analyzed simultaneously. The sampling ratio is 20 Hz, and a total of 180 s of data with sliding time window at 10 s are used to identify the mode frequency and damping ratio, which is shown in Fig. 13a. The identified output mode shapes of these nine power signals is presented in Fig. 13b.

Results in Fig. 13a shows that the dominant frequency is about 0.25 Hz and the damping ratio is about 0.022, which matches well with the results based on an earlier algorithm named Frequency-Domain Decomposition in [7].

7 Conclusions

The article demonstrates the feasibility of FDO method for estimating the dominant poorly damped oscillatory modes and their mode shapes from ambient data. Multiple signals have been analyzed simultaneously to improve the effectiveness of the estimation, and mode shape results have also been estimated. Results from a linear test system and a nonlinear two-area test power system show that the estimated modal parameters are reasonably close to the analytical values. Case studies from August 10, 1996 western American blackout are also shown in the article.

It has been pointed out that this method will be useful when the estimated modes are well separated and the modes are poorly damped. It has also shown that mode shape determination of two modes with close frequencies is difficult. However,

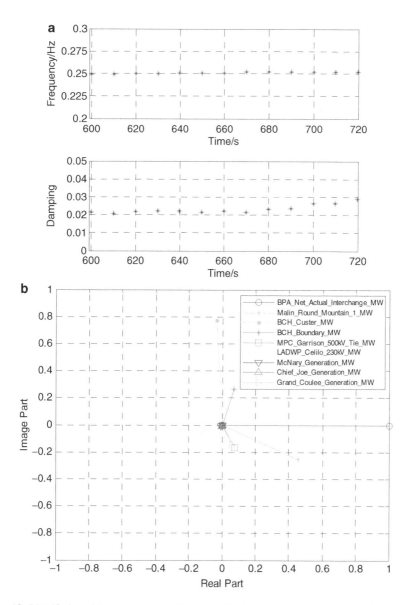

Fig. 13 Identified modal parameters. (**a**) Plots of estimated mode frequency and mode damping ratio; (**b**) Plot of averaged output mode shape

when suitable measurement signals are used, the mode parameters can still be identified. A good selection of output signals to be used in modal analysis is an important task for effective real-time modal identification.

Acknowledgments The authors gratefully acknowledge support from US Department of Energy, Power System Engineering Research Center, and Tennessee Valley Authority.

References

1. Hauer JF, Demeure CJ, Scharf LL (1990) Initial results in Prony analysis of power system response signals. IEEE Trans Power Syst 5(1):80–89
2. Zhang L, Brincker R, Andersen P (2005) An overview of operational modal analysis: major development and issues. In: Proceedings of the 1st international operational modal analysis conference, Copenhagen, Denmark, April 2005, pp 179–190
3. Pierre JW, Trudnowski DJ, Donnelly MK (1997) Initial results in electromechanical mode identification from ambient data. IEEE Trans Power Syst 12(3):1245–1251
4. Zhou N, Pierre JW, Hauer JF (2006) Initial results in power system identification from injected probing signals using a subspace method. IEEE Trans Power Syst 21(3):1293–1302
5. Zhou N, Trudnowski DJ, Pierre JW (2008) Electromechanical mode online estimation using regularized robust RLS methods. IEEE Trans Power Syst 23(4):1670–1680
6. Trudnowski DJ, Pierre JW, Zhou N (2008) Performance of three mode-meter block-processing algorithm for automated dynamic stability assessment. IEEE Trans Power Syst 23(2):680–690
7. Liu G, "Mani" Venkatasubramanian V (2008) Oscillation monitoring from ambient PMU measurements by frequency domain decomposition. In: Proceedings of IEEE international symposium on circuits and system, Seattle, USA, May 2008, pp 2821–2824
8. Zhang L, Wang T, Tamura Y (2005) A frequency-spatial domain decomposition (FSDD) technique for operational modal analysis. In: Proceedings of the 23rd international modal analysis conference (IMAC), Orlando, Florida, April 2005
9. Felber AJ (1993) Development of a hybrid bridge evaluation system. Ph.D. Dissertation, The University of British Columbia
10. Pintelon R, Guillaume P, Rolain Y et al (1993) Identification of Transfer Functions in the Frequency Domain, a Survey. In: Proceedings of the 32nd conference on decision and control, San Antonio, Texas, December 1993, pp 557–566
11. Hermans L, Van der Auweraer H, Guillaume P (1998) A frequency-domain maximum likelihood approach for the extraction of modal parameters from output-only data. In: Proceedings of the 23th international seminar on modal analysis, Leuven, Belgium
12. Kakimoto N, Sugumi M, Makino T et al (2006) Monitoring of interarea oscillation mode by synchronized phasor measurement. IEEE Trans Power Syst 21(1):260–268
13. Liu G, "Mani" Venkatasubramanian V, Carroll JR (2009) Oscillation monitoring system using synchrophasors. In: Proceedings of IEEE PES general meeting, Calgary, Canada, July 2009
14. Powertech Labs Inc., TSAT and SSAT user manuals. Power System Analysis Package (PSAPACK), Canada
15. Kundur P (1994) Power system stability and control. McGraw-Hill Inc.

Coherent Swing Instability of Interconnected Power Grids and a Mechanism of Cascading Failure

Yoshihiko Susuki, Igor Mezić, and Takashi Hikihara

Abstract We describe a dynamical mechanism of cascading failure in a system of interconnected power grids. This mechanism is based on the discovery of (Susuki Y et al. (2011) J Nonlinear Sci 21(3):403–439), an emergent and undesirable phenomenon of synchronous machines in a power grid, termed the Coherent Swing Instability (CSI). In this phenomenon, most of the machines in a sub-grid coherently lose synchronism with the rest of the grid after being subjected to a local and finite disturbance. By numerical analysis of a system of weakly interconnected power grids, we present a phenomenon of coupled swing dynamics in which the CSI happens for all of the power grids in a successive manner. We suggest that a small disturbance in one grid can grow, spill to the other grids, and cause the whole system to fail. This mechanism enables the development of dynamically relevant tools for monitoring and control of wide-area disturbances, which become feasible when the physical power network is overlaid with an information network, like the smart grid.

1 Introduction

Wide-area disturbances have been reported in large-scale power grids. Examples include the 2003 blackouts in North America and Europe [1] and the 2006 system disturbance in Europe [33]. Such disturbances are very costly to modern society and have reminded us of the importance of stable and reliable electricity supply.

Y. Susuki (✉) • T. Hikihara
Department of Electrical Engineering, Kyoto University, Katsura, Nishikyo,
Kyoto 615-8510 Japan
e-mail: susuki@ieee.org; hikihara@kuee.kyoto-u.ac.jp

I. Mezić
Department of Mechanical Engineering, University of California, Santa Barbara,
CA 93106-5070 USA
e-mail: mezic@engineering.ucsb.edu

A. Chakrabortty and M.D. Ilić (eds.), *Control and Optimization Methods for Electric Smart Grids*, Power Electronics and Power Systems 3,
DOI 10.1007/978-1-4614-1605-0_9, © Springer Science+Business Media, LLC 2012

Cascading failure is defined in [14] as a sequence of dependent failures of individual components that successively weakens the power grid. It is regarded as a major cause of disturbance propagation in a system of interconnected power grids. There is a large amount of past and current work on mathematical modeling and numerical simulations of cascading failures as reviewed in [14].

Understanding the dynamics of cascading failure is important for designing the smart grid operation. In [11], the term *Smart Grid* refers to a modernization of the electricity delivery system so it monitors, protects, and automatically optimizes the operation of its interconnected elements, and it will be characterized by a two-way flow of electricity and information to create an automated, widely distributed energy delivery network. This new technology implies the integration of information, communications, and power technologies. It is mainly driven by the large penetration of renewables such as solar and wind power, the aging grid infrastructure (especially in USA), and the emergence of wide-area disturbances as mentioned above. Thus, it is necessary to explore the dynamics of power grids for making the smart grid vision feasible.

In this chapter, we perform numerical analysis of coupled swing dynamics in power grids, based on the notion of power grid instability developed in [28]. In the previous paper, we studied a phenomenon in short-term[1] swing dynamics of multi-machine power grids which we termed the *Coherent Swing Instability* (CSI), based on [9, 10, 19]. This is an emergent and undesirable phenomenon of synchronous machines in a power grid, in which machines in a subset of the grid coherently lose synchronism with the rest of the grid after being subjected to a local disturbance. CSI is a nonlocal instability[2] occurring in a high dimensional dynamical system dominated by inertia in which one nonlinear mode is weak compared with many linear oscillatory modes. It is interpreted as an emergent transmission path of energy from the linear oscillatory modes to the nonlinear mode, which determines the spatially averaged motion of a power grid. The purpose of this chapter is, based on the result on CSI, to describe a dynamical mechanism of cascading failure in a system of interconnected power grids. Here, we use the dynamical systems approach to identify the mechanism: see related work [24, 30, 35].

The contribution of this chapter is a collection of data on CSI and cascading dynamics in a grid configuration of strong inner-connection and weak interconnection. Short-term swing dynamics are mainly studied using the nonlinear swing equations [16], which are a relatively simple dynamical system of ordinary differential equations. By numerical simulations of the swing equations, we analyze the CSI phenomenon in a system of weakly interconnected power grids and exploit the dynamics of CSI to identify the mechanism of cascading failure. The basic test

---

[1]Zero- to ten-seconds [16].

[2]The phenomenon we study here does not happen upon an infinitesimally small perturbation around an equilibrium of the dynamical system. However, it encompasses the situation when the system escapes a predefined set around the equilibrium. In this way, the notion of instability that we address here is nonlocal.

system used in this chapter is the New England (NE) power grid [3, 23] that we weakly interconnect into a larger power grid. The CSI phenomenon happens for all "unit" NE grids due to a swing wave propagating from other parts of the system and initiates the cascade of unit grid failures. Propagation of swing waves is studied in real power grids (Italy [2], west Japan [31], Texas [21], and Northeast America [5]). Swing dynamics have been reported as a cause of cascading failures in the 2003 blackouts in USA-Canada [34] and Italy [7]. We suggest that a small disturbance in one unit grid grows, spills to the other unit grid as a swing wave, and finally causes the whole system to fail. Throughout, we demonstrate that the CSI could be a part of the dynamical mechanism of cascading failure in power grids. Note that similar data on cascading dynamics in a system of nearly solvable ring power grids is reported in [29].

The rest of this chapter is organized as follows. In Sect. 2, we introduce the phenomenon of CSI using numerical simulations of short-term swing dynamics in the NE power grid. The contents are reported in [28]. In Sect. 3, we provide numerical analysis of a phenomenon of short-term swing dynamics in a system of weakly interconnected NE grids. The system consists of seven "unit" NE grids coupled via weak tie lines in series. In Sect. 4, we discuss topics related to the numerical result and conclude this chapter. Dynamics-based tools for monitoring and control of wide-area disturbances is addressed.

2 Introduction to Coherent Swing Instability

We introduce the phenomenon of CSI using numerical simulations of the NE 39-bus test system. The NE test system or grid is shown in Fig. 1 and is known as a benchmark system exhibiting coupled swing dynamics of synchronous machines [3]. The grid consists of ten generation units (equivalent ten synchronous generators), 39 buses, and lossy AC transmission lines. Most of the buses possess constant active and reactive power loads. The details of the grid, such as unit rating and line data, are available in [23].

2.1 Nonlinear Swing Equations

First of all, we introduce the equations of motion for the NE grid. Assume that generator 1 is the infinite bus[3] in order to explicitly represent the outside of the grid. The short-term swing dynamics of generators 2–10 are modeled by the nonlinear swing equations [16]:

---

[3] A voltage source of constant voltage and constant frequency.

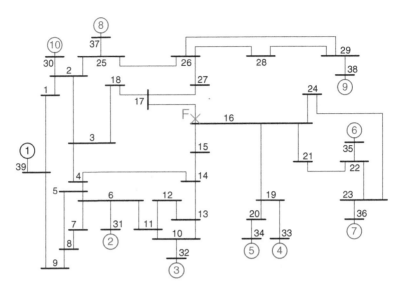

Fig. 1 New England (NE) power grid [3, 23]. The grid consists of ten generation units (equivalent ten synchronous generators, which are denoted as circled numbers), 39 buses, and lossy AC transmission lines

$$
\left.\begin{array}{l}
\dfrac{d\delta_i}{dt} = \omega_i, \\[2ex]
\dfrac{H_i}{\pi f_b} \dfrac{d\omega_i}{dt} = -D_i\omega_i + P_{mi} \\[2ex]
\qquad\qquad -V_i^2 G_{ii} - \displaystyle\sum_{j=1, j\neq i}^{10} V_i V_j \left\{ G_{ij} \cos(\delta_i - \delta_j) + B_{ij} \sin(\delta_i - \delta_j) \right\},
\end{array}\right\}
$$

$$(1)$$

where the integer label $i = 2, \ldots, 10$ denotes generator i. The variable δ_i represents the angular position of rotor in generator i with respect to bus 1 and is in radians (rad). The variable ω_i represents the deviation of rotor speed in generator i relative to system angular frequency $2\pi f_b = 2\pi \times (60\,\text{Hz})$ and is in radians per second (rad/s). We set the variable δ_1 to a constant, because bus 1 is assumed to be the infinite bus. The parameters f_b, H_i, D_i, P_{mi}, V_i, G_{ii}, G_{ij}, and B_{ij} are in per unit system except for f_b in Hertz (Hz) and for H_i and D_i in seconds (s). The mechanical input power P_{mi} to generator i and the internal voltage V_i of generator i are normally assumed to be constant for short-term swing dynamics [16]. The parameter H_i denotes the per unit time inertia constant of generator i, and D_i denotes its damping coefficient. The parameter G_{ii} denotes the internal conductance, and $G_{ij} + iB_{ij}$ (where i is the imaginary unit) denotes the transfer admittance between generators i and j. They are the parameters that change as the network topology changes. Electrical loads are modeled as passive admittances.

2.2 Numerical Simulations

We numerically simulate coupled swing dynamics of generators 2–10. All numerical simulations discussed in this chapter were performed using MATLAB: for example, the function ode45 was used for numerical integrations of (1). The voltage V_i and the initial condition $(\delta_i(0), \omega_i(0) = 0)$ for generator i are fixed using power flow computation. The inertia constant H_i is the same as in [23]. For the simulation, we use the following load condition: P_{mi} and constant power loads are 50% at their rating. The damping D_i is fixed at 0.005 s for each generator.[4] The elements G_{ii}, G_{ij}, and B_{ij} are calculated using the data in [23] and the result on power flow computation. Also, we use the following fault condition: each generator operates at a steady condition at $t = 0$ s. Then a three-phase fault happens at point F near bus 16 at $t = 1$ s $- 20/(60\,\text{Hz}) = 2/3$ s ≈ 0.67 s, and line 16–17 trips at $t = 1$ s. The fault duration is 20 cycles of a 60 Hz sinusoidal wave. The fault is modeled by adding a small impedance (10^{-7}i) between bus 16 and the ground.

Figure 2 shows the time responses of angular position δ_i and relative rotor speed ω_i of generator i. Before $t \approx 0.67$ s (this is the onset time of fault), each generator operates at a steady condition. In the fault duration from $t \approx 0.67$ to 1 s, all the generators 2–10 accelerate from their steady conditions. After the line trip at $t = 1$ s, they respond in an oscillatory manner. These oscillations are bounded during the period from $t = 1$ to 8 s. At about time 8 s, they begin to grow coherently and finally diverge. That is, every generator loses synchronism with the infinite bus at the same time. This corresponds to the growth of amplitude of inter-area mode oscillation between the NE grid and the infinite bus, namely, the outside of the grid. This is typical of the CSI phenomenon.

In the following, we demonstrate two methods for elucidating dynamical features of the phenomenon. First, we use the notion of *collective* variables for characterizing the spatially averaged motion of a power grid. The collective variables are well known as the COA (Center-Of-Angle) variables [3]. For the NE grid, the COA δ_{COA} and its time derivative ω_{COA} are defined as

$$\delta_{\text{COA}} := \sum_{i=2}^{10} \frac{H_i}{H}\delta_i, \quad \omega_{\text{COA}} := \frac{d\delta_{\text{COA}}}{dt} = \sum_{i=2}^{10} \frac{H_i}{H}\omega_i, \qquad (2)$$

where $H := \sum_{i=2}^{10} H_i$. The variables $(\delta_{\text{COA}}, \omega_{\text{COA}})$ describe the averaged motion of all the generators in the grid. Figure 3 plots the trajectory of (1) showing the CSI phenomenon in Fig. 2 on δ_{COA}–ω_{COA} plane. The trajectory starts near the origin at time 0 s, makes a couple of almost periodic loops around the initial point, and finally diverges.

---

[4]In the case that the relative rotor speed ω_i is in per unit system with base quantity $2\pi f_b$, the damping coefficient $D_i = 0.005$ s is equal to 1.88 in per unit system with its base quantity $1/(2\pi f_b)$.

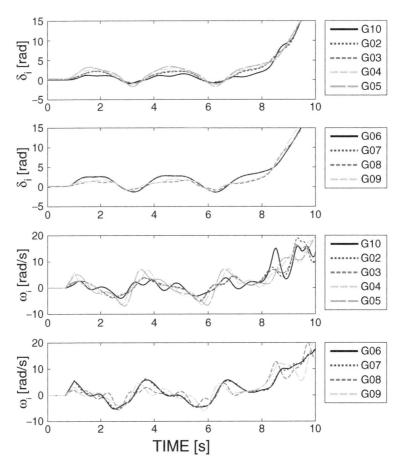

Fig. 2 Coupled swing dynamics in the NE power grid. The upper two plots are for angular positions δ_i of the nine generators, and the lower plots are for the relative rotor speed ω_i

Second, we use the *Proper Orthonormal Decomposition* (POD) in order to decompose the phenomenon in Fig. 2, which has been used in the context of power system analysis [17, 24]. POD provides a basis for the modal decomposition of an ensemble of functions, such as data obtained in the course of experiments, and provides energy-wise, the most efficient way of capturing the dominant components of the process [12, 13]. Consider finite simulation outputs of angular positions, $\{\delta_i(nT_s)\}$ ($i = 2, \ldots, 10, n = 0, \ldots, N_s - 1$), where T_s is the sampling period of outputs, and N_s is the number of samples. The outputs are represented by

$$\delta_i(nT_s) = \sum_{j=1}^{9} e_{ij} a_j(nT_s). \tag{3}$$

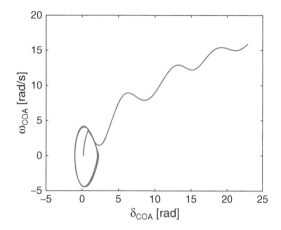

Fig. 3 Collective dynamics in the NE 39-bus test system. This plot corresponds to the projected trajectory onto the plane of the Center-Of-Angle (COA) variables $(\delta_{COA}, \omega_{COA})$ for the phenomenon observed in Fig. 2

We require the time-invariant basis vectors $\{e_{ij}\}$ $(i = 2, \ldots, 10)$ to be orthonormal and closest in energy norm to the output, and call them the *Proper Orthonormal Modes* (POMs). Every vector $\{e_{ij}\}$ is obtained by computing the correlation matrix R from $\{\delta_i(nT_s)\}$ and by finding the orthonormal eigenvectors of R: see [12] for details. The time-varying coefficient a_j $(j = 1, \ldots, 9)$ in the POD holds the following correlation property: $\langle a_j a_k \rangle = \langle a_j^2 \rangle$ (if $j = k$) or 0 (otherwise), where $\langle \bullet \rangle$ represents a time average of $\{\bullet\}$. POMs are ordered by $\langle a_j^2 \rangle \geq \langle a_{j+1}^2 \rangle$.

POMs are obtained using $N_s = 5341$ snapshots in the simulation outputs partially shown in Fig. 2. The time interval is [1 s, 90 s], and T_s is equal to 1/(60 Hz). Figure 4 shows the projection of the trajectory of (1) onto subspaces spanned by every POM. The projected trajectory $(a_j(nT_s), b_j(nT_s))$ for the j-th POM $(j = 1, \ldots, 9)$ is computed as

$$a_j(nT_s) = \sum_{i=2}^{10} e_{ij}\delta_i(nT_s), \quad b_j(nT_s) = \sum_{i=2}^{10} e_{ij}\omega_i(nT_s), \qquad (4)$$

where $n = 0, \ldots, N_s - 1$ (because of the smallness of the damping term we use the same modes for the angles and their time derivatives). In the first POM, the trajectory shows a transition from periodic motion to divergent one. The trajectory of the first POM coincides with the trajectory projected onto the COA plane in Fig. 3 by rotating it by 180° around the origin. On the other hand, in the other POMs, each trajectory shows a periodic or quasi-periodic motion. This is confirmed by looking at the results on power spectra (see [28]). The emergent phenomenon shown in Fig. 2 happens in the dynamical system with one *nonlinear* mode and many linear oscillatory modes.

These results enable us to explain the CSI phenomenon. In the dynamical system, the linear oscillator modes are strong because the interconnection term in (1) is strong (see [28]) due to the structure of interaction between generators, where

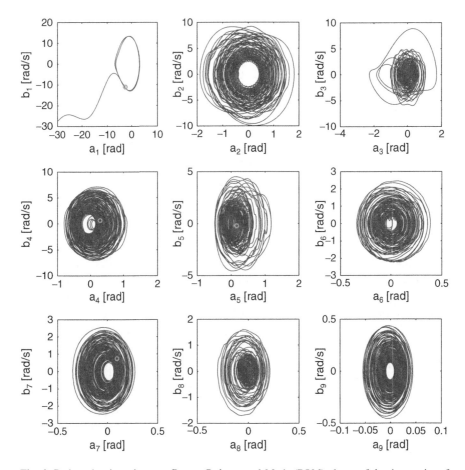

Fig. 4 Projected trajectories onto Proper Orthonormal Mode (POM) planes of the time series of coefficients a_j during [1 s, 90 s] with sampling frequency 60 Hz

many generators affect the dynamics of any single one. Compared with the linear modes, because the local term in (1), which represents the interaction between any generator and the infinite bus, is weak, the nonlinear mode is also weak. The linear oscillations then act as perturbations on the nonlinear mode. The perturbations are normally small due to the weakness of the coupling between the modes, but they become large if any linear mode and the nonlinear one satisfy a condition of internal resonance (see [10]). In this way, the amplitude of the first nonlinear POM in Fig. 4, or equivalently the projected COA trajectory in Fig. 3 can escape the region of bounded motions in the dynamical system of the first POM. This is the dynamical mechanism causing the CSI. It implies that the coupling of grid architecture and dynamics of the system matters the most.

Note that the NE grid is a test benchmark system, being a slight simplification of the real NE grid. Although the mathematical models are derived under reasonable assumptions for short-term rotor swing stability, they do not necessarily represent the true dynamics of the NE grid. Here, it is valuable to discuss whether the CSI in Fig. 2 can occur in a real power grid. The fault duration, which we set at 20 cycles in the simulation, is normally less than ten cycles. Such a long duration may imply the malfunction of protection systems, and hence the CSI in Fig. 2 may be regarded as a rare event in short-term swing dynamics. However, in Sect. 3 we will show that in a system of interconnected power grids, CSI is observed in the case of fault duration eight cycles. Furthermore, in the simulation we ignore the effect of load dynamics. The effect is normally negligible because it does not affect short-term swing dynamics [16] and will not cause any drastic change of simulation results. Thus, we suggest that CSI is a phenomenon that can occur for various configurations close to real power grids.

3   Dynamical Mechanism of Cascading Failures

In this section, we study a phenomenon of short-term swing dynamics in a system of Weakly Interconnected NE (WINE) grids. The system is shown in Fig. 5 and consists of the $N(= 7)$ NE grids (each of which we call the unit grid), the infinite bus, and weak interconnections. Each unit grid in Fig. 5 has equal specification of synchronous generators, loads, ac transmission lines, and network topology. The N unit grids are joined to each other in series via weak interconnections. Bus 24 in unit grid #i ($i = 1, \ldots, N - 1$) and bus 39 in unit grid #$(i + 1)$ are joined by a transmission line. We make the three assumptions: (i) generator 1 in unit grid #1 is the infinite bus; (ii) there is no generator 1 in the other unit grids; and (iii) the impedance of lines joining two different unit grids is three times as large as those of line joining buses 26 and 29. The reason why we chose line 26–29 is that it has the largest value of impedance in each unit grid. Thus, we *weakly* interconnect the seven unit grids, in each of these unit grids, nine synchronous machines are *strongly* connected.

Fig. 5 System of Weakly Interconnected New England (WINE) grids. The system consists of the $N(= 7)$ NE power grids and weak tie lines joining them. Generator 1 in unit grid #1 is assumed to be the infinite bus

3.1 Nonlinear Swing Equations

In the same manner as the single NE grid in Sect. 2, we use the nonlinear swing equations for modeling and analysis of coupled swing dynamics in the WINE system. The short-term swing dynamics of generator j in unit grid #i ($j = 2, \ldots, 10, i = 1, \ldots, N$) are represented by

$$
\left.
\begin{aligned}
\frac{\mathrm{d}\delta_{ij}}{\mathrm{d}t} &= \omega_{ij}, \\
\frac{H_j}{\pi f_{\mathrm{b}}} \frac{\mathrm{d}\omega_{ij}}{\mathrm{d}t} &= P_{\mathrm{m}ij} - D_j \omega_{ij} \\
&\quad - V_{ij} V_{11}\{G_{ij,11} \cos(\delta_{ij} - \delta_{11}) + B_{ij,11} \sin(\delta_{ij} - \delta_{11})\} \\
&\quad - V_{ij}^2 G_{ij} - \sum_{k=1, k \neq i}^{N} \sum_{l=2, l \neq j}^{10} V_{ij} V_{kl}\{G_{ij,kl} \cos(\delta_{ij} - \delta_{kl}) \\
&\quad + B_{ij,kl} \sin(\delta_{ij} - \delta_{kl})\}.
\end{aligned}
\right\} \tag{5}
$$

The variable δ_{ij} represents the angular position of rotor in generator j in unit grid #i with respect to the infinite bus and is in radians (rad). The variable ω_{ij} represents the deviation of rotor speed in generator i relative to system angular frequency $2\pi f_{\mathrm{b}}$ and is in radians per second (rad/s). The variable δ_{11} is the angular position of the infinite bus and becomes constant from its definition. The parameters H_j, $P_{\mathrm{m}ij}$, D_j, V_{ij}, $G_{ij,ij}$, and $G_{ij,kl} + \mathrm{i}B_{ij,kl}$ are in per unit system except for H_j and D_j in seconds (s). The constants H_j and D_j are introduced in Sect. 2.1. The constant $P_{\mathrm{m}ij}$ is the mechanical input power to generator j in unit grid #i, and V_{ij} is the internal voltage of generator j in unit grid #i. They are assumed to be constant in the same manner as in Sect. 2. The constant $G_{ij,ij}$ denotes the internal conductance of generator j in unit grid #i, and $G_{ij,kl} + \mathrm{i}B_{ij,kl}$ denotes the transfer admittance between generators j in unit grid #i and l in unit grid #k. The constant V_{11} is the voltage of the infinite bus, and $G_{ij,11} + \mathrm{i}B_{ij,11}$ is the transfer admittance between generator j in unit grid #i and the infinite bus. The admittance $G_{ij,kl} + \mathrm{i}B_{ij,kl}$ is the parameters that change as the network topology changes.

3.2 Numerical Simulations

We numerically simulate coupled swing dynamics of 63 generators. Basically, the system parameters are based on the data provided in [23] and in Sect. 2. The voltages V_{ij} and initial conditions $(\delta_{ij}(0), \omega_{ij}(0)) = (\delta_{ij}^*, 0\,\mathrm{rad/s})$ are fixed using power flow computation. The constant δ_{ij}^* is the value of angular position δ_{ij} at a steady operating condition before the fault. The parameter H_j is the same as in Sect. 2. The mechanical input power $P_{\mathrm{m}ij}$ and constant power loads are also the same as in [23].

The damping D_i is fixed at 0.01 s for each generator. The elements $G_{ij,ij}$, $G_{ij,kl}$, and $B_{ij,kl}$ are calculated using the data in [23]. We use the following fault condition: each generator operates at the steady condition at $t < 1\,\text{s} - 8/(60\,\text{Hz}) \approx 0.87\,\text{s}$, a three-phase fault happens near bus 39 in unit grid #1 at $t \approx 0.87\,\text{s}$, and line 1–39 trips at $t = 1\,\text{s}$. The fault duration is eight cycles of a 60 Hz sinusoidal wave. The fault is simulated by adding a small impedance (10^{-7}i) between bus 39 in unit grid #1 and the ground.

Figure 6 shows the time responses of angular positions δ_{ij}. The notation #i–Gj in the figure indicates that the corresponding colored line shows the time response of δ_{ij} for generator j in unit grid #i. The local fault happens in unit grid #1. The angular positions in unit grid #1 show swings at the onset time of fault $t \approx 0.87\,\text{s}$. On the other hand, the angular positions in unit grids #2 to #7 do not show any swings at the onset time and do remain at their steady conditions. After a while, the swings in unit grid #1 propagate through the WINE system and reach the last unit grid #7 at about time $t = 4\,\text{s}$. This propagation causes secondary swings in every unit grid. After reaching the last unit grid, it propagates through the grid in the backwards direction. During this initial swing propagation, the angular positions in each unit grid show coherent oscillations similar to that of the single NE grid in Sect. 2. The oscillations in unit grid #1 are damped due to both nonzero damping and dispersion effect caused by weak interconnections. The oscillations in the other unit grids are also damped and then show slight growth as time passes. Although the slight growth of angular positions is undesirable, it is still bounded and does not represent the loss of transient stability of synchronous generators in these unit grids. When the swings return to unit grid #1 at time $t = 9\,\text{s}$, the angular positions δ_{1j} begin to grow in a coherent manner and finally diverge. After the first divergence, the angular positions δ_{2j} in unit grid #2 next begin to diverge in a coherent manner. This cascade of coherent divergences continues up to the last unit grid. The coherent divergence in a unit grid corresponds to the loss of transient stability. This result indicates that a local disturbance in one unit grid grows, spills to the other grids, and finally causes the whole system to fail.

In the single NE grid, the motion of a hidden nonlinear mode was captured by projecting the full-system dynamics onto the phase plane of COA variables. Since the POD approach leads to the same nonlinear mode as the COA approach, we now investigate the cascade of unit grid failures using the notion of COA. Here, it is not effective to define the COA for the full-system dynamics of 63 generators, because it does not provide any insight of the interaction of different unit grids. In this case, we define the COA for each unit grid i $(i = 1, \dots, N)$ as

$$\delta_{i(\text{COA})} := \sum_{j=2}^{10} \frac{H_j}{H} \delta_{ij}, \quad \omega_{i(\text{COA})} := \sum_{j=2}^{10} \frac{H_j}{H} \omega_{ij}, \tag{6}$$

where $\delta_{i(\text{COA})}$ denotes the COA of unit grid #i and $\omega_{i(\text{COA})}$ its time derivative. The variables represent the spatially averaged motion of all the generators in unit grid #i. Figure 7 plots the trajectory of (5) showing the cascade in Fig. 6 on

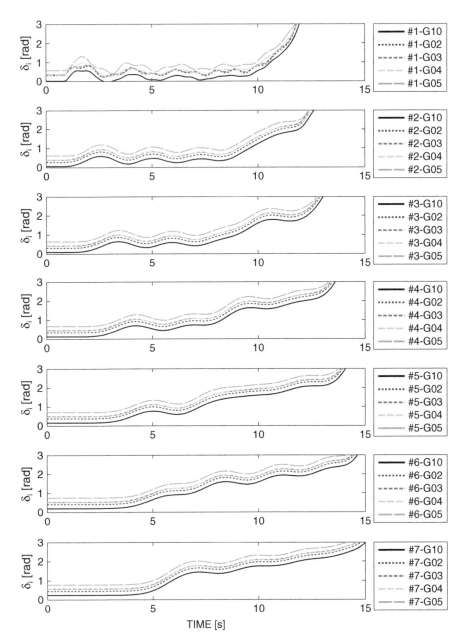

Fig. 6 Coupled swing dynamics in a system of WINE grids. The notation #i–Gj denotes the time response of angular position δ_{ij} of generator j in unit grid #i

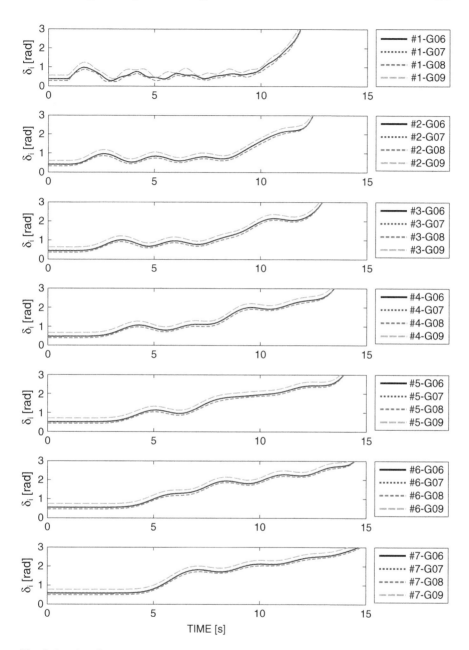

Fig. 6 (continued)

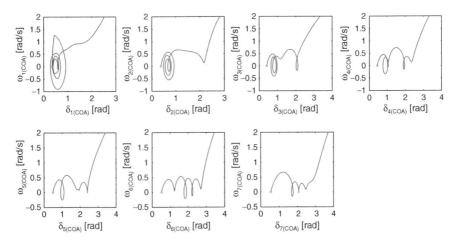

Fig. 7 Collective dynamics in a system of WINE grids. These plots correspond to the projected trajectories onto the planes of the COA variables (6) for the phenomenon shown in Fig. 6

$\delta_{i(\mathrm{COA})}-\omega_{i(\mathrm{COA})}$ planes. In unit grid #1, the trajectory executes damped oscillations for a while. However, the swings which return to unit grid #1 interrupt the trajectory and *kick* it. This kicking induces the divergence of trajectory in unit grid #1. In unit grid #2, the trajectory is kicked twice and finally diverges. The first kick is due to the swings propagating from unit grid #3, and the second kick is due to the divergence of unit grid #1. Similar behaviors for the trajectories of unit grids #3 and #4 are identified in Fig. 7. On the other hand, the trajectories of unit grids #5 to #7 show different behaviors. They do not show clear periodic motions and do drift to right when the swings kick them twice, and finally they diverge.

4 Summary and Discussion

In this chapter, we have studied the emergent instability phenomenon, termed the CSI, in power grids. Section 3 was devoted to numerical analysis of coupled swing dynamics in the WINE system shown in Fig. 5. The analysis suggests that the CSI phenomenon happens for all of the unit grids due to a swing wave propagating from other parts of the system and initiates the cascade of unit grid failures. In this section, we discuss several topics related to the result and close this chapter.

The analysis in Sect. 3 was performed by introducing the COA variables for each unit grid with coherent generators. The COA variables were useful for the dynamical analysis performed in this chapter, because the WINE system possessed a trivial set of strongly inner-connected grids, in each of which generators may exhibit coherent motions. Generally speaking, it is not easy to find such a set of coherent generators for real power grids. Identifying coherent generators is necessary for applying the developed theory to data on cascading failures in real power grids. Such identification techniques have been reported by many researchers, for example, the Lyapunov function method [22], the singular perturbation method [4, 6, 36], and the Koopman mode analysis (KMA) [27] that is introduced in the last of this section. These methods can identify a set of unit grids consisting of coherent generators, so that one can investigate the COA dynamics for each identified grid.

As introduced in Sect. 1, cascading failures are fairly complicated phenomena emerging in interconnected dynamical systems. It would be impossible to obtain a simple mechanism that can explain all dynamics and events in a cascading failure. For a power grid consisting of many sub-grids, the instability shown in Sect. 2 describes a failure of one sub-grid caused by the loss of transient stability. In this chapter, we show that a sequence of sub-grid failures is induced by a sequence of CSIs. Thus, we suggest that the CSI could be a part of the dynamical mechanism of cascading failure of large-scale power grids.

Also, it is often said that causes of cascading failures include correct/incorrect relay operations and hidden failures [32] that are discontinuous actions in the dynamics of power grids. This is naturally modeled by a hybrid dynamical system. In [30], the authors use a hybrid dynamical system for modeling and analysis of the cascading failure leading to the 2003 blackout in Italy. Analysis of the hybrid model shows that the swing dynamics, especially desynchronization of individual generators, result from a network switching with a simple relay feedback controller. On the other hand, in this chapter we describe another scenario of swing dynamics leading to desynchronization of individual generators *without* any network switching. This is a counterexample to the standard argument as stated in the beginning of this paragraph. Of course, the dynamic phenomenon that we studied in this chapter may be an extreme example for cascading failures. Our mechanism of cascading failure will need further research in more realistic test systems and for practical data on cascading failures.

It was suggested that cascading failures are partly due to the loss of transient stability based on real data, for example, the September 2003 blackout in Italy [7]. Our study suggests that such instability could happen via a swing wave propagation mechanism. To the best of our knowledge, there is yet no real data proving that the swing wave propagation is an initiation of cascading failure. In order to identify our mechanism in real power grids, a measurement system that can simultaneously monitor global dynamics of large-scale power grids is needed. This can be carried out with the emerging technology of wide-area measurement with the aid of synchronized PMUs[5] (see e.g. [8, 25]).

The cascading dynamics that we studied in Sect. 3 are, needless to say, an undesirable phenomenon of power grids and should be prevented by grid design or avoided by control. Figure 6 shows that the difference of COAs increases as time passes. This large difference may trigger the action of protection systems that are normally equipped with a tie line. A key point for control is how to detect the

---

[5]Phasor Measurement Units.

propagation of swing waves in a power grid. One possible solution is again the emerging wide-area measurement-based control and analysis (see e.g. [15]). Wide-area measurement is expected to stabilize spatiotemporal dynamics in large-scale power grids.

In the current analysis, we used the dynamical systems approach to elucidate a core cause of wide-area disturbances. As a next step, it is necessary to consider how to apply the dynamical perspective to monitoring and control of power grids. Currently, we are developing methodology and tools for monitoring of power grids based on the result. A key method in our development is the Koopman mode analysis (KMA) that is based on a fully nonlinear spectral theory and represents an extension of linear oscillatory mode analysis [18]. KMA is dynamically consistent with underlying (possibly nonlinear) physics and provides a new approach to model validation and reduction [18, 20]. We show that KMA provides a method for identifying coherent generators from sensor data [27] and defines a *precursor* to CSI with its computation method based on sensor data and mathematical modeling [26]. The precursor is based on the discovery of emergent transmission path of energy from many linear oscillatory modes to the nonlinear mode as mentioned in Sect. 2. Both the methods need data collected in a real power grid and computation based on a mathematical model. Hence, they would be suitable as tools implemented to the future smart grid in which the physical power network is overlaid with an information network.

Acknowledgments We are grateful to Professor Petar Kokotović and Professor Joe H. Chow for discussions and valuable suggestions. This work was supported in part by JSPS Postdoctoral Fellowships for Research Abroad, in part Grant-in-Aid for Global COE Program "Education and Research on Photonics and Electronics Science and Engineering," MEXT, Japan, and in part NICT Project ICE-IT (Integrated Technology of Information, Communications, and Energy). During this work, Y. Susuki was with the Department of Mechanical Engineering at the University of California, Santa Barbara.

References

1. Andersson G, Donalek P, Farmer R, Hatziargyriou N, Kamwa I, Kundur P, Martins N, Paserba J, Pourbeik P, Sanchez-Gasca J, Schulz R, Stankovic A, Taylor C, Vittal V (2005) Causes of the 2003 major grid blackouts in North America and Europe, and recommended means to improve system dynamic performance. IEEE Trans Power Syst 20(4):1922–1928
2. Arcidiacono V, Ferrari E, Saccomanno F (1976) Studies on damping of electromechanical oscillations in multimachine systems with longitudinal structure. IEEE Trans Power App Syst PAS-95(2):450–460
3. Athay T, Podmore R, Virmani S (1979) A practical method for the direct analysis of transient stability. IEEE Trans Power App Syst PAS-98(2):573–584
4. Avramović B, Kokotović PV, Winkelman JR, Chow JH (1980) Area decomposition for electromechanical models of power systems. Automatica 16:637–648
5. Chow JH (2010) Personal communication
6. Chow JH, Cullum J, Willoughby A (1984) A sparsity-based technique for identifying slow-coherent areas in large power systems. IEEE Trans Power App Syst PAS-103(3):463–473

7. Corsi S, Sabelli C (2004) General blackout in Italy Sunday September 28, 2003, h. 03:28:00. In: Proceedings of IEEE PES general meeting, Denver, USA, vol 2, pp 1691–1702

8. De La Ree J, Centeno V, Thorp JS, Phadke AG (2010) Synchronized phasor measurement applications in power systems. IEEE Trans Smart Grid 1(1):20–27

9. Du Toit P, Mezić I, Marsden J (2009) Coupled oscillator models with no scale separation. Physica D 238(5):490–501

10. Eisenhower B, Mezić I (2010) Targeted activation in deterministic and stochastic systems. Phys Rev E 81:026,603

11. Electric Power Research Institute (2009) Report to NIST on the Smart Grid Interoperability Standards Roadmap (Contract No. SB1341-09-CN-0031), June 17, 2009

12. Feeny BF, Kappagantu B (1998) On the physical interpretation of proper orthogonal modes in vibrations. J Sound Vib 211(4):607–616

13. Holmes P, Lumley JL, Berkooz G (1996) Turbulence, coherent structures, dynamical systems, and symmetry. Cambridge University Press, Cambridge

14. IEEE PES CAMS Task Force on Cascading Failure (2008) Initial review of methods for cascading failure analysis in electric power transmission. In: Proceedings of IEEE PES general meeting, Pittsburgh

15. Kamwa I, Grondin R, Hébert Y (2001) Wide-area measurement based stabilizing control of large power systems—A decentralized/hierarchical approach. IEEE Trans Power Syst 16(1):136–153

16. Kundur P (1994) Power system stability and control. McGraw-Hill, New York

17. Messina AR, Vittal V (2007) Extraction of dynamic patterns from wide-area measurements using empirical orthogonal functions. IEEE Trans Power Syst 22(2):682–692

18. Mezić I (2005) Spectral properties of dynamical systems, model reduction and decompositions. Nonlinear Dyn 41:309–325

19. Mezić I (2006) On the dynamics of molecular conformation. Proc Natl Acad Sci USA 103(20):7542–7547

20. Mezić I, Banaszuk A (2004) Comparison of systems with complex behavior. Physica D 197:101–133

21. Murphy RJ (1996) Disturbance recorders trigger detection and protection. IEEE Comput Appl Power 9(1):24–28

22. Ohsawa Y, Hayashi M (1981) Construction of power system transient stability equivalents using the Lyapunov function. Int J Electron 50(4):273–288

23. Pai MA (1989) Energy function analysis for power system stability. Kluwer Academic Publishers, Boston

24. Parrilo PA, Lall S, Paganini F, Verghese GC, Lesieutre BC, Marsden JE (1999) Model reduction for analysis of cascading failures in power systems. In: Proceedings of American Control Conference, San Diego, pp 4208–4212

25. Phadke AG (1993) Synchronized phasor measurement in power systems. IEEE Comput Appl Power 6(2):10–15

26. Susuki Y, Mezić I (2011) Nonlinear Koopman modes and a precursor to power system swing instabilities. IEEE Trans Power Syst (submitted for possible publication)

27. Susuki Y, Mezić I (2011) Nonlinear Koopman modes and coherency identification of coupled swing dynamics. IEEE Trans Power Syst 26(4):1894–1904

28. Susuki Y, Mezić I, Hikihara T (2011) Coherent swing instability of power grids. J Nonlinear Sci 21(3):403–439

29. Susuki Y, Mezić I, Hikihara T (2010) Coherent swing instability of power systems and cascading failures. In: Proceedings of IEEE PES general meeting, Minneapolis

30. Susuki Y, Takatsuji Y, Hikihara T (2009) Hybrid model for cascading outage in a power system: A numerical study. IEICE Trans Fund Electr E92-A(3):871–879

31. Tamura Y (1995) Possibility of parametric resonance in power system. Trans Electr Eng Jpn 112-B(8):657–663 (in Japanese)

32. Thorp JS, Phadke AG, Horowitz SH, Tamronglak S (1998) Anatomy of power system disturbance: importance sampling. Int J Electr Power 20(2):147–152

33. Union for the Coordination of the Transmission of Electricity (2007) Final Report on System Disturbance on 4 November 2006, January 2007
34. U.S.-Canada Power System Outage Task Force (2004) Final Report on the August 14, 2003, Blackout in the United States and Canada: causes and recommendations, April 2004
35. Venkatasubramanian VM, Li Y (2004) Analysis of 1996 Western American electric blackouts. In: Proceedings of Bulk Power System Dynamics and Control—VI, Cortina d'Ampezzo, Italy, pp 685–721
36. Winkelman JR, Chow JH, Bowler BC, Avramović B, Kokotović PV (1981) An analysis of interarea dynamics of multi-machine systems. IEEE Trans Power App Syst PAS-100(2): 754–763

Toward a Highly Available Modern Grid

N. Eva Wu and Matthew C. Ruschmann

Abstract The concept of fault-coverage and how it affects the availability of a dynamic grid is explained through a two-area power system represented by an aggregated swing model. Fault-coverage is intended to serve as a criterion for decisions in redundancy management to benefit system availability upon occurrence of a disturbance due to loss of equipment. The criterion allows the incorporation of formal measures of uncertainties associated with real-time fault diagnosis, as well as formal control performance measures. Also investigated is the effect of the availability of a modern grid's supporting structure on the availability of the grid, with focus on a network of measurement units. The focus stems from the recognition of a greater need for real-time diagnosis and control in a modern power grid. A redundancy architecture design problem is formulated based on a Markov model of a measurement network, and a solution is presented that minimizes the number of phasor measurement unit (PMU) restorations and the usage of communication links to a PMU while maintaining a prescribed data availability at any PMU. A 3-bus/ 3-PMU network is used as an example to explain the formulation and the solution of the redundancy architecture design problem.

1 Introduction

Restructuring the power grid [11] implies a higher degree of interconnectivity, including the partaking of intermittent energy sources, and a heavier reliance

N.E. Wu (✉) • M.C. Ruschmann
Department of Electrical and Computer Engineering, Binghamton University, Binghamton, NY 13902, USA
e-mail: evawu@binghamton.edu; mruschm2@binghamton.edu

A. Chakrabortty and M.D. Ilić (eds.), *Control and Optimization Methods for Electric Smart Grids*, Power Electronics and Power Systems 3,
DOI 10.1007/978-1-4614-1605-0_10, © Springer Science+Business Media, LLC 2012

on the information technology in the grid's supporting structure,[1] where, among many components, new control and monitoring devices are housed. Complexity, uncertainty, and vulnerability accompanying the new opportunities intensify the need for a formal availability study, which must also integrate the availability of the growing supporting structure, and incorporate the consequences of real-time decisions.

Despite the remarkable achievements of the past few decades in the effort to enhance the electric grids' security and reliability,[2] in 2003 alone, four major blackouts occurred in US-Canada (fifty million people affected), Chile (fifteen million), Sweden (four million), and Italy (fifty-six million) [23]. The author of *The Unruly Power Grid* [17] cited some advanced studies, which suggest that major blackouts are inevitable. One theory attributes such events to insufficient stability margins as the pursuit of maximal return on investment in the power grid prevails, and the other to the complexity of increasingly interconnected systems, which finds support in the established theory of large deviations in high-dimensional risks [14]. Both theories appeal to our intuitions: cost in building power systems that make rare events rarer is too high, and complexity is well recognized as being detrimental to reliability.

A variety of separate indices have been used in the analysis of reliability of the power grid for each of the generation, transmission, and distribution systems [24]. In addition, the notion of security has been used to indicate the ability of the electric power system to withstand large disturbances, due to, for example, unanticipated loss of system facilities [27]. A probabilistic framework has been used to provide a unified treatment of steady-state stability and dynamic security [29]. Some recent work [18] has included flexible ac transmission systems (FACTS) in the reliability analysis. However, the use of multiple indices and a separate notion of security indicate that the linkage between the transient behavior caused by faults in a dynamic system and the system reliability, which is apparently affected by its security, has not been adequately established. This linkage is completely absent when uncertain real-time information is involved in managing the equipment redundancy.

This chapter considers the issue of enhancing the security during the operation when major upsets occur, and quantifies the link between security and reliability in the face of two types of faults. The first type is associated with the grid's equipment, such as generating units, transformers, transmission lines, and loads. The ultimate objective is to accommodate this type of faults timely to the extent

---

[1] A supporting structure is comprised of the network where communication, computing, and control technologies are implemented and hosted, such as control and measurement devices, software, database and processing servers, routers and switches, wired and wireless links, and possibly human operators.

[2] The definitions on reliability and security, in traditional power engineering terms, are the ability of a power system to provide uninterrupted service and the ability of the system to withstand disturbances, respectively.

possible through exploitation and management of equipment redundancy, despite the presence of faults of a second type. The second type of faults is associated with the components residing in the grid's supporting structure, such as control, processing, and measurement devices. The growing presence of powerful and fast new devices in FACTS and high voltage dc transmission systems (HVDC) [22], and phasor measurement units (PMU) [25], has greatly enhanced the potentiality to exploit and manage the redundancy in the power grid. Fault-coverage [30] is to be used here as a criterion to guide the redundancy architecture design and online redundancy management.

Since the recovery of grid operation is always expected, steady-state availability,[3] rather than reliability, is adopted as a measure of the grid's performance. Steady-state availability is roughly the fraction of time in the long run that the system operates satisfactorily. The degree of satisfaction can be defined at multiple levels, and made to incorporate economic factors. Under a stochastic discrete state framework, availability can be computed as the sum of state probabilities for the states that are categorized as representing an operative grid of a given performance threshold.

A common security issue in a power system is associated with containing a swing, usually at a frequency of a fraction of a Hertz, of inter-area voltage phasor angles against one another as a result of large disturbances [12] due to, for example, equipment faults. The system recovery depends on whether the grid has been configured to tolerate a specified number of equipment loss. If so, protection devices, such as relays and circuit breakers [1], are relied on to remove the faulty equipment before a critical clearance time, beyond which the inter-area synchronization is lost.

It is noted, however, that disturbances during a fault may cause protection devices to trip intact equipment, and thus lead to cascading events, when the protection devices cannot distinguish the causes for the fluctuating voltages and currents. The current diagnostic support[4] in the grid is inadequate [26] to modify pre-set re-closure protocols in real time. Although new schemes that feedback measured or estimated information over a wide area in the modernizing grid are expected to be capable of performing more complex redundancy management in real-time, risks due to uncertainties in the measured or estimated information must be formally incorporated into the decision process. It is our objective to find a way to incorporate formal measures of uncertainties resulting from real-time fault diagnosis, so as to make the least risky decisions in redundancy management.

An appropriate measure of performance for a swinging power system is its ability to return to a desirable equilibrium, which can be assessed by determining whether a departing trajectory can be placed in a post-fault region of attraction [21] in the inter-area frequency-angle space at the time the fault is cleared. In this chapter, the

---

[3] Availability is the probability that a system is performing its required function at a given point in time when used under stated operating conditions [5].

[4] A large body of literature exists in the area of fault diagnosis, which provides the basis for fault-tolerant control [6].

computed fault coverage for the two-area system is shown to capture the effect of delayed control or management actions, the effect of uncertain diagnostic outcomes, and the effect of controllability.

The trend to acquire and to act upon real-time information in the operation of a modern power system places more critical reliance on the measurement devices. The addition of PMUs to the power grid has motivated studies on their placement to achieve a desired observability [3,7,15,19]. However, the placement schemes are all deterministic in nature, and the relation between the measurement data availability and the grid availability has not been established. In this chapter, an initial step is taken toward establishing the relationship, and furthermore toward optimizing the redundancy architecture of the measurement network necessary to support data availability in wide-area applications.

The design of redundancy architecture is approached by first modeling measurement data availability as a Markov process. A novel structure design problem for a measurement network is formulated as an optimization problem, where prices are assigned to system configurations and constraints on the data availability are enforced. The structural design problem is demonstrated through a 3-PMU network example. The solution dictates the need to place a PMU at a bus and the need to have communication links to a PMU for a required data availability at the PMU.

The chapter is organized as follows. Section 2 establishes the linkage between fault-coverage, which quantifies the security measure, to availability, which replaces the traditional reliability measure, for recoverable systems. Fault coverage is defined and evaluated for a two-area power system. The section also establishes the relation between the grid availability and that of the network of measurement devices on which the grid relies for its real-time operation. Section 3 formulates a redundancy architecture design problem for a network of measurement devices. The solution sought is to meet the required data availability with the least cost. The architecture design is detailed for a 3-PMU network. Section 4 concludes the chapter.

2 Redundancy Management in Dynamic Systems Based on Fault-Coverage

This section links the traditional notions of security and reliability defined for a power system through formalizing the relationship between fault-coverage and availability. It has been rationalized earlier that availability is a more appropriate measure for a recoverable system than reliability. The section also establishes that high fault-coverage is the key to achieving high availability in a dynamic system. Fault coverage is evaluated for a swing model of a two-area power system. The results in this section also show how uncertainty, controllability, and delay in control actions affect coverage.

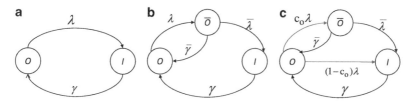

Fig. 1 Aggregated availability models of (**a**) an existing system, (**b**) an enhanced system with new redundancy configuration and management capabilities, and (**c**) the enhanced system with imperfect fault-coverage $\bar{c}_0 = 1 - c_0 > 0$, for which the average fractions of time the systems spend on state I (unavailability) are π_I^a, π_I^b, π_I^c, respectively, as expressed in (3)

2.1 Benefit of High Coverage in Redundancy Management

Consider a system of two states, an operative state O and an inoperative state I, as seen in Fig. 1a, where each state is formed by aggregating many states of the same category. In particular, state I can be formed by aggregating all inoperative states of varying root causes. Let us use a simplifying assumption for the moment that all state holding time distributions [20] are exponential, i.e., $Pr[T(S) \leq t] = 1 - e^{-\Lambda(S)t}$, where $\Lambda(S)$ is the sum of all outgoing transition rates from state S. Thus, $\Lambda(O) = \lambda$ and $\Lambda(I) = \gamma$. Recovery rate γ from state I, for example, implies that average recovery time from state I to state O is $1/\gamma$ units of time.

Viewed as a stochastic process, where randomness is caused by, for example, times of arrival of forced outages of equipment, unavailability is solved as the probability of being in state I. Highly available systems make use of direct or analytic redundancy to tolerate faults. Analytic redundancy [10] is provided by static and dynamic relations among system variables, which can enhance system tolerance to faults in a cost-effective manner beyond the exhaustion of direct redundancy. The addition of new control and measurement devices, such as those in FACTS and HVDC systems, and PMUs, can greatly enhance the ability to configure and manage analytic redundancy, and thus allows us to introduce a new intermediate aggregated state $\bar{O}$, as shown in Fig. 1b, at which the overall system is operative despite the presence of some faults.

To achieve fault-tolerance in a controlled power system, three elements must be present: (1) Redundancy is present in the distribution, transmission, and generation equipment with sufficient load reserve capacity[5]; (2) Monitoring devices work properly to help identify departures in systems' states in a timely manner, or root causes of disruptive events; (3) Control devices respond properly and timely to help accommodate the adverse effects of the disruptive events.

---

[5]Redundancy at the equipment level is determined during power system planning, which is outside of the scope of this chapter. Power system planning is made up of the electrical load forecast, generation planning, and network planning [28].

In Fig. 1c, coverage c_0 is introduced, which is the probability of reaching state $\bar{O}$ conditioned on transition out of state O [16]. Lack of perfect coverage ($c_0 < 1$) is attributed to missing some of the elements in above. Presence of uncertainties in the real-world almost always leads to $c_0 < 1$. Mathematically, coverage modifies the transition from O to $\bar{O}$ in Fig. 1b to that in Fig. 1c according to a decomposition property of a Poisson process [20], where the arrival of an event may lead to one of several next states with certain probabilities (c_0 and $1 - c_0$ in this example). Denote by π_I^a, π_I^b, and π_I^c the steady-state probabilities at state I for the ergodic chains in Fig. 1a, b, and c, respectively. The availability in each case is given by $1 - \pi_I^i$, where $i = a, b, c$.

Upon solving for the steady-state values of π_I^a, π_I^b, and π_I^c from the Chapman-Kolmogorov equation at steady-state

$$\begin{bmatrix} \dot{\pi}_O^i & \dot{\pi}_{\bar{O}}^i & \dot{\pi}_I^i \end{bmatrix} = \begin{bmatrix} 0 & 0 & 0 \end{bmatrix} = \begin{bmatrix} \pi_O^i & \pi_{\bar{O}}^i & \pi_I^i \end{bmatrix} \begin{bmatrix} -\lambda & c_0\lambda & \bar{c}_0\lambda \\ \bar{\gamma} & -\bar{\lambda} - \bar{\gamma} & \bar{\lambda} \\ \gamma & 0 & -\gamma \end{bmatrix}, \quad (1)$$

and $\pi_O^i + \pi_{\bar{O}}^i + \pi_I^i = 1$ with $\bar{c}_0 = 1 - c_0$, the following conclusions can be drawn.[6]

1. The benefit of redundancy management to availability is confirmed by

$$\pi_I^b = \frac{\lambda}{(\lambda + \gamma) + (\lambda + \bar{\gamma})(\gamma/\bar{\lambda})} < \frac{\lambda}{\lambda + \gamma} = \pi_I^a. \quad (2)$$

When $\bar{\gamma} \gg \bar{\lambda}$ holds, i.e., recovery from $\bar{O}$ to state O is much more likely to occur before the occurrence of cascading events leading to state I, the benefit of redundancy management is seen to be enormous. If in addition $\bar{\gamma} \gg \gamma$, i.e., recovery to state O from $\bar{O}$ is much faster than that from I, the benefit of redundancy management is even greater.

2. The effect of imperfect coverage is clearly revealed by

$$\pi_I^a = \frac{\lambda}{\lambda + \gamma} \geq \pi_I^c = \frac{\lambda + \bar{c}_0\bar{\gamma}\lambda/\bar{\lambda}}{\lambda + \gamma + (\lambda + \bar{\gamma})\gamma/\bar{\lambda} + \bar{c}_0(\bar{\gamma} - \gamma)\lambda/\bar{\lambda}}$$

$$\geq \pi_I^b = \frac{\lambda}{(\lambda + \gamma) + (\lambda + \bar{\gamma})\gamma/\bar{\lambda}}. \quad (3)$$

Inequalities in (3) state that availability worsens as coverage lowers. In fact

$$\frac{\partial \pi_I^c}{\partial \bar{c}_0} = \frac{(\lambda + \bar{\gamma})(1 + \gamma/\bar{\lambda})\gamma\lambda/\bar{\lambda}}{[\lambda + \gamma + (\lambda + \bar{\gamma})\gamma/\bar{\lambda} + \bar{c}_0(\bar{\gamma} - \gamma)\lambda/\bar{\lambda}]^2} > 0 \quad (4)$$

for $0 < \bar{c}_0 < 1$. Equivalently, $\partial \pi_I^c/\partial c_0 < 0$ for $0 < c_0 < 1$. Therefore, availability is a monotonically increasing function of coverage c_0.

---

[6]In the case of Fig. 1a, the state probability associated with $\bar{O}$ is removed, and so are the second column and row of the rate transition matrix in (1).

Fig. 2 Two different values
of coverage resulting from
integrals of $f_{(\hat{x},\hat{t}_0)}(t, x, t_0)$
over two sets of initial states,
corresponding to two control
laws, assuming t_0 is known

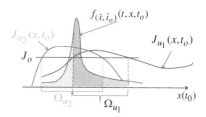

The above elucidates the role of coverage in a system's availability from an aggregated view. The discrete state set can be expanded to include $|\bar{O}| + 2$ total limp-along states. Corresponding to each state in the intermediate state category $\bar{O}$, there is a distinct fault-coverage parameter. The key qualitative conclusion is the monotonic dependence of availability on coverage.

2.2 Evaluation of Fault-Coverage for a Two-Area Swing Model

This section discusses how fault coverage is evaluated for a dynamic system. The initial definition of fault-coverage for a dynamic system is given in [30]. Let $f_{(\hat{x},\hat{t}_0)}(t, x, t_0)$ be a snapshot at t of the probability density function for the estimate of state x after the onset time t_0 of a discrete event. x can be generalized to include also a set of parameters describing the model of the system. Let $J_u(x, t_0)$ denote the performance afforded by control law $u(t) \in \mathcal{U}, t \geq t_0$, where $\mathcal{U}$ is an admissible set, taking into consideration of both transient and steady-state behaviors of the controlled system based on state $x(t)$, and J_0 is a specified threshold, above which the system performance is satisfactory. Define

$$c_{o,u}(t) = \int_{\Omega_u} f_{(\hat{x},\hat{t}_0)}(t, x, t_0) dx dt_0, \tag{5}$$

where $\Omega_u \equiv \{x(t) | J_u(x, t_0) \geq J_0, u \in \mathcal{U}\}$. $c_{o,u}(t)$ is the probability that the system enters state $\bar{O}$ with control law u applied, after the onset time t_0 of an upset event that triggers the outgoing transition from state O.

Figure 2 illustrates how two values of coverage are obtained as the areas under $f_{(\hat{x},\hat{t}_0)}(t, x, t_0)$ of the estimate at t, over Ω_{u_i}, where control performance $J_{u_i}(x, t_0) \geq J_0$ is met for $i = 1$ and 2, respectively. By inspection

$$c_{o,u_2}(t) = \int_{\Omega_{u_2}} f_{(\hat{x},\hat{t}_0)}(t, x, t_0) dx dt_0 \geq c_{o,u_1}(t) = \int_{\Omega_{u_1}} f_{(\hat{x},\hat{t}_0)}(t, x, t_0) dx dt_0. \tag{6}$$

To achieve a higher availability, redundancy management should choose to execute control law $u_2(t)$. Coverage has naturally linked the continuous state variable x, and discrete state set $S = \{O, \bar{O}, I\}$ in a way fundamentally different from the linkage commonly perceived in a hybrid system [2].

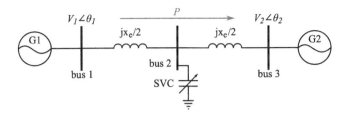

Fig. 3 Two-area power system with a midpoint static var compensator (SVC) shown

To show how fault-coverage quantifies the notion of security traditionally defined as the ability of a power system to withstand disturbances, an aggregated two-area power system is considered. Our presentation on the swing model follows that in [12]. Synchronous generator $G1$ supplying P MW of power to synchronous generator $G2$ is connected to $G2$ by a set of transmission lines with an equivalent reactance x_e, as shown in Fig. 3. Two cases are considered. One with and one without the voltage support by a static var compensator (SVC) in midpoint.

For simplicity the machine voltage magnitudes and angle differences are considered to be the same as the corresponding quantities extracted at the buses. Define the difference in bus voltage angles by $\delta = \delta_2 - \delta_1$, where δ_1 and δ_2 are the rotor angles of the generators. Each machine can be viewed as an aggregate of all the slow-coherent generators. Assuming constant internal voltages V_1 and V_2 at the aggregated generators, the dynamics of the two-machine system are

$$\dot{\delta} = \Omega\omega, \tag{7}$$

$$2H\dot{\omega} = P_m - \frac{V_1 V_2 \sin(\delta)}{x_e}, \text{without SVC}, \tag{8}$$

$$2H\dot{\omega} = P_m - \frac{V_1 V_2 \sin(\delta/2)}{x_e}, \text{with SVC}, \tag{9}$$

where $\Omega = 2\pi f$, with f the nominal system frequency in Hertz, $\omega = \omega_2 - \omega_1$ is in per unit (pu), with ω_1 and ω_2 the machine speeds for $G1$ and $G2$, respectively, $H = H_1 H_2/(H_1 + H_2)$ is the equivalent inertia, with H_1 and H_2 the inertias for $G1$ and $G2$, respectively, D is the damping coefficient, and $P_m = (H_2 P_{m1} - H_1 P_{m2})/(H_1 + H_2)$ is the equivalent mechanical input power. The set of nominal parameters used for our computation are $V_1 = V_2 = 1$ pu, $x_e = 0.0770$ pu, $D = 5$, $H = 129$ pu, and $P_m = 10.4397$ pu. The system is operating at the stable equilibrium point at $\delta_0 = 53.5$ degrees, and $\omega_0 = 0$ pu.

The space defined for coverage evaluation in this example involves only the state-space (δ, ω) of the swing model. The distribution function is produced in the process of fault diagnosis. More specifically, suppose an unbiased estimate of (δ, ω)

at current time along with an error covariance that defines a normal distribution is provided.[7]

Standard definitions on the qualifiers pre-fault ($t < t_0$), during fault (or fault-on, $t_0 \leq t \leq t_0 + \varepsilon$), and post-fault ($t > t_0 + \varepsilon$) assume three distinct system dynamics, and control actions involve only correct removal of faulty equipment at $t = t_0 + \varepsilon$ by traditional protective devices [27]. Interval ε is typically in the millisecond range whenever adequate redundant equipment exists following a prompt identification of the faulty equipment to allow the establishment of a post-fault stable equilibrium. A post-fault system is in discrete state $\bar{O}$. Although secure, a speedy restoration to pre-fault state O must ensue to maintain high system availability, according to the earlier discussion.

Security assessment using coverage permits more general operation settings expected of a modernizing grid, where measured information is used for more sophisticated redundancy management due to a wider class of possible control actions and performance measures. In the two-area system, among many choices of control performance, characteristic functions supported on regions of attraction are considered, because the major concern is whether the swing state trajectory can settle into an acceptable equilibrium after a control action is taken.

Consider the sum of the potential energy and kinetic energy for the two-area system [11], assuming known equilibrium and system parameters

$$V_e(\delta, \omega) = \frac{V_1 V_2}{x_e} [\cos(\delta_0) - \cos(\delta) + \sin(\delta_0)(\delta_0 - \delta)] + H \Omega \omega^2. \qquad (10)$$

This energy function satisfies the conditions that define a quasi-stability region [9] of the system. In this case, the level set of the energy function that intersects the closest unstable equilibrium point not contained in the quasi-stability region is an estimate of the boundary of the region of attraction. For the two-area model, assuming that $0 \leq \delta_0 \leq \frac{\pi}{2}$, the stable equilibrium point under consideration is $\delta = \delta_0, \omega = 0$ pu, and the closest unstable equilibrium is $\delta = \pi - \delta_0, \omega = 0$ pu.

Suppose a fault occurs due to a shorted transmission line to cause a drop in the bus voltages at t_0. The region of attraction shrinks suddenly to exclude the current system state at the onset of the fault, and the state trajectory continues to depart until an action is taken to remove the faulty line within a critical clearance time $\varepsilon < t_c$, provided that adequate equipment redundancy exists. The post-fault region of attraction is defined as the set of states attractive to a post-fault equilibrium established by the control action taken.

Real-time assessment of security regions has been a focused research in power systems for many years by many researchers [27]. Figure 4 depicts a pre-fault, a during fault, and two post-fault (with and without SVC) estimates of the regions of attraction, applying the algorithm for quasi-stability region estimation from [9]. The

---

[7]The discussion on diagnosis is omitted for it is beyond the scope of this chapter.

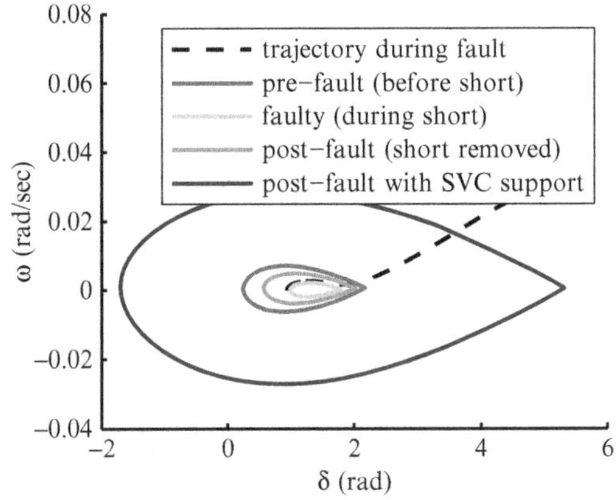

Fig. 4 Estimates of the
regions of attraction using the
Chiang algorithm [9] for the
pre-fault, during-fault, and
post-fault (with and without
SVC) two-area swing model;
the departing state trajectory
during fault

Fig. 5 Distributions of state
estimation error as time
progresses from the onset of
the transmission line short
circuit fault, and control
performance measures for
inaction u_1 and action u_2
(removal of faulty line)
defined on the regions
of attraction

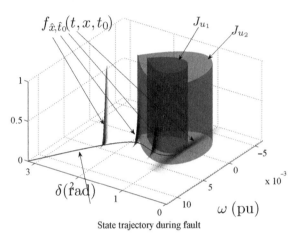

State trajectory during fault

largest region in Fig. 4 corresponds to post-fault region of attraction estimate with
an SVC support.

Figure 5 illustrates all necessary elements in the evaluation for the two-area
system: control performance threshold $J_0 = 1$, control performance measures
J_{u_1} of inaction u_1, and J_{u_2} of action u_2 with the supports being their respective
regions of attraction, as well as distribution $f_{(\hat{x},\hat{t}_0)}(t, x, t_0)$ of state estimation error
at four different points of time along a departing state trajectory. The equivalent line
reactance of the post-fault system increases to $x_e = 0.0846$ pu, and the post-fault
equilibrium swing angle increases to $\delta_0 = 62$ degrees, when u_2 is applied at $t + \varepsilon$,
where ε is less than the critical clearance time $t_c = 1.22$ s.

Rather than a discrete set of control actions, the admissible set for this example
is more properly characterized by $\mathcal{U} = \{1(t - t_0 - \tau), \tau \geq 0\}$, where $1(t - t_0 - \tau)$
is a step function of varying time delays.

It is seen in Fig. 5, as the state estimate improves over time, as indicated by the
decreasing error covariance of the distribution function, the risk of running beyond
the critical clearance time also increases. This suggests that there is an optimal time

Fig. 6 Calculated coverage
of inaction $u(t) \equiv 0$, and
coverage as a function of
delay in control action
$u(t) = 1(t - t_0 - \tau)$ with and
without the midpoint SVC

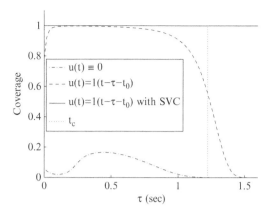

$0 < \tau_{opt} < \varepsilon < t_c$ for a decision on the control action, which can be seen in Fig. 6, where the horizontal axis should be interpreted as the time of a control action since the inception of the fault. Due to uncertainty, coverage is generally nonperfect, and reduces to an unacceptable level before the critical clearance time. Thus, waiting for uncertainty to reduce can be beneficiary, but a delayed action toward the critical clearance time is detrimental.

Omitted in the above consideration is the uncertainty associated with the removal of wrong equipment. In general, multiple values of fault coverage for a range of control actions are computed. Since the onset time of a fault is as hard to determine as state and parameter estimates in a dynamic system [4], coverage-based optimal redundancy management amounts to taking the least risky control action (highest coverage) as soon as the risk of the action is deemed acceptable.

When the midpoint SVC is in place, the calculated coverage is nearly perfect (≈ 1) for the two-area example because of the large region of attraction seen in Fig. 4. In addition, critical clearance time is significantly extended because the faulty state trajectory is allowed to depart farther. Thus increased controllability, in this case, the presence of the SVC, has greatly improved fault-coverage.

2.3 Need for a Resilient Supporting Structure

It has been discussed that control and measurement devices, such as FACTS, HVDC link systems, and PMUs, in a modern supporting structure can potentially establish a set of fast-recoverable limp-along states, aggregated as $\bar{O}$ in Fig. 1, and thus greatly enhance the grid's availability. In this section, the assumption that a supporting structure is always intact is removed to reveal the need for a resilient supporting structure.

Define a functional unit as a subsystem of a particular functionality that is necessarily available for the supporting structure to be available. The supporting structure can then be decomposed into a series of functional units. Each functional

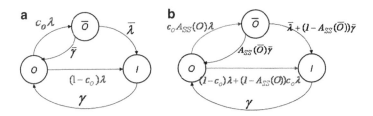

Fig. 7 (a) Rate transition diagram of the controlled grid with a perfect supporting structure, i.e. $A_{SS}(O) = A_{SS}(\bar{O}) = 1$; (b) rate transition diagram of the controlled grid with imperfect supporting structure $A_{SS}(O)$, $A_{SS}(\bar{O}) < 1$

Table 1 Effect of imperfect supporting structure, i.e. $A_{SS}(O)$, $A_{SS}(\bar{O}) < 1$, on state transition rates		Rate with imperfect supporting structure	Rate with perfect supporting structure
	State transition		
	$O \rightarrow \bar{O}$	$A_{SS}(O)c_0\lambda <$	$c_0\lambda$
	$O \rightarrow I$	$\bar{c}_0\lambda + \bar{A}_{SS}(O)c_0\lambda >$	$\bar{c}_0\lambda$
	$\bar{O} \rightarrow O$	$A_{SS}(\bar{O})\bar{\gamma} <$	$\bar{\gamma}$
	$\bar{O} \rightarrow I$	$\lambda + \bar{A}_{SS}(\bar{O})\bar{\gamma} <$	λ

unit has an up state, and a down state. Let $A_{SS}(x)$ denote the availability of the supporting structure, and $A_i(S)$ the availability of the ith functional unit. Then the availability of a structure with N functional units at current grid state S is

$$A_{SS}(x) = A_1(S) \times A_2(S) \times \cdots \times A_N(S), \tag{11}$$

where $S = O$, $\bar{O}_i$, I, and $\bar{O}_i$ is one of the $|\bar{O}| + 2$ grid states. The dependence on grid state S is deliberately shown to emphasize the fact that different grid state requires different support.

An example of a functional decomposition can be a three-unit supporting structure ($N = 3$): a functional unit that supports information acquisition, one that supports information processing and decision making, and finally one that supports execution of the decisions. Each of these functional units is a complex interconnection of many subsystems, some of which are self-contained control systems. Faults considered now are those impeding the flow of information from acquired data to decisions to be executed for controlling the grid.

To grasp the crux of the matter in terms of the impact of the supporting structure availability, let us revisit Fig. 1c, which is now renamed Fig. 7a, and compare the transition rates with those in Fig. 7b. Again, the decomposition property of Poisson processes [20] is invoked.

It is seen from Table 1 that an imperfect supporting structure always decreases the rates for desirable transitions and increases rates for undesirable transitions. Thus, a supporting structure with low availability can be a liability rather than an asset. The next section is to focus on redundancy architecture design of a supporting structure that satisfies a prescribed availability requirement. This requirement can be directly imposed by grid availability $\pi_O + \pi_{\bar{O}}$ through the rates shown in Table 1.

3   Redundancy Architecture Design for a PMU Network

The example in the last section reveals that reliable diagnostic outcomes, such as accurate state estimates, extracted from measurements, are essential in coverage-based redundancy management in the modernizing grid. This section focuses on the discussion of redundancy architecture design for the functional unit comprised of a network of PMUs (referred to as the PMU network hereafter) in the grid's supporting structure. The objective is to maintain a prescribed availability of the synchrophasors at the buses. Here, synchrophasors are narrowly referred to the GPS synchronized voltage phasors at the buses in the grid, though the conclusions drawn are applicable to more general data forms obtained through more general measurement devices.

3.1   Synchrophasor Availability at a Bus

Suppose for every bus that contains a PMU, a level of synchrophasor availability is imposed. A standard Markov chain is now used to model the overall availability of an n-PMU network. A state X_j is named by an n digit integer $B_1 \cdots B_n$. Each B_i, $i = 1, \ldots, n$, can take one of three values, $B_i \in \{0, 1, 2\}$. $B_i = 1$ indicates that the synchrophasor at bus i is supplied by a local PMU at the bus. Absence of local synchrophasor data at a bus will be simply described by "PMU failure", though it can be attributed to either the PMU hardware failure or intermittent data loss of various root causes. $B_i = 2$ indicates that the synchrophasor at bus i can be inferred from synchrophasors at some other buses, which implies that a certain redundancy configuration is in place. $B_i = 0$ indicates that there is no inferred synchrophasor at bus i. For convenience of bookkeeping, the subscript j in X_j is a decimal integer determined by rewriting the base 3 integer $B_1 \cdots B_n$ in base 10. Thus, the resulting state space $\mathscr{X} = \{X_0, \ldots, X_{n^3-1}\}$.

Two types of events are considered: PMU failure and PMU restoration. In a Markov chain, a random event life, T_e, comes from an exponential distribution, $f_{T_e}(t_e, \lambda) = \lambda e^{-\lambda t_e}$, where λ is the event occurrence rate. Let $\lambda = \rho_i$ be the failure rate of the PMU at bus i, and $\lambda = \gamma_i$ be the restoration rate of the PMU at bus i. Let Λ be the set of all possible transition rates.

When the ith digit, B_i, in state X_j flips from a 1, it either becomes a 2, which causes a state transition to X_y at rate $c_{j,y}\rho_i$, indicating redundancy is utilized to calculate the synchrophasor at bus i, or becomes a 0, which causes state transition to X_z at rate $(1 - c_{j,y})\rho_i$, indicating redundancy is not utilized. $c_{j,y} \in \{0, 1\}$ is called a decision variable, which modifies a transition rate through decomposing the associated Poisson process. A second kind of decision variable, denoted by $u_{y,j} \in \{0, 1\}$ is introduced to enable restoration of failed PMU at bus i. The decision $u_{y,j}$ is attached to the restoration rate γ_i. Decision variables are introduced to design redundant architecture in the PMU network.

Let $\pi_j(t)$ be the probability of being in state X_j at time t. Then the state probability distribution at t is given by the vector $\pi(t) = [\, \pi_0(t) \; \pi_1(t) \; \cdots \; \pi_{n^3-1}(t) \,]$. Evolution of the state-probability distribution over time can be solved from the Chapman–Kolmogorov equation [8, 20] $\dot{\pi}(t) = \pi(t)Q(c, u)$ for a given initial distribution, where $Q(c, u)$ is the transition rate matrix fully defined by the transition rates in Λ, and vectors c, u of decision variables. Steady-state probability distribution $\bar{\pi} = \lim_{t\to\infty} \pi(t)$ can be solved from $0 = \bar{\pi} Q(c, u)$ and $1 = \sum_j \bar{\pi}_j$, provided that the Markov chain is ergodic.

Synchrophasor availability is defined as the steady-state probability that the synchrophasor at bus i can be either estimated in the presence of local measurements or inferred remotely in the absence of local measurements. Recall that B_i is the ith digit for state X_j, which has state probability $\pi_j(t)$. Thus, the synchrophasor at bus i is available when $B_i \geq 1$. The steady-state availability of the synchrophasor at bus i, A_i is

$$A_i(\pi) = \sum_{j \in \mathscr{A}_i} \bar{\pi}_j, \tag{12}$$

where $\mathscr{A}_i$ is the set of all j for which X_j has $B_i \geq 1$.

3.2 Redundancy Architecture Design Under Availability Constraints

Subject to $A_i(\pi) \geq a_i$ for a prescribed set of synchrophasor availability a_i at bus i, $i = 1, \cdots, n$, a redundancy architecture design problem is formulated in this section to minimize a total cost associated with PMU failures. It includes the cost of a communication link, such as an optical fiber, between buses to enable inferred synchrophasors, and the cost of restoration of the failed PMUs. A cost is incurred whenever a decision variable in c or u is set to unity. With a linear cost function, the architecture design problem is formulated as a mixed-boolean convex optimization problem, which can be solved using branch and bound [13]. If the steady-state solution asserts the restoration of a particular PMU unnecessary, the PMU can be spared at a bus in the first place. If the steady-state solution asserts the utilization of redundant synchrophasors at a particular bus unnecessary, the communication links to the PMU can be spared.

Let the price of utilizing redundancy to infer a synchrophasor be k_c, and the price of restoring a failed PMU be k_p. Let $\mathscr{X}_0$ and $\mathscr{X}_2$ be the sets of state names containing at least a 0, and at least a 2, respectively. The average cost at steady-state for utilizing redundancy to infer synchrophasors is defined as

$$K_C = \sum_{x \in \mathscr{X}_2} k_c \bar{\pi}_x. \tag{13}$$

The average cost for restoring failed PMUs is similarly defined as

$$K_P = \sum_{x \in \mathscr{P}_0 \cup \mathscr{P}_2} \sum_{y \in \mathscr{X}} k_p u_{x,y} \bar{\pi}_x, \qquad (14)$$

where $u_{x,y} = 0$ if the transition $x \to y$ is not related to a PMU restoration. The term $u_{x,y}$ appears in (14) because a cost is only incurred in state x if a PMU needs to be restored. Using this cost function, structure with a higher probability of use accumulates a larger cost.

An availability constrained redundancy architecture optimization problem is stated as follows:

$$\text{minimize} \quad \sum_{x \in \mathscr{P}_R} k_c \bar{\pi}_x + \sum_{x \in \mathscr{P}_0 \cup \mathscr{P}_2} \sum_{y \in \mathscr{X}} k_p u_{x,y} \bar{\pi}_x, \qquad (15)$$

$$\text{subject to:} \quad \sum_{j \in \mathscr{A}_i} \bar{\pi}_j \geq a_i \text{ for } i = 1, \ldots, n, \qquad (16)$$

$$\bar{\pi} Q(c, u) = 0, \quad \sum_j \bar{\pi}_j = 1, \quad \bar{\pi} \geq 0, \qquad (17)$$

$$c \in \{0, 1\}, \quad u \in \{0, 1\}. \qquad (18)$$

Theoptimization variables are vectors $\bar{\pi}$, c, and u. (16) enforces the availability requirement a_i at each bus i. Equation (17) constrains solution $\bar{\pi}$ to be a steady-state distribution. Equation (18) constrains the decision variables to be Boolean variables.

Although the problem is a linear program for fixed values of the Boolean variables, the relaxed problem when each Boolean variable is allowed to be in interval [0, 1] is nonconvex. This issue can be resolved by introducing new variables $v_{x,y} = c_{x,y} \bar{\pi}_x$ and $\mu_{x,y} = u_{x,y} \bar{\pi}_x$ into the first equality of (17), with $0 \leq v_{x,y} \leq \bar{\pi}_x$ and $0 \leq \mu_{x,y} \leq \bar{\pi}_x$. The variable substitution renders the optimization problem in (15)–(18) a mixed-Boolean convex program. It contains both continuous variables ($\bar{\pi}$, $v_{x,y}$, $\mu_{x,y}$) and Boolean variables (c, u), which can be solved using branch and bound [13].

3.3   A 3-Bus/3-PMU Example

An optimal redundancy architecture design is performed for the 3-bus/3-PMU power system in Fig. 3. The states are named based on whether the synchrophasors at each bus is calculated in the presence of local PMU measurements, or inferred in the absence of local PMU measurements. Thus, the state space with unreachable

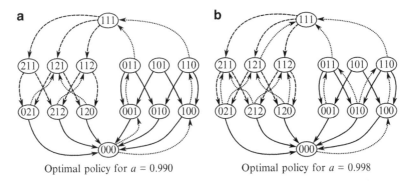

Optimal policy for $a = 0.990$ Optimal policy for $a = 0.998$

Fig. 8 The optimal redundancy architecture is depicted for two different synchrophasor availability requirements, $a = 0.99$ and $a = 0.998$. *Dashed arrows* indicate that $c_{x,y} = 1$. *Solid arrows* indicate that $c_{x,y} = 0$. *Dotted arrows* indicate that $u_{x,y} = 1$. Transitions with zero rate are omitted from the diagram. As a increases, the optimal architecture makes more aggressive use of redundancy and restoration, resulting a higher cost (**a**) Optimal policy for $a = 0.990$ (**b**) Optimal policy for $a = 0.998$

states removed is $\mathscr{X} = \{X_0 : 000, X_2 : 001, X_4 : 010, X_5 : 011, X_7 : 021, X_9 : 100, X_{10} : 101, X_{12} : 110, X_{13} : 111, X_{14} : 112, X_{15} : 120, X_{16} : 121, X_{22} : 211, X_{23} : 212\}$, as seen in Fig. 8. States such as $X_{19} : 201$ are removed from the state space because the voltage phasor at bus 1 cannot be calculated when voltage and current synchrophasors are not available at bus 2, which also means that the corresponding decision variable $c_{x,y} = 0$. To simplify, state $X_{21} : 210, X_5 : 012$, and $X_{23} : 212$ are aggregated as a single state. Another simplification is the use of equal failure rates and restoration rate for all PMUs, i.e., $\rho_1 = \rho_2 = \rho_3 = \rho$ and $\gamma_1 = \gamma_2 = \gamma_3 = \gamma$, though this need not be the case.

Without loss of generality, availability requirements for all synchrophasor are set equal, i.e., $a_1 = a_2 = a_3 = a$, which is not necessary in general. The set of states in which a voltage synchrophasor is available at bus 1 is any state where $B_1 \geq 1$, $\mathscr{A}_1 = \{9, 10, 12, 13, 14, 15, 16, 22, 23\}$. The sets $\mathscr{A}_2$ and $\mathscr{A}_3$ are found similarly. The set $\mathscr{X}_2 = \{7, 14, 15, 16, 22, 23\}$ is all states with any $B_i = 2$ for $i = 1, 2, 3$.

The optimal redundancy architecture is solved for the PMU network in Fig. 3 using the branch and bound method. The optimal structure is investigated for a range of parameter values. The nominal PMU failure rate is assumed to be $\rho = 0.01$ and restoration rate is $\gamma = 0.99$. This corresponds to a single PMU uptime of $\frac{\gamma}{\gamma + \rho} = 0.99$ if the failed PMU is restored ($u_{x,y} = 1$). $a \geq 0.990$ is enforced, based on the nominal uptime of a single PMU.

The optimal policies for $a = 0.990$ and $a = 0.998$ are depicted in Fig. 8. Although for the small system no PMUs or communication links can be spared, it is clear that less redundancy utilization and less restoration are required to meet $a = 0.990$ in Fig. 8a than to meet $a = 0.998$ in Fig. 8b. In fact, the optimal policies in Fig. 8a do not have any links from $X_{22} : 211, X_{16} : 121$, or $X_{14} : 112$ to $X_{13} : 111$, which shows that there is no need to restore all three PMUs when redundancy is used. In Fig. 8b, one of these links, along with several links elsewhere, has been

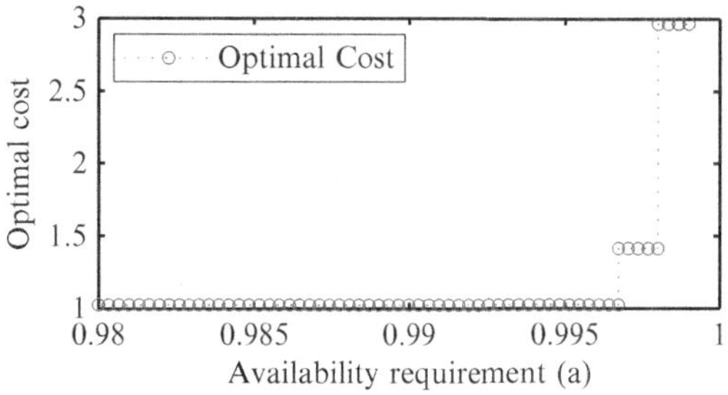

Fig. 9 Optimal cost across
a range of values of required
synchrophasor availability.
The cost jumps where
structural changes occur

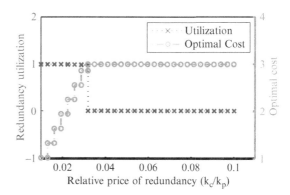

Fig. 10 Optimal cost and
redundant PMU utilization
across a range of price ratios
on utilizing redundancy to
restoring PMUs, k_c / k_p. The
discontinuities in the cost are
where structural changes
occur. For $k_c / k_p \geq 0.03$, the
optimal policy is to not use
any redundancy

enabled in order to meet the more stringent availability requirement. From state
$X_0 : 000$ following up the right sides of the diagrams in Fig. 8, the system prefers
not to use any redundancy until the system has returned to as good as new, $X_{13} : 111$.
Although these actions seem counterintuitive, they are optimal due to the very small
probability of $X_0 : 000$ and all subsequent states on the right.

The minimum cost is plotted over a range of synchrophasor availability requirements in Fig. 9, confirming that cost increases as the availability requirement
increases. Discontinuities in the cost indicate changes in the optimal architecture,
which are reflected in Fig. 8. Figure 10 demonstrates the policy changes as the
price of PMU interconnection changes relative to the price of PMU restoration. As
the price of PMU interconnection increases, the usage of redundant synchrophasor
data decreases. Eventually, avoidance of using the redundant synchrophasor data
causes the steady-state cost to plateau as shown in Fig. 10. The effects of increasing
restoration rate relative to the failure rate are observed in Fig. 11. Faster PMU
restoration reduces the overall cost, because probability of being in unavailable
states decreases. The utilization of redundant synchrophasors increases with the
PMU restoration rate as observed in Fig. 11 because redundant synchrophasors are
more readily available, and the necessity to restore to as good as new reduces (π_{13}
becomes small).

The current model does not distinguish the root causes leading to PMU outages.
It in fact only captures outages due to PMU failures properly. Intermittent loss
of synchrophasors need not invoke restoration. Thus, more detailed modeling is

Fig. 11 Optimal cost and
redundant synchrophasor
utilization across a range
of values on the ratio of the
restoration rate to PMU
failure rate, γ/ρ. As the
restoration rate increases,
the usage of redundancy
increases

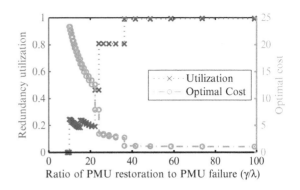

needed, which inevitably requires a larger state-space. Despite the larger size,
the problem formulation and the method of solution remain the same. Redundant
information may also be provided by traditional state estimation and SCADA data.

4   Conclusions and Future Work

The essence of the concepts presented in this chapter has to do with exploiting
analytical redundancy potentially affordable by new control and measurement
devices to help create a set of limp-along states from which significantly faster
restoration to normal state is implementable in comparison with the restoration rates
from blackouts. Thus, a highly available grid can be expected in a cost-effective
manner. An example of a two-area swing model has been used to explain how fault-
coverage can guide the management of redundancy in equipment with minimum
risk. A 3-bus/3-PMU example has been used to explain how data availability
requirement can be met through a method of redundancy architecture design of
measurement networks.

The two examples solved, however, have the smallest nontrivial sizes, and serve
mainly to explain the concepts and methods introduced. An immediate extension of
this work is to apply the concepts and methods to power systems and measurement
networks of realistic scales. It must be recognized, on the other hand, that the
problem formulations for both the redundancy management and the redundancy
architecture design are applicable to selected portions of the power grid, or of a
measurement network, for which the computational demands are manageable, and
the solutions can still provide significant benefits.

Other aspects of development being pursued by the authors include the em-
ployment of state and parameter estimation schemes based on the raw PMU
measurements as an approach to fault diagnosis, the interpretation of the outcomes
of the architecture design of a measurement network as a solution to the placement
of measurement devices, and the utilization of a Markov decision process as an
alternative framework for the redundancy architecture design.

Acknowledgements The first author gratefully acknowledges the enormous values of the insights provided by Dr. Joe H. Chow of RPI, Dr. Aranya Chakrabortty of NCSU, and Dr. Xiaohong Guan of XJTU in separate sessions of discussion on these topics with her, as well as the financial support from the XJTU during her two-month visit last year, where she started to explore this area.

References

1. Anderson PM, (1998) Power system protection. IEEE Press, New Jersey
2. Antsaklis PJ, (ed) (2000) Hybrid systems: theory and applications, Special Issue. Proc IEEE 88:879–1130
3. Baldwin T, Mili L, Boisen M (1993) Power system observability with minimal phasor measurement placement. IEEE Trans Power Syst 8:707–715
4. Basseville M, Nikiforov IV (1993) Detection of abrupt changes: theory and application. Prentice-Hall, NJ
5. Barlow RE, Proschan F (1996) Mathematical theory of reliability. Society for industrial mathematics, New York
6. Blanke M, Kinnaert M, Lunze J, Staroswiecki M (2003) Diagnosis and fault-tolerant control. Springer-Verlag, Berlin
7. Brueni D, Heath L (2005) The PMU placement problem. SIAM J Discrete Math 19:744
8. Cassandras C, Lafortune S (1999) Introduction to discrete event systems, Springer, New York
9. Chiang H, Fekih-Ahmed L (1996) Quasi-stability regions of nonlinear dynamical systems I: fundamental theory and applications. IEEE Transact Circuit Syst 43:627–635
10. Chow EY, Willsky AS (1984) Analytical redundancy and the design of robust detection systems. IEEE Trans Automat Contr 29:603–614
11. Chow JH, Wu FF, Momoh JA (ed) (2005) Applied mathematics for restructured electric power systems: optimization, control, and computational intelligence. Springer, USA
12. Chow JH, Chakrabortty A, Arcak M, Bhargava B, Salazar A (2007) Synchronized phasor data based energy function analysis of dominant power transfer paths in large power Systems. IEEE Trans Power Syst 22:727–734
13. Davis R, Kendrick D, Weitzman M (1971) A Branch-and-bound algorithm for zero-one mixed integer programming problems. Oper Res 19:1036–1044
14. Dembo A, Zeitouni O (1998) Large deviations techniques and applications. Applications of mathematics, 2nd edn., 38 Springer-Verlag, Berlin
15. Donmez B, Abur A (2010) A computationally efficient method to place critical measurements. IEEE Trans Power Syst 26:924–931.
16. Dugan JB, Trivedi KS (1989) Coverage modeling for dependability analysis of fault tolerant systems. IEEE Trans Comput 38:775–787
17. Fairley P (2004) The unruly power grid. IEEE Spectrum 41:22–27
18. Faza AZ, Sedigh S, McMillin BM (2007) Reliability modeling for the advanced electric power grid. Comput Saf Reliab Secur 4680:370–383
19. Huang J, Wu NE (2011) . Fault-tolerant sensor placement based on control reconfigurability. In: Proceedings of the 18th IFAC World Congress, Milano, Italy
20. Kao EPC (1997) An introduction to stochastic processes. Duxbury Press, Chicago
21. Khalil HK (2002) Nonlinear systems, 3rd ed. Prentice Hall, Upper Saddle River, New Jersey
22. Kundur P (1994) Balu NJ, Lauby MG (ed) Power system stability and control. McGraw-Hill, New York
23. Madani V, Novosel D (2005) Getting a grip on the grid. IEEE Spectrum 42:42–47
24. Mountford JD, Austria RR (1999) Keeping the lights on. IEEE Spectrum 36:34–39
25. Phadke AG, Thorp JS (2008) Synchronized phasor measurements and their applications, Springer, Berlin

26. U.S.-Canada Power System Outage Task Force (2004) Final Report on the August 14 2003 Blackout in the U.S. and Canada: Causes and Recommendations https://reports.energy.gov Cited 15 April 2011
27. Varaiya P, Wu FF, Chen R (1985) Direct methods for transient stability analysis of power systems: recent results. Proc IEEE 73: 1703–1715
28. Wang X, McDonald JR (ed) (1993) Modern power system planning, McGraw-Hill, New York
29. Wu FF, Tsai Y (1983) Probabilistic dynamic security assessment of power systems: Part I—basic model. IEEE Trans Circuit Syst 30: 148–159
30. Wu NE (2004) Coverage in fault-tolerant control. Automatica 40:537–548

Models and Control Strategies for Data Centers in the Smart Grid

Luca Parolini, Bruno Sinopoli, and Bruce H. Krogh

Abstract This chapter discusses the challenges and the possibilities offered by controlling a data center as a node of the smart-grid. The communication between the grid and the data center takes the form of a time-varying and power-consumption-dependent electricity price. A model that considers both the computational and the physical characteristics of a data center, as well as their interactions, is proposed. Two control approaches are discussed. The first control approach does not consider the interaction between the computational and the thermal characteristics of a data center. We call such a control approach *uncoordinated*. The second controller considers the coupling between the computational and the thermal characteristics. We call such a control approach *coordinated*. Simulation results, discussed at the end of the chapter, show that the coordinated control approach can lead to larger income for data center operators than the uncoordinated approach.

1 Introduction

The number of data centers has significantly increased throughout the world, fueled by the increasing demand for remote storage and cloud computing services. Power consumption of data centers has also significantly increased. A report of the U.S. Environmental protection agency (EPA) published in August 2007 shows that data center power consumption doubled from 2000 to 2006, reaching a value of 60 TWh/yr (terawatt hour/year) [22]. Efficiently powering and cooling data centers have also become a challenging problem. Large-scale data centers are tailored for peak power consumption of dozens of megawatts and have operational costs as high

L. Parolini (✉) • B. Sinopoli • B.H. Krogh
Carnegie Mellon University, 5000 Forbes Avenue, Pittsburgh, PA 15213, USA
e-mail: lparolin@ece.cmu.edu; brunos@ece.cmu.edu; krogh@ece.cmu.edu

A. Chakrabortty and M.D. Ilić (eds.), *Control and Optimization Methods for Electric Smart Grids*, Power Electronics and Power Systems 3, DOI 10.1007/978-1-4614-1605-0_11, © Springer Science+Business Media, LLC 2012

as \$5.6 M/year [5]. Energy consumed for computation and cooling is dominating data center run-time management and operating costs and even a small percentile of power reduction can have large impact in the economics of data center operation.

A measure of the efficiency of a data center, typically used in industry, is the *power usage effectiveness* (PUE) [20]. The PUE of a data center is defined as the ratio between the total data center power consumption and the power consumption of the information technology (IT). A PUE equal to 1.0 indicates that all of the data center power consumption is due to the IT. The PUE values collected over 60 different data centers are discussed in [20]. The average PUE value is 2.03, eight data centers have PUE values lower than 1.5, whereas six data centers have PUE values larger than 2.75. The majority of the data centers have PUE values between 1.5 and 2 [20]. One of the shortcomings of the PUE index is that it does not indicate how efficiently the IT is used.

From the power-grid perspective, the peculiarities of the data centers are their high value of power consumption per squared meter and the timescale at which power consumption can be controlled, e.g., on the minute timescale. A low-density data center can have a peak power consumption of 800 W/m^2 (75 W/sq.ft.), whereas a high-density data center can reach 1.6 KW/m^2 (150 W/sq.ft.) [6, 16, 18]. In a deregulated electricity market, the price of electricity can change over time and also over geographic locations. A viable technique for reducing the variability of the electricity cost is to require to some costumers to cap their power consumption upon request from the grid. Such a technique, already applied by some independent system operators (ISOs), is called *demand response program* (DRP).

As in the work of Parolini et al. [14], we consider the case where a data center controller can leverage over two *service level agreements* (SLAs). The first SLA regulates the income of the data center based on the *quality of service* (QoS) provided to the users. The second SLA regulates the cost of the electricity. We consider the case where the data center participates into a DRP with the power-grid. The communication between the power-grid and the data center takes form of a time-varying and power-consumption-dependent electricity price. As long as the average power consumption of a data center is kept below a time-varying threshold, the data center buys electricity at a reduced price. When the average data center power consumption exceeds the time-varying threshold, the additional energy is provided at a higher cost. Average values of the data center power consumption are computed over a given time window.

2 A Thermal-Computational Model for Data Centers

Data center technology can be roughly classified into three groups: IT, CT, and support technology [13, 14]. The IT comprises servers, storage devices, and network-related components such as switches, firewalls, and routers. The CT comprises components such as *computer room air conditioners* (CRACs) and

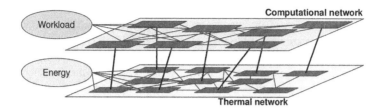

Fig. 1 Networked model of a data center (© IEEE 2011), reprinted with permission

fans [2,3,15,17,23]. The support technology class includes devices such as batteries, backup power generators, uninterruptible power supply (UPSs), power distribution units (PDUs), etc.

Data centers are large-scale systems with timescales ranging from milliseconds to tens of minutes. The actuators that can be exploited by the control algorithms are heterogeneous in nature and their effects are relevant at different time and spatial scales. For example, *dynamic voltage and frequency scaling* (DVFS) techniques operate at the millisecond level and influence the power consumption of a single server [1, 4, 8]. At the rack scale, virtual machine (VM) migration affects the power consumption of multiple servers and operate on the minute timescale [7, 9]. In such a scenario, a hierarchical control architecture is a desirable control approach. As in the work of Parolini et al. [13, 14], we consider a hierarchical-distributed approach to the data center control. At the highest level of the hierarchy, we consider a centralized controller called *data center controller*. The data center controller provides a collection of bounds and optimal set points to lower-level controllers that then operate independently from each other [13]. Individual racks, or collections of racks, are modeled as single IT components called *zones*. In this chapter, we focus on the highest level of the hierarchy.

A coupled networked model is considered. The first network is called *computational network* and it defines the relationship between the workload execution, the QoS, and the amount of power that each zone consumes. The second network is called *thermal network* and it defines the heat exchanged among the devices in the data center. Figure 1 provides a graphical representation of the proposed model. Zones have both computational and physical characteristics. These characteristics are represented by two coupled nodes. One node models the computational characteristics of a zone and another node models the physical characteristics of the zone. The first node belongs to the computational network, whereas the second node belongs to the thermal network. In order to simplify the notation, we assume that the two coupled nodes representing a zone have the same index in their respective networks, i.e. the ith node of the computational network is coupled with the ith node of the thermal network. Among the devices in the CT, we focus on the CRAC units. Other devices such as fans can be controlled at lower levels of the hierarchy. A CRAC unit has only physical characteristics, which are represented by a single node of the thermal network. We discuss in this chapter a simplified scenario, where the power consumption of the devices in the support class and their effects to the workload execution are negligible.

In the rest of the chapter, the variables k and v are elements of $\mathbb{Z}$ with $v \geq k$ and the variable τ is an element of $\mathbb{R}$. With $\hat{x}(v|k)$, we denote the expected value of the variable $x(v)$ based on the information available up to the beginning of the kth interval.

2.1 Computational Network

Let us consider a fluid approximation of the workload execution process and define N as the number of nodes in the computational network. Due to the fluid approximation, the workload arrival rate and the workload departure rate can be described by differentiable functions.

Let $\lambda^W(\tau)$ denote the workload arrival rate to the data center at time τ. The relative amount of workload assigned to the ith node at time τ is denoted by $s_i(\tau)$. The variables $\{s_i(\tau)\}$ represent the workload scheduling (or allocation) action. The rate at which workload migrates from the ith computational node to the jth computational node at time τ is denoted by $\delta_{j,i}(\tau)$. With $\mu_i(\tau)$ we denote the workload execution rate of the ith node at time τ. The variables $\{\mu_i(\tau)\}$, $\{s_i(\tau)\}$, and $\{\delta_{j,i}(\tau)\}$ are controllable variables. The amount of workload in the ith node at time τ is denoted by $l_i(\tau)$. The workload evolution at the ith computational node can be written as

$$\dot{l}_i(\tau) = \mathscr{I}_i \left[\lambda^W(\tau)s_i(\tau) + \sum_{j=1}^{N} \delta_{i,j}(\tau) - \left(\mu_i(\tau) + \sum_{j=1}^{N} \delta_{j,i}(\tau) \right) \right], \quad (1)$$

where the function $\mathscr{I}_i$ is defined as

$$\mathscr{I}_i = \begin{cases} 1 \text{ if } l_i(\tau) > 0 \text{ or} \\ \quad \left(\lambda^W(\tau)s_i(\tau) + \sum_{j=1}^{N} \delta_{i,j}(\tau) \right) > \left(\mu_i(\tau) + \sum_{j=1}^{N} \delta_{j,i}(\tau) \right) \\ 0 \text{ otherwise.} \end{cases} \quad (2)$$

Let us define the $N \times 1$ vectors $l(\tau) = \left[l_1(\tau) \ldots l_N(\tau) \right]^T$, $s(\tau) = \left[s_1(\tau) \ldots s_N(\tau) \right]^T$, $\mu(\tau) = \left[\mu_1(\tau) \ldots \mu_N(\tau) \right]^T$, and the $N^2 \times 1$ vector $\delta(\tau) = \left[\delta_{1,1}(\tau) \delta_{1,2}(\tau) \ldots \delta_{N,N} \right]^T$. The vector $l(\tau)$ is the state of the computational network. The variable $\lambda^W(\tau)$ is the uncontrollable input, whereas the vectors $\mu(\tau)$, $s(\tau)$, and $\delta(\tau)$ are the controllable inputs. The fluid model discussed in this chapter can be extended to include multiple workload classes, hardware requirements, and the interaction among the workload allocated to different nodes [12–14].

2.2 Thermal Network

Let M denote the number of nodes in the thermal network. The heat exchanged among the nodes of the thermal network is described by means of temperature variations of the nodes. Two temperatures are defined for every node: the *input* temperature and the *output* temperature. With $T_{in,i}(\tau)$ we denote the input temperature of the ith node at time τ. The variables $\{T_{in,i}(\tau)\}$ represent the recirculation and the cooling effects due to all of the nodes in the thermal network. With $T_{out,i}(\tau)$ we denote the output temperature of the ith node at time τ. The variables $\{T_{out,i}(\tau)\}$ represent the amount of heat contained in the ith thermal node at time τ.

As discussed in the work of Tang et al. [19], we consider a linear relationship between the input and the output temperatures of every node

$$T_{in,i}(\tau) = \sum_{j=1}^{M} \psi_{i,j} T_{out,j}(\tau), \qquad \text{for all } i = 1, \ldots, M. \tag{3}$$

The coefficients $\{\psi_{i,j}\}$ are non-negative and $\sum_{j=1}^{M} \psi_{i,j} = 1$ for all $i = 1, \ldots, M$. An experimental method to estimate the values of the coefficients $\{\psi_{i,j}\}$ is discussed in [19]. Let us collect the $\{\psi_{i,j}\}$ coefficients in the $M \times M$ matrix Ψ and the input and the output temperatures of every node in the $M \times 1$ vectors $T_{in}(\tau)$ and $T_{out}(\tau)$, respectively. The relationship between the output and the input temperature vectors can now be rewritten as

$$T_{in}(\tau) = \Psi T_{out}(\tau). \tag{4}$$

As discussed in [12, 14, 22] and in the references therein, the thermal constraints of the IT devices are formulated in terms of their inlet temperatures. In the proposed model, the constraints can be stated as

$$\underline{T_{in}} \leq T_{in}(\tau) \leq \overline{T_{in}}, \qquad \text{for all } \tau \in \mathbb{R}, \tag{5}$$

where the vectors $\underline{T_{in}}$ and $\overline{T_{in}}$ represent the lower and the upper limits of the input temperature vector.

Let us consider a thermal node modeling a zone. The evolution of the output temperature of the node is given by

$$\dot{T}_{out,i}(\tau) = -k_i T_{out,i}(\tau) + k_i T_{in,i}(\tau) + c_i p_i(\tau), \tag{6}$$

where $\frac{1}{k_i}$ is the time constant of the ith node, c_i is the coefficient that maps power consumption into output temperature variation, and $p_i(\tau)$ represents the power consumption of the ith at time τ [11, 14]. The power consumption of the ith node depends on the workload execution rate of the coupled node in the computational network. We consider the following relationship

$$p_i(\tau) = 1_{\{l_i > 0\}} \alpha_i \mu_i(\tau), \tag{7}$$

where α_i is a non-negative coefficient. The choice of the linear relationship, for the case $l_i(\tau) > 0$, stems from the assumption that lower level controllers, such as the one proposed by Tolia et al. [21], can be used to make the power consumption of a zone proportional to the amount of workload processed by the zone. The proposed power consumption model can be extended to include more complicated functions, which may account for the ON-OFF state of every servers.

Let us focus on a thermal node representing a CRAC unit. The evolution of its output temperature is given by

$$\dot{T}_{\text{out},i}(\tau) = -k_i T_{\text{out},i}(\tau) + k_i \min\{T_{\text{in},i}(\tau), T_{\text{ref},i}(\tau)\}, \tag{8}$$

where $T_{\text{ref},i}(\tau)$ represents the reference temperature of the CRAC node i and it is a controllable variable [11, 12, 14]. The min operator in (8) forces the output temperature of the node to be always lower than or equal to its input temperature. As discussed in the work of Moore et al. [10], the power consumption of a CRAC node can be modeled as

$$p_i(\tau) = \begin{cases} c_i \dfrac{T_{\text{in},i}(\tau) - T_{\text{out},i}(\tau)}{COP(T_{\text{out},i}(\tau))} & T_{\text{in},i}(\tau) \geq T_{\text{out},i}(\tau) \\ 0 & T_{\text{in},i}(\tau) < T_{\text{out},i}(\tau), \end{cases} \tag{9}$$

where c_i is a coefficient that depends on the amount of air that flows through the CRAC unit every second and on the specific heat capacity of the air. The function $COP(T_{\text{out},i}(\tau))$ represents the *coefficient of performance* (COP) of the CRAC unit modeled by the ith node.

Let us collect the power consumption of the thermal nodes representing a zone in the $N \times 1$ vector $\mathbf{p}_N(\tau)$. The power consumption values of the thermal nodes representing a CRAC are collected in the vector $\mathbf{p}_C(\tau)$. The vector $\boldsymbol{T}_{\text{ref}}(\tau)$ collects all of the reference temperatures of the CRAC nodes. The state of the thermal network is the vector $\boldsymbol{T}_{\text{out}}(\tau)$, the controllable input of the thermal network is the vector $\boldsymbol{T}_{\text{ref}}(\tau)$ and the uncontrollable input of the thermal network is the vector $\mathbf{p}_N(\tau)$. The outputs of the thermal network are the vector $\mathbf{p}_C(\tau)$ and the vector of node input temperatures $\boldsymbol{T}_{\text{in}}(\tau)$. The vector $\boldsymbol{T}_{\text{in}}(\tau)$ is function of the network state (4) and therefore it is an output of the network. However, when we look at a single thermal node, the input temperature becomes an uncontrollable input of the node. In this sense, the input vector is an output of the thermal network and, at the same time, it is an uncontrollable input for each of the thermal node.

3   Control Strategies

Two control strategies are considered. The first controller treats the computational management problem and the thermal management problem separately. We call such a controller *uncoordinated*. The second controller treats the computational management problem and the thermal management problem jointly. We call such a controller *coordinated*. Both controllers are based on a model predictive control (MPC) approach. The goal of the two controllers is to minimize the expected run-time cost of operating the data center. The expected run-time cost is composed of three parts: (a) the expected cost of powering the data center, (b) the expected cost of the chosen control action, and (c) the expected QoS cost. The uncoordinated controller and the coordinated controller use different approaches to estimate the powering cost of the data center.

The uncoordinated controller is based on a two-step optimization problem. In the first step, the expected cost of powering the data center accounts only for the expected power consumption of the servers. In the second step, the expected cost of powering the data center accounts only for the expected power consumption of the CRAC units. Both of the powering costs are based on the reduced (time-varying) electricity price. Let $p_{\mathcal{N}}(k)$ denote the average power consumption of the servers during the kth interval. The expected cost of powering the data center, in the first step of the optimization problem, can be stated as

$$\hat{c}_{e_{U,1}}(v|k) = \hat{\alpha}_e(v|k)\hat{p}_{\mathcal{N}}(v|k), \tag{10}$$

where $\alpha_e(k)$ represents the reduced electricity price during the kth interval. Let $p_{\mathcal{C}}(k)$ denote the average power consumption of the CRAC units during the kth interval. The expected cost of powering the data center, in the second step of the optimization problem, can be stated as

$$\hat{c}_{e_{U,2}}(v|k) = \hat{\alpha}_e(v|k)\hat{p}_{\mathcal{C}}(v|k). \tag{11}$$

The expected cost of the chosen control action at time v, based on the information available up to the beginning of the kth time interval, is given by

$$\hat{c}_{\delta,s}(v|k) = \hat{\delta}(v|k)^{\mathsf{T}}D\hat{\delta}(v|k) - s(v|k)^{\mathsf{T}}Ss(v|k), \tag{12}$$

where the matrices D and S are positive semidefinite. The fist term in (12) penalizes the migration of workload among nodes, whereas the second term in (12) penalizes the drop of the workload. The expected QoS cost at time v, based on the information available up to the beginning of the kth time interval, is given by

$$\hat{c}_{\text{QoS}}(v|k) = l(v|k)^{\mathsf{T}}C_Q l(v|k), \tag{13}$$

where the matrix C_Q is positive semidefinite.

The expected run-time cost considered by the uncoordinated controller, in the first step of its optimization problem, is

$$J_{U,1}(v|k) = \hat{c}_{e_{U,1}}(v|k) + \hat{c}_{\delta,s}(v|k) + \hat{c}_{\text{QoS}}(v|k). \tag{14}$$

The expected run-time cost considered by the uncoordinated controller, in the second step of its optimization problem, is

$$J_{U,2}(v|k) = \hat{c}_{e_{U,2}}(v|k).$$ (15)

The controllable variables in the computational and in the thermal network are subject to constraints. The vector of workload execution rate is bounded by

$$\mathbf{0} \le \mu(k) \le \overline{\mu},$$ (16)

where $\mathbf{0}$ is the zero vector, i.e. the vector whose elements are all equal to 0, and the inequalities are applied componentwise. As in [14], the constraints on the workload migration rates can be written as

$$\mathbf{0} \le \Delta^{-}\delta(k) \le l(k),$$ (17)

$$\delta_{i,j}(k)\delta_{j,i}(k) = 0, \quad \text{for all} \quad i, j = 1, \dots, N.$$ (18)

The constraints in (17) force the controller to not reallocate more workload than the amount available in a zone at the beginning of the kth interval. The inequality in (18) prevent the controller to migrate workload to and from the same couple of zones. The scheduling control action is bounded by

$$\mathbf{0} \le s(k) \le \mathbf{1},$$ (19)

$$\mathbf{1}^{\mathsf{T}}s(k) \le \mathbf{1}.$$ (20)

The constraint in (20) allows the data center to drop a certain amount of workload. The constraints on the reference temperature vector are given by

$$\underline{T}_{\text{ref}} \le T_{\text{ref}}(k) \le \overline{T_{\text{ref}}}.$$ (21)

Let $\mathsf{T} \in \mathbb{N}$ denote the horizon for the optimization problems solved by the uncoordinated controller and by the coordinated controller, i.e. the number of future intervals over which the evolution of the data center is considered. Let us define the sets $M = \{\hat{\mu}(k|k), \dots, \hat{\mu}(k + \mathsf{T}|k)\}$, $S = \{\hat{s}(k|k), \dots, \hat{s}(k + \mathsf{T}|k)\}$, $\mathscr{D} = \{\hat{\delta}(k|k), \dots, \hat{\delta}(k + \mathsf{T}|k)\}$, and $\mathscr{T}_{\text{ref}} = \{\hat{T}_{\text{ref}}(k|k), \dots, \hat{T}_{\text{ref}}(k + \mathsf{T}|k)\}$. The two-step optimization problem, solved by the uncoordinated controller at the beginning of the kth interval, can now be written as

1.

$$\min_{M,S,\mathscr{D}} \sum_{v=k}^{k+\mathsf{T}} J_{U,1}(v|k)$$

s.t.

computational constraints

control constraints. (22)

2.

$$\min_{\mathscr{T}_{ref}} \sum_{v=k}^{k+T} J_{U,2}(v|k)$$

s.t.

thermal constraints

control constraints.   (23)

The optimization problem in (23) is instantiated based on the solution obtained for the optimization problem in (22).

Let us focus now on the coordinated controller. Let $p(k)$ denote the average total data center power consumption during the kth interval, i.e. $p(k) = p_{\mathscr{N}}(k) + p_{\mathscr{C}}(k)$. Let $\bar{p}(k)$ be the expected average power consumption threshold during the kth interval. The coordinated controller computes the expected powering cost of the data center as

$$\hat{c}_e(v|k) = \begin{cases} \hat{\alpha}_e(v|k)\hat{p}(v|k) & \hat{p}(v|k) \le \hat{\bar{p}}(v|k) \\ \hat{\alpha}_e(v|k)\hat{\bar{p}}(v|k) + \hat{\beta}_e(v|k)\left(\hat{p}(v|k) - \hat{\bar{p}}(v|k)\right) & \hat{p}(v|k) > \hat{\bar{p}}(v|k), \end{cases}$$
(24)

where $\alpha_e(k)$ represents the reduced price of the electricity and $\beta_e(k)$ represents the higher price at which electricity is bought when the total average power consumption is greater than $\bar{p}(k)$. The expected run-time cost considered by the coordinated controller is

$$J_C(v|k) = \hat{c}_e(v|k) + \hat{c}_{\delta,s}(v|k) + \hat{c}_{QoS}(v|k).$$
(25)

The optimization problem solved by the coordinated controller can now be written as

$$\min_{M,S,\mathscr{D},\mathscr{T}_{ref}} \sum_{v=k}^{k+T} J_C(v|k)$$

s.t.

computational constraints

thermal constraints

control constraints.   (26)

4   Simulation Results

Let us consider the data center in Fig. 2. The data center comprises 32 racks and 4 CRAC units. Each rack contains 42 servers, which are cooled through a raised-floor architecture. Racks are grouped in eight zones. Two kinds of servers are placed

Fig. 2 An example of a data center layout. (© IEEE 2011), reprinted with permission

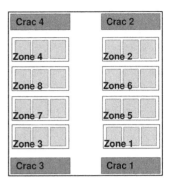

in the data center. The first kind is located in zones 1–4 and the latter is located in zones 5–8. Under the same workload, servers in zones 5–8 consume 10% less power than the other servers, but they are not cooled as efficiently as the servers in zones 1–4. CRAC units are identical and their efficiency increases quadratically with their output temperature. A local controller is embedded into each zone and it forces the zone to behave as postulated in the model discussed in Sect. 2. Simulations were developed using the TomSym language and KNITRO was used as numerical solver.[1] Data for the simulation are taken from [14].

Two cases are considered in the simulations. In the first case we compare the performance of the coordinated controller against the performance of the uncoordinated controller. For both of the controllers, we choose a high cost of dropping workload and a high cost for processing the workload with low QoS. In this a case, the two controllers are forced to process all of the incoming workload and also, they are forced to maximize the QoS. Therefore, the two control strategies only differ on the workload allocation strategy. In the second case, we consider the performance of two coordinated controllers. The first coordinated controller has a very large cost of dropping jobs and a large cost of processing the workload with low QoS. The second coordinated controller has a small cost of dropping jobs, whereas the cost of processing the workload with low QoS is still high. In this case, the main difference between the two controllers resides in the possibility, for the second coordinated controller, to drop jobs.

Figure 3 shows the workload arrival rate at the data center over time. Workload arrival rate represents a scaled version of the request rate arrived to an EPA server on August 30th 1995.[2] Figure 4 shows the reduced cost and the higher cost of the electricity over time. These data are used both for the first simulation case and for the second simulation case.

Let us focus on the first simulation case. Figure 5 shows the total data center power consumption obtained by the coordinated controller and by the uncoordinated controller. The time-varying power consumption threshold is also shown in Fig. 5.

---

[1]http://tomsym.com/ and http://www.ziena.com/knitro.html.

[2]Source: The Internet Traffic Archive http://ita.ee.lbl.gov/.

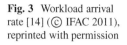
Fig. 3 Workload arrival rate [14] (© IFAC 2011), reprinted with permission

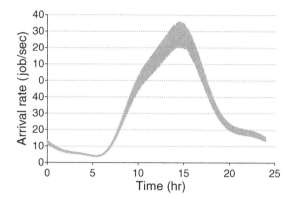

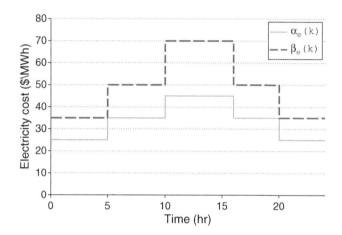

Fig. 4 Reduced and higher electricity costs [14] (© IFAC 2011), reprinted with permission

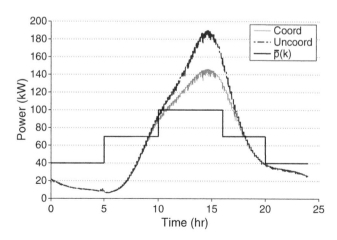

Fig. 5 Total data center power consumption [14] (© IFAC 2011), reprinted with permission

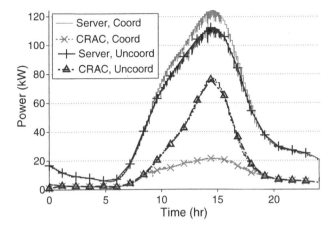

Fig. 6 Server and CRAC power consumption [14] (© IFAC 2011), reprinted with permission

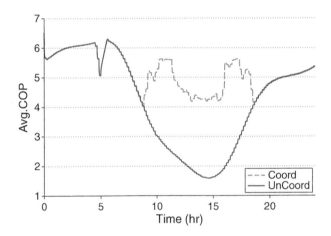

Fig. 7 Average coefficient of performance of the CRAC units

When the workload arrival rate is reduced, i.e. before about time 8 h and after about time 17 h, the two controllers obtain the same total data center power consumption. Also, during those periods of time, both of the two controllers allocate the workload on the energy-efficient servers, i.e. on the servers in zones 5–8. When the workload arrival rate increases, the coordinated controller tends to use the efficiently cooled servers. In such a case, the increase in power consumption, due to the use of the nonenergy-efficient servers, is compensated by the savings in the cooling power consumption. Figure 6 shows the total server power consumption and the total cooling power consumption obtained by the uncoordinated controller and by the coordinated controller. The average efficiency of the CRAC units over time is shown in Fig. 7. It can be seen that a little before time 5 h both the coordinated and the

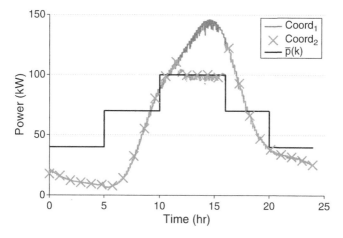

Fig. 8 Total data center power consumption [14] (© IFAC 2011), reprinted with permission

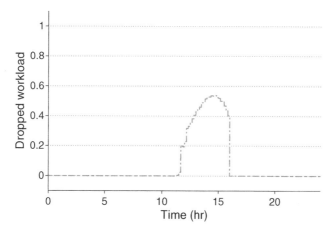

Fig. 9 Relative amount of dropped workload [14] (© IFAC 2011), reprinted with permission

uncoordinated controller have a decrease in the cooling efficiency. This happens because both of the controllers over-cool the data center just before the increase in the electricity cost, which happens at time 5 h. Over-cooling the data center allows both controllers to turn-off the CRAC units for the following 30 min. In the simulation, both of the controllers were able to maximize the QoS of the incoming workload and also they did not drop the incoming workload. Computational and thermal constraints were enforced by both controller throughout the simulation.

Let us focus on the second simulation case. Figure 8 shows the total data center power consumption obtained by the two coordinated controllers. Figure 9 shows the relative amount of workload dropped by the second coordinated controller. As shown in Fig. 8, the actions of the two controller only differ when the cost of the current is over a certain value. Furthermore, the second coordinated controller does not drop the workload as soon as the electricity prices increases, but it caps the power consumption of the data center only when both the average workload arrival rate and the electricity price exceed certain values.

5 Discussion

In this chapter, we discussed a model that includes both the computational and the physical characteristics of a data center, as well as their interactions. Additional elements of the computational or of the physical characteristics can be included in the model [13, 14]. Two cases have been considered in the simulation section. The first simulation compares the total data center power consumption obtained by an uncoordinated and by a coordinated controller, when they are forced to process all of the incoming workload at the highest QoS. The second simulation compares the behavior of two coordinated controllers. The two controllers differ on the way their cost function is formulated. The first controller has a large cost for dropping the workload, whereas the second controller has a reduced cost. The second simulation case shows that the choice about whether to drop or not the workload depends on the couple (electricity cost, workload arrival rate) and not uniquely on the electricity cost. As discussed in [14], further research is necessary to understand how an SLA with the power-grid should be stipulated in order to allow the data center to become an efficient smart node of the power-grid.

Acknowledgements The authors thank the research group at the Parallel Data Lab of Carnegie Mellon University Pittsburgh, PA, and the researchers at the Sustainable Ecosystem Research Group of HP Labs, Palo Alto, CA, for the helpful discussions and useful advices on modeling and control of data centers.

This material is based upon work partially supported by the National Science Foundation under Grant ECCS 0925964. Any opinions, findings, and conclusions or recommendations expressed in this material are those of the author(s) and do not necessarily reflect the views of the National Science Foundation.

References

1. Aydin H, Zhu D (2009) Reliability-aware energy management for periodic real-time tasks. IEEE Trans Comput 58(10):1382–1397
2. Bash CE, Patel CD, Sharma RK (2006) Dynamic thermal management of air cooled data centers. In: Proc. of the 10th Intersociety Conf. Thermal and Thermomechanical Phenomena in Electronic Systems (ITherm), San Diego, CA, pp 445–452
3. Breen TJ, Walsh EJ, Punch J, Shah AJ, Bash CE (2010) From chip to cooling tower data center modeling: Part I influence of server inlet temperature and temperature rise across cabinet. In: Proc. 12th IEEE Intersociety Conf. Thermal and Thermomechanical Phenomena in Electronic Systems (ITherm), Las Vegas, NV, pp 1–10

4. Choudhary P, Marculescu D (2009) Power management of voltage/frequency island-based systems using hardware-based methods. IEEE Trans Very Large Scale Integrat (VLSI) Syst 17(3):427–438
5. Fan X, Weber WD, Barroso LA (2007) Power provisioning for a warehouse-sized computer. In: International Symposium on Computer Architecture, San Diego, CA, pp 13–23, June 2007
6. Greco RA (2003) High density data centers fraught with peril. Slides, EYP Mission Critical Facilities Inc.
7. Jin H, Deng L, Wu S, Shi X, Pan X (2009) Live virtual machine migration with adaptive, memory compression. In: Proc. IEEE Int. Conf. Cluster Computing and Workshops CLUSTER, pp 1–10
8. Kim J, Yoo S, Kyung CM (2011) Program phase-aware dynamic voltage scaling under variable computational workload and memory stall environment. IEEE Transactions on Computer-Aided Design of Integrated Circuits and Systems 30(1):110–123
9. Ma F, Liu F, Liu Z (2010) Live virtual machine migration based on improved pre-copy approach. In: Proc. IEEE Int Software Engineering and Service Sciences (ICSESS) Conf, Beijing, pp 230–233
10. Moore J, Chase J, Ranganathan P, Sharma R (2005) Making scheduling "Cool": temperature-aware workload placement in data centers. In: USENIX Annual Technical Conference, Anaheim, California, pp 61–75, April 10–15
11. Parolini L, Sinopoli B, Krogh BH (2009) A unified thermal-computational approach to data center energy management. In: Fifth International Workshop on Feedback Control Implementation and Design in Computing Systems and Networks, San Francisco, California, USA, April 16, 2009
12. Parolini L, Garone E, Sinopoli B, Krogh BH (2010) A hierarchical approach to energy management in data centers. In: Proc. of the 49th IEEE Conference on Decision and Control, Atlanta, Georgia, USA, pp 1065–1070
13. Parolini L, Tolia N, Sinopoli B, Krogh BH (2010) A cyber-physical systems approach to energy management in data centers. In: First international conference on cyber-physical systems, Stockholm, Sweden, pp 168–177
14. Parolini L, Sinopoli B, Krogh BH (2011) Model predictive control of data centers in the smart grid scenario. In: Proc. of the 18th International Federation of Automatic Control (IFAC) World Congress, Vol 18, Part1, Milano, Italy
15. Patel CD, Sharma RK, Bash C, Beitelmal M (2002) Thermal considerations in cooling large scale high compute density data centers. In: ITHERM, San Diego, CA, USA, pp 767–776
16. Patterson M, Costello D, Grimm PF, Loeffler M (2007) Data center TCO; a comparison of high-density and low-density spaces. White paper, Intel Corporation
17. Patterson MK, Fenwick D (2008) The state of data center cooling. White paper, Intel Corporation
18. Sharma RK, Bash CE, Patel CD, Friedrich RJ, Chase JS (2005) Balance of power: Dynamic thermal management for internet data centers. IEEE Internet Comput 9:42–49
19. Tang Q, Mukherjee T, Gupta SKS, Cayton P (2006) Sensor-based fast thermal evaluation model for energy efficient high-performance data centers. In: Intelligent Sensing and Information Processing, Bangalore, India, pp 203–208
20. The Green Grid (2007) The green grid data center power efficiency metrics: PUE and DCiE. White paper, Technical Committee
21. Tolia N, Wang Z, Ranganathan P, Bash C, Marwah M, Zhu X (2009) Unified power and cooling management in server enclosures. In: InterPACK, San Francisco, California, USA, pp 721–730
22. US Environmental Protection Agency (2007) Report to congress on server and data center energy efficiency. Tech. rep., ENERGY STAR Program
23. Walsh EJ, Breen TJ, Punch J, Shah AJ, Bash CE (2010) From chip to cooling tower data center modeling: Part II influence of chip temperature control philosophy. In: Proc. 12th IEEE Intersociety Conf. Thermal and Thermomechanical Phenomena in Electronic Systems (ITherm), Las Vegas, NV, pp 1–7

Electrical Centrality Measures for Power Grids

Zhifang Wang, Anna Scaglione, and Robert J. Thomas

Abstract This chapter investigates measures of centrality that are applicable to power grids. Centrality measures are used in network science to rank the relative importance of nodes and edges of a graph. Here we define new measures of centrality for power grids that are based on its functionality. More specifically, the coupling of the grid network can be expressed as the algebraic equation $YU = I$, where U and I represent the vectors of complex bus voltage and injected current phasors; and Y is the network admittance matrix which is defined not only by the connecting topology but also by the network's electrical parameters and can be viewed as a complex-weighted Laplacian. We show that the relative importance analysis based on centrality in graph theory can be performed on power grid network with its electrical parameters taken into account. In the chapter the proposed electrical centrality measures are experimented with on the NYISO-2935 system and the IEEE 300-bus system. The centrality distribution is analyzed in order to identify important nodes or branches in the system which are of essential importance in terms of system vulnerability. A number of interesting discoveries are also presented and discussed regarding the importance rank of power grid nodes and branches.

1 Introduction

The electric power grid is one of the most critical infrastructures. The interconnectivity of the power grid enables long-distance transmission for more efficient system operation; however, it also allows the propagation of disturbances in the

Z. Wang (✉) • A. Scaglione
University of California, One Shields Ave, Davis, CA 95616, USA
e-mail: zfwang@ucdavis.edu; ascaglione@ucdavis.edu

R.J. Thomas
Cornell University, Ithaca, NY 14853, USA
e-mail: rjt1@cornell.edu

A. Chakrabortty and M.D. Ilić (eds.), *Control and Optimization Methods for Electric Smart Grids*, Power Electronics and Power Systems 3,
DOI 10.1007/978-1-4614-1605-0_12, © Springer Science+Business Media, LLC 2012

network. The nondecreasing frequency of large cascading blackouts in the United States reveals the existence of intrinsic weakness in the large electric power grids. Studies on the power grid system structures and vulnerability analysis have attracted many research efforts in the past years (see [1–6]).

It has been observed that the electric power grid network has a distinct topology. [7] provided a systematic investigation of the topological and electrical characteristics of power grid networks based on both available real-world and synthetic power grid system data: first, power grids have salient "small-world" properties, since they feature much shorter average path length (in hops) and much higher clustering coefficients than that of Erdös-Rényi random graphs with the same network size and sparsity[8]; second, their average node degree does not scale as the network size increases, which indicates that power grids are more complex than small-world graphs; in particular, it is found that the node degree distribution is well fitted by a mixture distribution coming from the sum of a truncated Geometric random variable and an irregular Discrete random variable. [4] highlighted that the topology robustness of a network is closely related to its node degree distribution. [6] investigated the deviation of the node degree distribution of power grids from a pure Geometric distribution and concluded that it substantially affects the topological vulnerability of a network under intentional attacks. That is, compared to a network with a pure Geometric node degree distribution, the power grid appears to be more vulnerable to intentional attacks when nodes with large degrees become first targets of the attack. Another less explored but equally important aspect that characterizes a power grid network is its distribution of line impedances, whose magnitude exhibits a heavy-tailed distribution, and is well fitted by a clipped double-Pareto-logNormal (dPlN) distribution [7].

With recent advances in network analysis and graph theory, many researchers have applied centrality measures to complex networks in order to study network properties and to identify the most important elements of a network. Various centrality measures have been defined and used to rank the relative importance of nodes and edges in a graph. [9] investigated the property of community structure in many types of networks in which network nodes are joined together in tightly knit groups, between which there are only looser connections. It also proposed a method for detecting such communities based on a generalized centrality measure of "edge betweenness" and experimented with the proposed algorithm on a collaboration network and a food web network. A different centrality measure of vertex by net flow of random walkers, which does not flow along the shortest paths was proposed in [10]. This centrality is known to be particularly useful for finding vertices of high centrality that do not happen to lie on the shortest paths and shown to have a strong correlation with degree and betweenness centrality.

Hines, et al. provided some insights into the topological and electrical structures of electrical power grids, pointing out the differences of the topology of power grids from that of Erdös-Rényi random graphs, Watts-Strogatz "small-world" networks [8], or "scale-free" networks (see [11, 12]). An "electrical centrality measure" was proposed to be calculated based on the impedance matrix Z^{bus} and used their centrality measure to explain why in power grids a few number of highly connected bus failures can cause cascading effects, which was referred to as "scale-free network" vulnerability. However, as shown in Sect. 4, the proposed electrical centrality measure was incorrectly defined and the corresponding analysis on the vulnerability was in fact misleading.

Rajasingh et al. [13] developed a formula to compute the betweenness centrality for a regular grid network. Torres and Anders [14] discussed different methods of graph theory, mainly the topology centralities, for ranking the relative importance of substations in a power grid and illustrated the procedure on a synthetic 5-node test system . Gorton et al. [15] proposed a new method for contingency selection based on the concept of graph edge betweenness centrality, which can be used for the contingency analysis of large-scale power grids. And [16] improved the model of power flow distribution in the power grid network. That is, the flows are not concentrated only along the shortest paths; instead, they are randomly distributed on all the paths between nodes, as random walks. A centrality measure was defined accordingly for the transmission network analysis and was applied to the IEEE 14-bus system.

In this chapter, we investigate the measures of centrality that are applicable to power grids and their meanings. We define new measures of centrality for a power grid that are based on its functionality. More specifically, the coupling of the grid network can be expressed as the algebraic equation $YU = I$, where U and I represent the bus voltage and injected current vectors; and Y is the network admittance matrix, which is defined not only by the connecting topology but also its electrical parameters and can be viewed as a complex-weighted Laplacian. A simple transformation allows us to compute a weighted adjacency matrix from the weighted Laplacian Y. Therefore, the relative importance analysis based on centrality in graph theory can be performed on power grid network with its electrical parameters taken into account. In this chapter, we present and discuss some interesting discoveries on the importance rank of the power grid nodes and branches which are obtained from the experiments on the NYISO system and the IEEE 300-bus system. It is found that when electrical parameters are incorporated into the centrality definition, the distribution of some centrality measures becomes very different from the original ones, which were based on the topological structure alone; and with some proposed electrical centrality measures, a large amount of system centrality can reside in a small number of nodes in the system. These findings will help us to identify the electrically critical components of the system for the vulnerability analysis and to search for ways of enhancing system robustness.

2   System Model

The power network dynamics are coupled by its network equation

$$YU = I, \tag{1}$$

where U and I represent the bus voltage and injected current vectors; and Y is the network admittance matrix, which is determined not only by the connecting topology but also its electrical parameters. Given a network with n nodes and m links (which may also be referred to as "buses and branches (or lines)" in power grid analysis; or "vertices and edges" in graph theory and network analysis), each link $l = (s,t)$ between nodes s and t has a line impedance $z_{pr}(l) = r(l) + jx(l)$, where $r(l)$ is the resistance and $x(l)$ the reactance. Usually for high-voltage transmission network, $x(l) \gg r(l)$, i.e., its reactance, with $l = 1, 2, \ldots, m$, dominates. The line admittance is obtained from the inverse of its impedance, i.e.,

$$
\begin{aligned}
y_{pr}(l) &= g(l) + jb(l) \\
&= 1/z_{pr}(l).
\end{aligned}
\tag{2}
$$

Assume that a unit current flows along the link $l = (s,t)$ from node s to t; then the caused voltage difference between the ends of the link equals $\Delta u = U(s) - U(t) = z_{pr}(l)$ or equivalently $\Delta u = 1/y_{pr}(l)$. Therefore, $z_{pr}(l)$ can be interpreted as the "electrical" distance between node s and t and $y_{pr}(l)$ reflects the "coupling" strength between the two end nodes.

The line-node incidence matrix A of the network A, with size $m \times n$, can be written as

$$
A : \begin{cases} A(l,s) = & 1 \\ A(l,t) = & -1 \\ A(l,k) = & 0, \text{ with } k \neq s \text{ or } t. \end{cases}
\tag{3}
$$

The Laplacian matrix L of the network, with size $n \times n$, can be obtained as

$$
L = A^{T} A
\tag{4}
$$

with

$$
L(s,t) = \begin{cases} -1, & \text{if there exists link } s - t, \quad \text{for } t \neq s \\ k, & \text{with } k = -\sum_{t \neq s} L(s,t), \quad \text{for } t = s \\ 0, & \text{otherewise}, \end{cases}
\tag{5}
$$

with $s,t = 1, 2, \ldots, n$.

The network admittance matrix Y of the network, with size $n \times n$, can be obtained as

$$
Y = A^{T} diag(\mathbf{y}_{pr}) A,
\tag{6}
$$

where $\mathbf{y}_{pr}$ is the line admittance vector. The entries in Y are as follows:

$$
\begin{cases} Y(s,t) = -y_{pr}(s,t), & \text{link } s - t \text{ exists, for } t \neq s \\ Y(s,s) = \sum_{t \neq s} y_{pr}(s,t), & \text{for } t = s \\ Y(s,t) = 0, & \text{otherewise.} \end{cases}
\tag{7}
$$

with $s,t = 1, 2, \cdots, n$.

A close comparison of the matrix structures of L and Y uncovers some interesting discoveries. It is known that the Laplacian matrix L fully describes the topology of a network; while the network admittance matrix Y not only contains information about the system topology but also contains information about its electrical coupling. The off-diagonal entries of Y, $Y(s,t)$ equals the line admittance of the link between node s and t (with a "−" sign), whose magnitude reflects the coupling strength between the two nodes. The diagonal entries of the Laplacian L represent the total number of links connecting each node with the rest of the network. Whereas a diagonal entry of Y represents the total coupling capability one node has with the rest of the network. Therefore the network admittance matrix Y can be viewed as a complex-weighted Laplacian; and the Laplacian L can be equivalent to a "flat" network admittance matrix, which assumes all the links in the network have the same line impedance (with a common proportional factor). These analogies are very important in the sense that, as shown in the next section, they will help extend the centrality measures which were originally defined on a network topological structure to be more appropriately defined on the electrical structure.

3   Centrality Definitions and Extensions

Centrality measures are used in network science to rank the relative importance of vertices and edges in a graph. Within graph theory and network analysis, there are various measures of the centrality of a vertex or an edge. In the following subsections, we examine the definitions of four widely used measures of centrality, i.e., degree centrality, betweenness, closeness, and eigenvector centrality. Then we discuss how to extend the definitions to corresponding "electrical" measures of centrality for power grids.

3.1   Degree Centrality

The simplest centrality for a vertex is its node degree, i.e. , the total number of edges incident upon a node. This centrality represents the connectivity of a node to the rest of the network and reflects the immediate chance for a node to exert its influences to the rest of the network or to be exposed to whatever is flowing through the network, such as disturbances, shared information, power or traffic flows, or even a virus. For a graph $G := (V, E)$ wit n vertices, where V represents the set of vertices and E the set of edges, given its Laplacian L, the degree centrality of a vertex or node v is defined as

$$C_d(v) = \frac{deg(v)}{n-1} = \frac{L(v,v)}{n-1}, \tag{8}$$

where $n - 1$ is used as a normalization factor.

For a node in the power grid network, its connectivity or "coupling" with the rest of the network is not only related to how many links it connects but also related to the connecting strength of each link; and the admittance of each link just reflects this coupling strength. Therefore, by using the analogy between the Laplacian L and the network admittance matrix Y, we define the electrical degree centrality $C_{d_Y}(v)$ as

$$C_{d_Y}(v) = \frac{\|Y(v, v)\|}{n - 1}. \tag{9}$$

3.2 Eigenvector Centrality

Eigenvector centrality is a measure of the importance of a node in a network according to its adjacency matrix.

Given a network $G := (V, E)$, its adjacency matrix $\mathscr{A}$, one eigenvalue λ, and the corresponding eigenvector x satisfy

$$\lambda x = \mathscr{A} x. \tag{10}$$

The centrality of a node v is defined as the v-th entry of the eigenvector x corresponding the largest eigenvalue $\lambda_{\max}$:

$$C_e(v) = x_v = \frac{1}{\lambda_{\max}} \sum_{j=1}^{n} \mathscr{A}(v, j) x_j. \tag{11}$$

Clearly, the centrality of node v is proportional to the sum of the centralities of all its neighboring nodes. The definition chooses the eigenvector corresponding to the largest eigenvalue $\lambda_{\max}$ in order to guarantee all the centrality scores, which are all the entries in the eigenvector, to be positive (see Perron-Frobenius Theorem [17]).

As stated in Sect. 2, the off-diagonal entries in the network admittance matrix Y can be viewed as the connectivity strength between neighboring nodes in the network. Therefore, just as we extract the adjacency matrix from the Laplacian, $\mathscr{A} = -L + D(L)$, we can retrieve the complex-weighted electrical adjacency matrix as

$$\mathscr{A}_Y = -Y + D(Y), \tag{12}$$

where $D(\cdot)$ represent the diagonal matrix retrieved from the original matrix.

After performing the eigen-analysis of $\mathscr{A}_Y$, we take the magnitude of the entries of the eigenvector as the centrality measure.

$$C_{e_Y}(v) = \|x_v\| = \left\| \frac{1}{\lambda_{\max}} \sum_{j=1}^{n} \mathscr{A}_Y(v, j) x_j \right\|. \tag{13}$$

3.3 Closeness Centrality

Compared to degree centrality, the definition of closeness centrality is more sophisticated. It is the mean geodesic distance (i.e., the shortest path length in hops) between a vertex v and all the other vertices reachable from it:

$$C_c(v) = \frac{\sum_{t \in V \setminus v} d_G(v, t)}{n - 1}, \tag{14}$$

with $d_G(v, t)$ being the shortest path length between vertices v and t. Obviously, definition (14) in fact measures how "far away" a node is from the rest of the network instead of its closeness. Therefore, some researchers define closeness to be the reciprocal of this quantity, to make the name more appropriate [18], that is

$$C_c(v) = \frac{n - 1}{\sum_{t \in V \setminus v} d_G(v, t)}. \tag{15}$$

The shortcoming of the definition of closeness centrality is that it does not properly reflect how vulnerable is a network to becoming disconnected. In fact, the shortest path length $d_G(v, t)$ between vertices v and t turns out to be infinity if the network is disconnected and there is not a path between the two vertices. As a result, the definitions of (14) and (15) can only be applied to connected networks. In order to incorporate the disconnectivity and to more conveniently measure the network vulnerability, Danalchev modified in [19] the definition of closeness to be:

$$C_c(v) = \sum_{t \in V \setminus v} 2^{-d_G(v,t)}. \tag{16}$$

Because the power grid networks we are interested in are connected topologies, undirected with neither multiple links nor self-loops, definition (15) above is still suitable for our purposes.

In all definitions above, the distance along a path from vertex v to t is measured as in "hops". That is, $d_G(v, t)$ equals the total number of hops along the path:

$$d_G(v, t) = \sum_{(i,j) \in E \cap path(v \rightarrow t)} 1. \tag{17}$$

To adapt the definition of closeness centrality to nodes in a power grid network, we define the "electrical distance" between the nodes as $d_Z(v, t)$, which is counted in "electrical hops" as

$$d_Z(v, t) = \left\| \sum_{(i,j) \in E \cap path(v \rightarrow t)} Z_{pr}(i, j) \right\|, \tag{18}$$

where $Z_{pr}(i, j)$ is the line impedance of the link (i, j). Therefore, the corresponding closeness centrality based on electrical distance is defined as

$$C_{cz}(v) = \frac{n-1}{\sum_{t \in V \backslash v} d_Z(v, t)}.$$ (19)

It is worth noting that the line impedance Z_{pr} in a power grid is a complex number, i.e., $Z_{pr} = R + jX$, where R is the resistance and X the reactance. Therefore, the electrical distance $d_Z(v, t)$ is in fact a complex number though one could take the magnitude of $d_Z(v, t)$ to make a more "real" distance measure. According to the definition in (18), the line impedance of each link is in fact used as the edge weight in the search of the shortest path between v and t. However, this will complicate the shortest-path search algorithm because one cannot compare or add up two complex-number weights so straightforward as with real-number weights. On the other hand, it is known that for the high-voltage transmission network in a power grid the reactance X is usually the dominant component of a line impedance, whereas R only takes a trivial value which in many cases can even be neglected. Therefore for the purpose of simplicity, we can only take the reactance X as the edge weights and hence the distance evaluation can be approximated by real numbers.

3.4 Vertex and Edge Betweenness Centrality

Vertex Betweenness is one of the most widely used centrality measure. It was first suggested by Freeman (1977) in [20]. This measure reflects the influence of a node over the flow of information between other nodes, especially in cases where information flow over a network primarily follows the shortest available path.

Given a undirected graph $G(V, E)$, the betweenness of a node v is defined as the number of shortest paths between pairs of other vertices that run through v:

$$C_b(v) = \frac{\sum_{s \neq v \neq t \in V} \sigma_{st}(v)/\sigma_{st}}{(n-1)(n-2)/2},$$ (20)

where σ_{st} the number of shortest paths from s to t and $\sigma_{st}(v)$ is the total number from the mentioned paths that pass through vertex v.

In order to find which edges in a network are most between other pairs of vertices, Girvan and Newman [9] generalize Freemans betweenness centrality to edges and define the edge betweenness of an edge as the number of shortest paths between pairs of vertices that run along it. If there is more than one shortest path between a pair of vertices, each path is given equal weight such that the total weight of all of the paths is unity. Note that the normalization factor of edge betweenness is different from that of vertex betweenness.

$$C_b(e) = \frac{\sum_{s \neq t \in V} \sigma_{st}(e)/\sigma_{st}}{n(n-1)/2}.$$ (21)

Obviously, vertices or edges that occur on many shortest paths have higher betweenness than those that do not. It is found that removal of the nodes or edges with larger betweenness will put the network at higher risk to become disconnected.

Definitions (20) and (21) are based on the shortest path counted in hops. Using the shortest electrical path counted in electrical hops as (18), we can define the electrical betweenness for nodes and edges in power grid networks, which are denoted as $C_{bZ}(v)$ and $C_{bZ}(e)$, respectively.

4   Some Discussion on the Electrical Centrality Proposed in [11]

Hines and Blumsack (2008) proposed an "electrical centrality measure," which is calculated based on the Z^{bus} matrix [11]. The Z^{bus} matrix is the inverse of the Y matrix, which, unlike the Y matrix, is a nonsparse (dense) matrix. That is, $Z^{bus} = Y^{-1}$. This centrality measure has also been adopted by other researchers [14].

The principle of assigning this centrality measure is restated as follows (see [11, 12]):

It was claimed that the equivalent electrical distance between node k and l is thus given by the magnitude of the (k, l) entry in the Z^{bus} matrix. Smaller $\|Z^{bus}_{k,l}\|$ corresponds to a shorter electrical distances and a stronger coupling between these node hence a larger propensity for power to flow between these nodes. From the nonzero off-diagonal entries in the Z^{bus} matrix, select only the same number of "links" (k, l) as that in Y, which have the shortest electrical distances (corresponding to the entries with the least $\|Z^{bus}_{k,l}\|$'s). With these newly selected "links", one has a different "electrical" topology from the original network. Then node degrees, counted on the newly formed electrical topology, are defined as the electrical betweenness, or the electrical degrees.

With further analysis based on the electrical degrees, it is found that although a power grid topologically does not have the properties of a scale-free network[1], electrically, there exist a number of highly connected nodes in the electrical topology (obtained from Z^{bus}) similar to what would be expected from a scale-free network. That is, the electrical node degree distribution has a "fat" tail. This finding, according to [11], explains the relatively high vulnerability of power grid to failures at some "hub" buses (i.e., intentional attacks to high-degree nodes).

However, a deeper analysis of the impedance matrix Z^{bus} reveals that the definition of the electrical betweenness (degree) and the corresponding analysis in [11, 12] are potentially misleading. The problem is that the entry of the impedance matrix, $Z^{bus}_{k,l}$, does not represent the mutual electrical distance between node k and l.

---

[1] i.e., its node degree distribution does not have a "fat" power-law tail.

In fact, from the network equation $U = Z^{bus} I$, we can learn the physical meaning of the entry of $Z_{k,l}^{bus}$: when only node k has a unit injected current, the voltage increases occurring at node l and k are $Z_{k,l}^{bus}$ and $Z_{k,k}^{bus}$, respectively. That is

$$\Delta U(l) = Z_{k,l}^{bus},$$

$$\Delta U(k) = Z_{k,k}^{bus}. \tag{22}$$

And usually $\|Z_{k,k}^{bus}\| > \|Z_{k,l}^{bus}\|$, which means the Z^{bus} matrix is diagonally dominant. Only if $Z_{k,l}^{bus}$ is close to $Z_{k,k}^{bus}$, which happens when $Z_{k,l}^{bus}$ has a relatively large magnitude, there exists a strong coupling between the two nodes. Whereas the mutual electrical distance between node k and l equals the voltage difference caused by an injected unit current at node k and an output unit current from node l. That is

$$Z_{k \leftrightarrow l} = Z_{k,k}^{bus} + Z_{l,l}^{bus} - 2Z_{k,l}^{bus}, \tag{23}$$

which also implies that a large entry of $Z_{k,l}^{bus}$, not a small one as shown in [11], gives a small mutual electrical distance and a strong coupling between the two nodes.

On the other hand, for a connected power grid network with grounding branches, its network admittance matrix Y is nonsingular therefore the inverse of Y, which is Z^{bus}, exists. The network equation $YU = I$ and $U = Z^{bus} I$ are equivalent to each other. That is, the matrices Y and Z^{bus} in fact describe the same electrical and topological structure of the power grid. It is, however, misleading to interpret Z^{bus} as something that reveals a structurally different "new" topology with denser connections or with a different node degree distribution.

5   Experiment Results

Based on defined centrality measures in Sect. 3, we perform experiments on the NYISO-2935 system and the IEEE-300 system. The NYISO-2935 system is a representation of the New York Independent System Operator's transmission network containing 2,935 nodes and 6,567 links, with an average node degree $\langle k \rangle = 4.47$ and an average shortest path length $\langle l \rangle = 16.43$ (in hops). The IEEE-300 system is a synthesized network from the New England power system and has a topology with 300 nodes and 409 links, with $\langle k \rangle = 2.73$ and $\langle l \rangle = 9.94$. We evaluate the relative importance for the nodes and lines according to different centrality measures and normalize the results to make sure the sum of all the vertex or edge centralities in a system equals 1.0. Then we analyze their distribution and correlation, and identify the most "significant" nodes in each system.

Figure 1 compares the distribution of degree centrality and eigenvector centrality for the nodes in the NYISO system and shows the correlation between different centrality measures. It can be seen that when electrical parameters are incorporated into the centrality definition, the distribution of the degree centrality and the

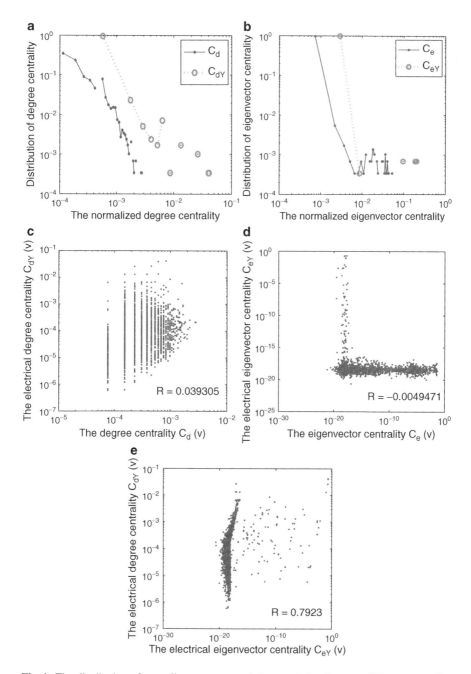

Fig. 1 The distribution of centrality measures and the correlation between different centrality measures – the NYISO-2935 System: (**a**) degree centrality, C_d and C_{dY}; (**b**) eigenvector centrality, C_e and C_{ey}; (**c**) correlation between C_d and C_{dY}; (**d**) correlation between C_e and C_{ey} (**e**) correlation between C_{dY} and C_{eY}

eigenvector centrality become very different from the original ones, which are based on the topological structure alone. The correlation between the "topology" centralities and the corresponding electrical centralities is very weak. However, there is a strong correlation between the two electrical centralities C_{d_Y} and C_{e_Y}.

Table 1 ranks the first ten most important nodes in the NYISO system according to different type of centrality measures. The bottom row gives the total centrality of each group of most important nodes. It shows that the degree centrality distributes quite "flatly" among the nodes, because the first ten most important nodes sum only to 2% of system's total centrality. However, when the electrical parameter is taken into account, a large amount of weight can shift into a small number of nodes in the system, e.g., the first ten most important nodes based on the electrical eigenvector centrality take more than 99.2% of the system's total centrality. This reveals that the eigenvector centrality vector is in fact a very sparse one. It also shows that C_{d_Y} and C_{e_Y} are kind of consistent with each other in the sense that they are able to locate a very similar group of most important nodes in the system (a 60% overlap) which, however, is quite different from the group identified by the topological centralities C_d and C_e.

Figure 2 compares the distribution of different types of centrality for the nodes in the IEEE-300 system. It can be seen that including electrical parameters causes a large change in the distribution of the degree centrality and the eigenvector centrality (see (a) and (b)). However, the effect of electrical parameters is not so evident in the distribution of the closeness and betweenness centrality (see (c), (d) and (e)). Figure 3 displays the correlation between each pair of topological and electrical centrality measures. It shows that strong correlations exist between the electrical and topological closeness centrality and between the electrical and topological betweenness centrality as well.

Table 2 ranks the first ten most important nodes in the IEEE-300 system according to eigenvector centrality and closeness centrality. The bottom row gives the total centrality of each group of most important nodes. The same concentration of the eigenvector centrality can be observed here as is observed with the NYISO system. It is interesting to notice that both the topological and electrical closeness centrality distribute very "flatly" among the nodes and the groups of important nodes located by the pair have a 60% overlap. Table 3 ranks the first ten most important nodes and branches in the IEEE-300 system according to betweenness centrality. The bottom row gives the total centrality of each group of most important nodes or branches. Similarly, we see that both the topological and electrical betweenness centrality distribute quite "flatly" among the nodes or edges and each pair of centrality measures are very consistent with each other. This indicates that the inclusion of electrical parameters in the betweenness centrality does not cause much difference in identifying the most important nodes or branches. In fact, there are 80% overlap between the nodes groups and 70% overlap between the branch groups, respectively, according to the topological and electrical betweenness measures. The reason for this strong consistence in identification of most important components or the strong correlations observed in the closeness or betweenness centrality distribution can be interpreted as follows: both closeness and betweenness centrality

Table 1 The list of nodes in the NYISO-2935 system that carry the most significant centralities

Ranking order	C_d	C_e	C_{dy}	C_{ey}
1	2,773	2,622	9	9
2	2,622	2,614	8	8
3	2,516	2,606	1,312	15
4	2,511	2,619	17	17
5	2,894	2,605	15	11
6	2,728	2,613	234	12
7	2,435	2,608	233	84
8	2,614	2,601	12	29
9	2,481	2,610	11	27
10	2,409	2,609	1,518	26
Total centrality	0.021471	0.40277	0.22631	0.99227

measures are defined based on the shortest path count; the transmission network of power grids is sparsely connected, therefore the shortest path between any two nodes tends to include a large number of hops (e.g., on average about 10 hops in the IEEE-300 system and 16 hops in the NYISO system); as a result, the differences among individual line impedances average out in the evaluation of centrality measure based on the shortest path count.

6 Conclusions

This chapter investigates measures of centrality that are applicable to power grids. New measures of centrality are defined for a power grid that are based on its electrical characteristics rather than just its topology. More specifically, the coupling of the grid network can be expressed as the algebraic equation $YU = I$, where U and I represent the bus voltage and injected current vectors; and Y is the network admittance matrix which is defined not only by the connecting topology but also by its electrical parameters and can be seen as a complex-weighted Laplacian. We show that the relative importance analysis based on centrality in graph theory can be performed on power grid network with its electrical parameters taken into account.

Based on defined centrality measures, experiments are performed on the NYISO-2935 system and the IEEE-300 system and obtained some interesting discoveries on the importance rank of the power grid nodes and lines. It has been found that when the electrical parameters are incorporated into the centrality definition, the distribution of the degree centrality and the eigenvector centrality become very different from the original ones, which are based on the topological structure alone. With the electrical degree centrality and the electrical eigenvector centrality a large amount of centrality can reside in a small number of nodes in the system and help locate of a quite different group of important nodes. From the

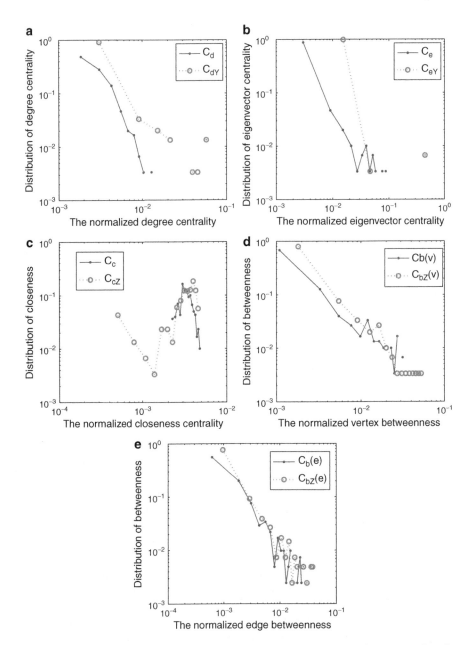

Fig. 2 The distribution of centrality measures – the IEEE-300 system: (**a**) degree centrality, C_d and C_{dY}; (**b**) eigenvector centrality, C_e and C_{eY}; (**c**) closeness centrality, C_c and C_{cZ}; (**d**) vertex betweenness centrality, $C_b(v)$ and $C_{bZ}(v)$; (**e**) edge betweenness centrality, $C_b(e)$ and $C_{bZ}(e)$

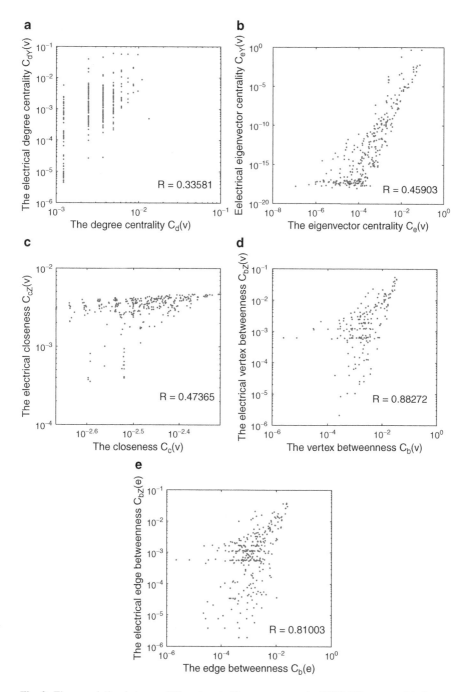

Fig. 3 The correlation between different centrality measures – the IEEE-300 system: (**a**) degree centrality, C_d and C_{dY}; (**b**) eigenvector centrality, C_e and C_{eY}; (**c**) closeness centrality, C_c and C_{cZ}; (**d**) vertex betweenness centrality, $C_b(v)$ and $C_{bZ}(v)$; (**e**) edge betweenness centrality, $C_b(e)$ and $C_{bZ}(e)$

Table 2 The list of nodes in the IEEE-300 system that carry the most significant centralities

Ranking order	C_e	C_{eY}	C_c	C_{cZ}
1	31	266	36	36
2	35	31	40	40
3	32	270	16	16
4	15	35	39	33
5	43	32	4	28
6	27	34	35	4
7	75	43	15	3
8	74	75	3	7
9	34	15	68	129
10	44	74	31	39
Total centrality	0.51401	0.99818	0.046687	0.046551

Table 3 The List of nodes and branches in the IEEE-300 system that carry the most significant betweenness centralities

Ranking order	$C_b(v)$	$C_{bZ}(v)$	$C_b(e)$	$C_{bZ}(e)$
1	3	36	68→40	40→36
2	40	40	40→36	36→16
3	68	3	16→4	16→4
4	36	16	4→3	4→3
5	16	4	129→3	68→40
6	31	68	129→109	129→3
7	109	109	266→31	129→109
8	4	129	36→16	7→3
9	266	7	52→39	36→35
10	129	35	173→68	35→31
Total centrality	0.27490	0.36570	0.20520	0.29115

experimental results of IEEE-300 system, it is shown that the effect of including electrical parameters is not so evident in changing the distribution of the closeness and betweenness centrality. And strong correlations exist between the electrical and topological closeness centrality; and between the electrical and topological betweenness centrality as well. The proposed electrical centrality measures can locate a quite different group of "important" nodes or vertices in the system from the topological centrality measures. However, more tests and analysis need to be done in order to validate the proposed measures, to further understand their physical meaning, and to apply the findings to search for ways of enhancing system robustness.

References

1. Carreras BA, Lynch VE, Dobson I, Newman DE (2002) Critical points and transitions in an electric power transmission model for cascading failure blackouts. Chaos 12(4):985–994
2. Albert R, Albert I, Nakarado GL (2004) Structural vulnerability of the north american power grid. Phys Rev E 69:1–4
3. Crucittia P, Latorab V, Marchioric M (2004) A topological analysis of the italian electric power grid. Physica A 338:92–97
4. Rosas-Casals M, Valverde S, Solé R (2007) Topological vulnerability of the european power grid under erros and attacks. Intl J Bifurcat Chaos 17(7):2465–2475
5. Wang Z, Thomas RJ, Scaglione A (2008) Generating random topology power grids. In: 41st Annual Hawaii International Conference on System Sciences (HICSS-41), Big Island, Hawaii
6. Wang Z, Scaglione A, Thomas RJ (2010) The node degree distribution in power grid and its topology robustness under random and selective node removals. In: 1st IEEE International Workshop on Smart Grid Communications (ICC'10SGComm), Cape Town, South Africa
7. Wang Z, Scaglione A, Thomas RJ (2010) Generating statistically correct random topologies for testing smart grid communication and control networks. IEEE trans on Smart Grid 1(1):28–39
8. Watts DJ, Strogatz SH (1998) Collective dynamics of 'small-world' networks. Nature 393:440–442
9. Girvan M, Newman MEJ (2002) Community structure in social and biological networks. PNAS 99(12):7821–7826
10. Newman MEJ (2005) A measure of betweenness centrality based on random walks. Soc Network 27:3954
11. Hines P, Blumsack S (2008) A centrality measure for electrical networks. In: the 41st Hawaii International Conference on System Sciences, Big Island, Hawaii
12. Hines P, Blumsack S, Sanchez EC, Barrows C (2010) The topological and electrical structure of power grids. In: the 43rd Hawaii International Conference on System Sciences, Kauai, Hawaii
13. Rajasingh I, Rajan B, D FI (2009) Betweeness-centrality of grid networks. In: International Conference on Computer Technology and Development, Kota Kinabalu, Malaysia
14. Torres A, Anders G (2009) Spectral graph theory and network dependability. In: Fourth International Conference on Dependability of Computer Systems, Brunow, pp 356–363
15. Gorton I, Huang Z, Chen Y, Kalahar B, Jin S, Chavarra-Miranda D, Baxter D, Feo J (2009) A high-performance hybrid computing approach to massive contingency analysis in the power grid. In: International Conference on e-Science and Grid Computing, Oxford, UK, pp 277–283
16. Zio E, Piccinelli R (2010) Randomized flow model and centrality measure for electrical power transmission network analysis. Reliab Eng Syst Saf 95(4):379–385
17. Horn R, Johnson C (1990) Matrix Analysis, chapter 8. Cambridge University Press, Cambridge
18. Sabidussi G (1966) The centrality index of a graph. Psychometrika 31(4):581–603
19. Dangalchev C (2006) Residual closeness in networks. Phisica A 365(2):556–564
20. Freeman LC (1977) A set of measures of centrality based on betweenness. Sociometry 40(1):35–41

Part III
Communications and Control

Optimal Charging Control for Plug-In Electric Vehicles

Zhongjing Ma, Duncan Callaway, and Ian Hiskens

Abstract This chapter discusses strategies to coordinate charging of autonomous plug-in electric vehicles (PEVs). The chapter briefly reviews the state of the art with respect to grid level analyses of PEV charging, and frames PEV coordination in terms of whether they are centralized or decentralized and whether they are optimal or near-optimal in some sense. The bulk of the chapter is devoted to presenting centralized and decentralized cost-optimizing frameworks for identifying and coordinating PEV charging. We use a centralized framework to show that "valley filling" charge patterns are globally optimal. Decentralized electricity cost minimizing frameworks for PEV charging can be framed in the context of non-cooperative dynamic game theory and are related to recent work on mean field and potential games. Interestingly, in this context it can be difficult to achieve a Nash equilibrium (NE) if electricity price is the sole objective. The decentralized algorithm discussed in this chapter introduces a very small penalty term that damps unwanted negotiating dynamics. With this term, the decentralized algorithm takes on the form of a contraction mapping and, in the infinite system limit, the NE is unique and the algorithm will converge to it under relatively loose assumptions.

Z. Ma
School of Automation, Beijing Institute of Technology, and the Key Laboratory of Complex System Intelligent Control and Decision (Beijing Institute of Technology),
Ministry of Education, Beijing, China
e-mail: mazhongjing@bit.edu.cn

D. Callaway (✉)
Energy and Resources Group, University of California, Berkeley, CA, USA
e-mail: dcal@berkeley.edu

I. Hiskens
Department of Electrical Engineering and Computer Science, University of Michigan, MI, USA
e-mail: hiskens@umich.edu

A. Chakrabortty and M.D. Ilić (eds.), *Control and Optimization Methods for Electric Smart Grids*, Power Electronics and Power Systems 3,
DOI 10.1007/978-1-4614-1605-0_13, © Springer Science+Business Media, LLC 2012

1 Introduction

This chapter addresses the intersection of two trends in the modernization of power systems: vehicle electrification and flexible loads. Pure electric vehicles and plug-in hybrids—referred to generally as plug-in electric vehicles or PEVs—are being offered or developed by a number of manufacturers, and depending on market conditions they could become prevalent in a short period of time. PEVs hold appeal for a number of reasons, including operating cost, environmental impact, energy security, and power system economics. However, electric power system operation may become more challenging if they are large in number and not properly coordinated. For example, if a large number of PEVs began charging around the time most people finish their evening commute, a new demand peak could result, possibly requiring substantial new generation capacity and ramping capability [1].

However, PEV charging is potentially very flexible, because the time required to charge a vehicle will in many cases be much less than the time the vehicle is plugged in and available to charge. Therefore, with the appropriate coordination schemes, undesirable demand peaks may be avoidable. In this chapter, we will cover recent efforts to study how to coordinate PEVs to achieve charging patterns that are manageable or optimal at the system level. Although the main approach we describe does not explicitly require a system operator or central decision-maker, it is necessary to aggregate from and disseminate information to PEVs so that they can locally compute their desired charging patterns in the context of anticipated system-level conditions.

PEV coordination strategies can be divided into two categories:

1. In *centralized strategies*, a central operator dictates precisely when and at what rate every individual PEV will charge. Decisions could be made on the basis of system-level considerations only, or they could factor vehicle-level preferences, for example desired charging interval, final state of charge (SOC), and budget. These strategies could be further distinguished by whether they attempt to identify charge patterns that are in some way optimal, or instead follow rules-of-thumb that seek to achieve aggregate charging patterns that are reasonably close to optimal. Examples of papers that focus on centralized strategies include [2–4].
2. *Decentralized or distributed strategies* allow individual PEVs to determine their own charging pattern. Vehicle charging decisions could, for example, be made on the basis of time-of-day or electricity price. The outcome of a decentralized approach may or may not be optimal, depending on the information and methods used to determine local charging patterns. Care must be taken to ensure charging strategies cannot inadvertently synchronize the responses of large numbers of PEVs, as the resulting abrupt changes in aggregate demand could potentially destabilize grid operations [5]. Examples of papers that focus on decentralized strategies include [6, 7].

A number of recent studies have explored the potential impacts of high penetrations of PEVs on the power grid [8–11]. In general, these studies assume that

PEV charging patterns "fill the valley" of night-time demand. However, none of the aforementioned studies address the issue of how to coordinate PEV charging patterns to achieve the valley-filling pattern. The authors' recent efforts [12–14], discussed throughout this chapter, focus on this coordination problem. The work applies concepts from optimal control and game theory to formally understand the nature of "optimal" PEV population charging patterns and the ability of decentralized charging algorithms to create optimal patterns.

This chapter has strong connections to the broader control and game theoretic literature. In particular, we will study problems that fall within the class of *potential games* identified by Monderer and Shapley [15]. In fact, the work is mathematically equivalent to routing and flow control games in telecommunications, where networks of parallel links are congested [16]. From this point of view, PEV charging games are conceptually similar to network games [17]. There is an extensive body of literature in this area, including research on *centralized mechanisms* to produce Nash equilibria [18, 19] and *decentralized or distributed mechanisms* [20–23]. Noncooperative game theory is commonly used to understand imperfect competition among generators in electricity markets [24–27]. Some game theoretic work has been done to understand demand-side behavior in the face of dynamic pricing tariffs [28, 29].

Although strongly connected, this chapter is distinct from the references in the previous paragraph in that (a) it examines multi-period demand-side behavior with a local energy constraint applied to the total energy consumed across all periods, and (b) it analyzes the problem from a decentralized perspective. In fact, the work is strongly related to the Nash certainty equivalence principle (or mean-field games), proposed by Huang et al. [30, 31] in the context of large-scale games for sets of weakly coupled linear stochastic control systems. Furthermore, in the infinite agent limit, the results we will present correspond to the classic Wardrop equilibrium [32].

2 Models for PEV Charging and Electricity Markets

2.1 PEV Charging Model and Centralized Control

We consider charging control of a significant PEV population of size N over charging horizon $\mathscr{T} \triangleq \{0, ..., T-1\}$, where T denotes the terminal charging instant. The population of PEVs is denoted $\mathscr{N} \triangleq \{1, ..., N\}$. For an individual PEV n, we adopt the notation of Table 1. The SOC of PEV n at instant t is given by $0 \leq x_{nt} \leq 1$, and the SOC dynamics are described by the simplified model,

$$x_{n,t+1} = x_{nt} + \frac{\alpha_n}{\beta_n} u_{nt}, \quad t \in \mathscr{T}, \tag{1}$$

with initial SOC x_{n0}, battery size $\beta_n > 0$, and charger efficiency $\alpha_n \in (0, 1]$.

Table 1 List of key symbols

$\mathcal{N}$	PEV populations
$\mathcal{T}$	PEV charging interval
x_{nt}	SOC of PEV n at instant t
u_{nt}	Charging rate of PEV n at instant t
β_n	Battery capacity of PEV n
α_n	Charging efficiency of PEV n
$\mathbf{u}$	Collection of charging controls of PEV population
$\Sigma(\mathbf{u}_n)$	Total charging energy delivered to PEV n
$\bar{\mathbf{u}}_t$	Average charging rate of PEV population at instant t
$\mathbf{u}_{-n}$	Collection of local charging controls except PEV n
$\mathcal{U}_n$	Set of admissible charging controls of PEV n
d_t	(Inelastic) non-PEV demand at instant t
$\mathbf{u}_n^*(\mathbf{z})$	Optimal charging control of PEV n w.r.t. $\mathbf{z}$

The charging control trajectory of PEV n, denoted $\mathbf{u}_n \equiv (u_{nt}; t \in \mathcal{T})$, is an *admissible charging control*, if it belongs to the set

$$\mathcal{U}_n(\gamma_n) \triangleq \{\mathbf{u}_n : u_{nt} \geq 0 \text{ and } \Sigma(\mathbf{u}_n) = \gamma_n\}, \tag{2}$$

where $\Sigma(\mathbf{u}_n) \triangleq \sum_{t \in \mathcal{T}} u_{nt}$, $\gamma_n \triangleq \frac{\beta_n}{\alpha_n}(1 - x_{n0})$.

For simplicity, we have omitted upper capacity constraints on individual PEV charge rates, for the reason that in all optimal strategies we observe, charging levels are well below any reasonable constraint. Moreover, we are assuming that transmission network constraints are nonbinding. The set of admissible charging control for the entire PEV population is denoted by $\mathcal{U}$. It follows from the SOC dynamics described in (1) that $x_{nT} = 1$ for any admissible control $\mathbf{u}_n \in \mathcal{U}_n(\gamma_n)$.

Subject to an admissible charging control, we define the cost associated with delivering the total system demand by,

$$\mathbb{J}(\mathbf{u}) = \sum_{t \in \mathcal{T}} p(r_t^N(\mathbf{u}_t)) \left(D_t^N + \sum_{n \in \mathcal{N}} u_{nt} \right), \tag{3}$$

where $\mathbf{u}_t \equiv (u_{nt}; n \in \mathcal{N})$ denotes the collection of PEV charging rates at time t, $p(r_t^N(\mathbf{u}_t))$ is the electricity charging price at instant t, and D_t^N is the total inelastic non-PEV demand at instant t. Because we are neglecting spatial constraints, the price $p(r_t^N(\mathbf{u}_t))$ is independent of location. We further assume that price is determined by the ratio between total demand and total generation capacity, so

$$r_t^N(\mathbf{u}_t) \triangleq \frac{1}{C^N} \left(D_t^N + \sum_{n \in \mathcal{N}} u_{nt} \right), \tag{4}$$

where C^N denotes the total generation capacity. The importance of the dependence of C and D_t on N is discussed later in this section.

This definition of price is consistent with a "real-time pricing" tariff. Because the generator unit commitment process is dynamic and depends on demand in all periods of the optimization horizon, price in any single hour will in practice be influenced by demand in all hours, and the price function could vary from one hour to another. In contrast, the formulation in (3) assumes price is independent of time and demand in other time steps; we are working to relax this assumption in ongoing research.

In what follows we will examine asymptotic properties in the large N limit. To preserve key properties at that limit, we assume that non-PEV demand and total generation capacity vary with the number of PEVs and make the following asymptotic assumptions as PEV population size approaches infinity,

$$\lim_{N \to \infty} \frac{D_t^N}{N} = d_t, \qquad \lim_{N \to \infty} \frac{C^N}{N} = c. \tag{5}$$

Equation (5) implies that larger power systems, with greater capacity and base demand, are required to support large numbers of PEVs. Direct substitution into (4) gives, $r_t \triangleq \lim_{N \to \infty} r_t^N(\mathbf{u}_t) = \frac{1}{c}(d_t + \bar{\mathbf{u}}_t) \triangleq r_t(\bar{\mathbf{u}}_t)$, where $\bar{\mathbf{u}}_t \triangleq \lim_{N \to \infty} \frac{1}{N} \sum_{n \in \mathcal{N}} u_{nt}$.

We define valley-filling charging as follows:

$$\bar{\mathbf{u}}_t^{\text{vf}} = \max\{0, \vartheta - d_t\}, \tag{6}$$

where ϑ is a constant determined by the total energy required to charge the PEV population. In words, in hours when $\vartheta > d_t$ this strategy chooses total PEV demand such that system-wide demand is equal to ϑ; otherwise, PEV demand is zero. The value of ϑ uniquely determines the total energy supplied for charging.

Ma et al. [14] establishes that if $p(r_t(\bar{\mathbf{u}}_t))$ is convex and increasing on $\bar{\mathbf{u}}_t$, then given non-PEV demand d_t, the optimal charging strategy $(\bar{\mathbf{u}}_t^*; t \in \mathcal{T})$ is valley-filling. As we will discuss later, the decentralized negotiation algorithm in Sect. 3 achieves this same minimum-cost strategy in the case of a homogeneous population of PEVs.

2.1.1 Spatially Distributed Charging

The valley-filling charging strategy given by (6) is optimal when all loads experience a common price. However, when network constraints become binding, for example when a transmission line encounters its flow limit, price differences will often develop between regions. Under such circumstances, valley filling is no longer optimal. We note though that normally the network is relatively lightly loadedovernight, when charging occurs. Flow limits are therefore least likely to be encountered during this period. It follows that valley filling remains a good approximation for optimal charging.

2.2   Decentralized Charging Control

In the rest of the chapter, we will study a decentralized game-based charging control strategy for large PEV populations. Consider the local cost function $\mathscr{J}_n$ for an individual PEV n subject to a collection of charging controls $\mathbf{u}$,

$$\mathscr{J}_n(\mathbf{u}) \triangleq \sum_{t \in \mathscr{T}} p\big(r_t(\overline{\mathbf{u}}_t)\big) u_{nt}. \tag{7}$$

The locally optimal charging control problem with respect to a fixed collection of controls $\mathbf{u}_{-n}$ is given by the minimization,

$$\mathbf{u}_n^*(\mathbf{u}_{-n}) \triangleq \underset{\mathbf{u}_n \in \mathscr{U}_n(\gamma_n)}{\operatorname{argmin}} \mathscr{J}_n(\mathbf{u}_n; \mathbf{u}_{-n}), \tag{8}$$

where $\mathbf{u}_{-n} \triangleq \{\mathbf{u}_m; m \in \mathscr{N}, m \neq n\}$, in other words $\mathbf{u}_{-n}$ denotes the collection of control strategies of all PEVs except the n-th. If a minimizing function exists, it will be referred to as an optimal control law for the local charging control problem.

A collection of PEV strategies $\{\mathbf{u}_n^*; n \in \mathscr{N}\}$ is a *Nash equilibrium (NE)* if each PEV n cannot benefit by unilaterally deviating from its individual strategy $\mathbf{u}_n^*$, i.e.,

$$\mathscr{J}_n(\mathbf{u}_n^*; \mathbf{u}_{-n}^*) \leq \mathscr{J}_n(\mathbf{u}_n; \mathbf{u}_{-n}^*), \quad \text{for all } \mathbf{u}_n \in \mathscr{U}_n, \text{ and all } n \in \mathscr{N}.$$

Assuming technical conditions of the price function p and inequality (11) are satisfied, $\overline{\mathbf{u}}^*(\mathbf{z})$ is a contraction mapping with respect to $\mathbf{z}$ [14]. This result motivates the following *iterative algorithm* for computing the NE associated with the decentralized charging control system:

(S1)   A utility (or system operator) broadcasts a non-PEV demand forecast ($d_t; t \in \mathscr{T}$) to all PEVs.
(S2)   Each PEV proposes a charging strategy that is cost minimizing with respect to an aggregate PEV demand, broadcast by the utility.
(S3)   The utility collects all the optimal charging strategies proposed in (S2), and updates the aggregate PEV demand, which is then rebroadcast to all PEVs.
(S4)   Repeat (S2) and (S3) until the optimal strategies proposed by all PEVs no longer change.   □

At convergence, the collection of optimal charging strategies is an NE. When the charging period occurs, each PEV implements its proposed optimal strategy.

At each iteration, every PEV optimizes its strategy relative to $\mathbf{u}$ determined in the previous iteration. As we will show in later in Fig. 3, charging intervals with high PEV demand at one iteration tend to induce low PEV demand at the following iteration, and vice versa. This occurs because PEVs move their charging requirements from expensive to inexpensive intervals; the resulting changes in demand reduce the marginal electricity price in the previously expensive intervals and raise the price in previously inexpensive intervals. This establishes an oscillatory pattern from one iteration to the next, preventing convergence to any NE.

To mitigate these oscillations, we modify the local cost function (7) to include a quadratic term that penalizes the deviation of an individual control strategy from the population (or "mass") average:

$$\mathscr{J}_n(\mathbf{u}) \triangleq \sum_{t \in \mathscr{T}} \left(p\left(\frac{1}{c}(d_t + \bar{\mathbf{u}}_t) \right) u_{nt} + \delta(u_{nt} - \bar{\mathbf{u}}_t)^2 \right), \tag{9}$$

where δ determines the magnitude of the penalty for deviating from the mass average. It will be shown that the presence of the δ-term ensures convergence to a unique collection of locally optimal charging strategies that is an NE. This NE only coincides with the globally optimal strategy (6) when all PEVs are identical (homogeneous). Nevertheless, we will see that the cost added due to this term can be quite small compared with the electricity price p.

3 Implementation for Infinite Systems

This section presents conditions for which Algorithm (S1–S4) yields an NE when (9) is used in Step S2. Although we assume $N \to \infty$ to obtain the results, numerical examples in Sect. 5 show that the results apply to finite systems as well.

Theorem 1. *A collection of charging strategies* $\mathbf{u} \in \mathscr{U}$ *for an infinite population of PEVs is an NE, if (i) for all* $n \in \mathscr{N}$, $\mathbf{u}_n$ *minimizes the cost function,*

$$J_n(\mathbf{u}_n; \mathbf{z}) = \sum_{t \in \mathscr{T}} \left(p\left(\frac{1}{c}(d_t + z_t) \right) u_{nt} + \delta(u_{nt} - z_t)^2 \right) \tag{10}$$

with respect to $\mathbf{z}$, *and (ii)* $z_t = \bar{\mathbf{u}}_t$, *for all* $t \in \mathscr{T}$, *i.e.,* $\mathbf{z}$ *is the average of each individual PEV's optimal strategy. A proof is presented in [14].*

As mentioned earlier, PEV charging games are related to the Nash certainty equivalence principle (also known as mean-field games). The key similarity is that individual agents are not influenced by other individuals, rather they only consider the so-called "mass average". In the case of PEV charging, the mass average influences individuals via the electricity price, which we specify as a function of the average charging trajectory $\bar{\mathbf{u}}_t$.

We now turn to identifying conditions that the system converges a unique NE for infinite population systems following (S1–S4) as below.

Theorem 2. *Assume $p(r)$ is continuous on r; then there exists an NE for the infinite population charging control system. Moreover following (S1–S4) the system converges to a unique NE if $p(r)$ is continuously differentiable and strictly increasing on r, and*

$$\frac{1}{2c} \max_{r \in [r_{min}, r_{max}]} \frac{dp(r)}{dr} \leq \delta \leq \frac{a}{c} \min_{r \in [r_{min}, r_{max}]} \frac{dp(r)}{dr} \tag{11}$$

for some a in the range $\frac{1}{2} < a < 1$, where r_{min} and r_{max} denote, respectively, the minimum and maximum possible r over the charging interval $\mathcal{T}$, subject to the admissible charging control set $\mathcal{U}_n(\gamma_n)$. A proof is presented in [14].

The theorem given above establishes a sufficient condition for a range of values of δ for which the system will converge to a unique NE. It may be difficult to satisfy this condition over a large demand range $[r_{min}, r_{max}]$, especially if the higher demand value approaches the capacity limits of the system (the supply curve is usually very steep there). However, for overnight charging this is less likely to be a binding factor. Moreover, as we will show using a numerical example in Sect. 5, convergence remains possible if condition (11) is slightly violated.

Having proven the implementation of the NE, here we establish that the associated NE is *nearly valley filling*. The following key points capture the various cases:

1. For any pair of charging instants, the one with the smaller non-PEV base demand is assigned a larger charging rate (for individual PEVs as well as for the average over all PEVs), and possesses an equal or lower total demand.
2. The total demand, consisting of aggregate PEV charging load together with non-PEV demand, is the same for all charging subintervals with strictly positive PEV charging rates. This is also true of the demand obtained by summing non-PEV demand with the charging load of any individual PEV.

These properties are formalized in the following theorem.

Theorem 3. *[14] Suppose that the set of charging trajectories $\mathbf{u}^* \equiv \{\mathbf{u}_n^*; n \in \mathcal{N}\}$ is an NE and $p(r)$ is strictly increasing on r; then for all $\delta > 0$, $\mathbf{u}_n^*$ and $\overline{\mathbf{u}}^*$ satisfy:*

(i) $\overline{u}_t^* \geq \overline{u}_s^*, \ d_t + \overline{u}_t^* \leq d_s + \overline{u}_s^*, \ u_{nt}^* \geq u_{ns}^*,$ *when $d_t \leq d_s$, with $t, s \in \mathcal{T}$*
$$\tag{12a}$$

(ii) $d_t + \overline{u}_t^* = \theta, \ d_t + u_{nt}^* = \theta_n,$ \hspace{2em} *for some $\theta, \theta_n > 0$, with $t \in \widehat{\mathcal{T}}$,*
$$\tag{12b}$$

where $\widehat{\mathcal{T}} \equiv \{t \in \mathcal{T}; u_{nt}^* > 0 \quad \text{for all } n \in \mathcal{N}\}$. A proof is presented in [14].

Remark. In case of homogeneous PEV populations, each of the individual optimal strategies $\mathbf{u}_n^*$ is coincident with their average strategy $\bar{\mathbf{u}}^*$. It follows that the properties of the NE specified in (12) are equivalent to,

$$\bar{\mathbf{u}}_t^* = \max\{0, \theta - d_t\}, \quad \text{for some } \theta > 0, \tag{13}$$

which is the normalized form of the optimal valley-filling strategy given in (6). In other words, in the case of a homogeneous PEV population, the NE coincides with the charging strategy given by centralized control, and is therefore globally optimal.

4 Extensions

4.1 Generalized Price Function and User Preferences

The local objective functions in (7) and (9) can be extended to include a number of important factors, including (a) an electricity price that depends on time as well as instantaneous demand and (b) user-specific charging preferences. The price function could vary with time as a result of changes in generator availability. The charging preferences could relate to desired times to initiate and complete charging, or battery state of health considerations. For example, in the latter case, battery chemistry may be such that it is best to avoid full SOC for as long as possible. A generalized local objective function can be written as follows:

$$\mathscr{J}_n(\mathbf{u}) \triangleq \sum_{t \in \mathscr{T}} \left(p_t \left(\frac{1}{c}(d_t + \bar{\mathbf{u}}_t) \right) u_{nt} + \delta(u_{nt} - \bar{\mathbf{u}}_t)^2 + \phi_{nt} u_{nt} + \pi_{nt} x_{nt} \right), \tag{14}$$

where ϕ_{nt} and π_{nt} reflect user-specific preferences on charging time and battery SOC as a function of time. By (1), $x_{n,t+1} = x_{nt} + \frac{\alpha_n}{\beta_n} u_{nt}$, and therefore $\pi_{nt} x_{nt}$ can be written in terms of $u_{n\tau}, \tau < t$ and constant parameters. Consequently, a broad range of user preference can be accomodated with a user- and time-specific parameter multiplied by u_{nt}. In this case, one can show that Theorems 1 and 2 still hold and therefore (S1–S4) will still converge to an NE. However, it is not possible to prove that the NE is a valley-filling result and in general it will not be.

4.2 Terminal Cost

User preferences could also be used to justify relaxing the "fully charged" constraint. This is because users may be willing to drive fewer miles or, in the case of plug-in *hybrid* electric vehicles, use a liquid fuel instead of electricity for a portion of their mileage. In this case, (9) becomes

$$\mathscr{J}_n(\mathbf{u}) \triangleq \sum_{t \in \mathscr{T}} \left(p \left(\frac{1}{c}(d_t + \overline{\mathbf{u}}_t) \right) u_{nt} + \delta(u_{nt} - \overline{\mathbf{u}}_t)^2 \right) + h_n(|u_n|_1), \qquad (15)$$

where $\mathbf{u}_n \in \widehat{\mathscr{U}}_n(\gamma_n) \triangleq \{\mathbf{u}_n : u_{nt} \geq 0 \text{ and } \Sigma(\mathbf{u}_n) \leq \gamma_n\}$, which is different from (2), and h_n is its local terminal cost function. The decentralized optimization problem now becomes more difficult to solve analytically; whereas the optimal total demand θ_n in (13) can be determined from the "shadow value" of the constraint γ_n, a slightly more involved process, which we omit here for the sake of space, is required in (15).

4.3 Adapting to Forecast Errors

The decentralized charging control strategy presented in Sect. 2.2 computes optimal control trajectories with respect to non-PEV demand predicted at the beginning of the charging interval. Adjustments cannot be made for subsequent disturbances or changes in the forecast. As a consequence, depending on the accuracy of the forecast at the beginning of the interval, the charging strategy $\mathbf{u}^*$ attained at convergence of (S1)–(S4) may actually result in suboptimal charging cost for each PEV. In order to improve the robustness of the decentralized charging strategy, the control algorithm can be revised in a manner that is consistent with model predictive control (MPC) methods. The modifications basically entail repetition of (S1)–(S4) as the charging interval progresses, with an updated demand forecast being used at each subsequent time step. In accordance with MPC, the control signals calculated for the first time instant are implemented by the PEVs, and the process repeats. Further details, including a comparison between the open-loop process and the MPC formulation, can be found in [13].

5 Numerical Examples

A range of examples will be used in this section to illustrate the main results of the chapter. In particular, we will consider the conditions for convergence of Algorithm 1, and explore the nature of valley filling.

The examples use a non-PEV demand profile, see solid curve marked with dots in Fig. 1, which shows the load of the Midwest ISO region for a typical summer day during 2007. It is assumed that the total generation capacity is 1.2×10^8 kW. Furthermore, the simulations are based on assuming $N = 10^7$, which corresponds to roughly 30% of vehicles in the MISO footprint. This gives $c^N \triangleq \frac{1}{N} C^N = 12$ kW, where the superscript N indicates finite population size. Also, we define $d_t^N \triangleq \frac{1}{N} D_t^N$.

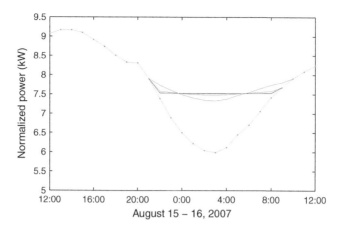

Fig. 1 Convergence of Algorithm 1 for a homogenous PEV population, with $\delta = 0.012$ (© IEEE 2010), reprinted with permission

The following PEV population parameters will be used for all the examples. All PEVs have an initial SOC of 15%, i.e., $x_{n0} = 0.15$ for all n, and 85% charging efficiency, i.e., $\alpha_n = 0.85$ for all n. The charging interval $\mathscr{T}$ covers the 12-hour period from 8:00 pm on one day to 8:00 am on the next. The continuously differentiable and strictly increasing price function

$$p(r) = 0.15r^{1.5} \text{ \$/kWh} \tag{16}$$

is used in all cases.

Other parameters, such as PEV battery size β_n and the tracking cost parameter δ, are specified within each of the examples.

5.1 Homogeneous PEV Populations

This section considers the computation of the NE for a homogeneous population of PEVs, each of which possesses an identical battery size of $\beta^n = 10\,\text{kWh}$. First, it may be verified from Fig. 1 that $r_{min} = \min_{t \in \mathscr{T}} \{d_t^N\}/c^N \approx 0.5$. To determine r_{max}, we assume the entire energy requirement γ_n is delivered over a single time step, so $r_{max} = (\max_{t \in \mathscr{T}} \{d_t^N\} + \gamma_n)/c^N \approx 1.5$. Referring to (16), this gives

$$\frac{1}{2c^N} \max_{[r_{min}, r_{max}]} \frac{\mathrm{d}p(r)}{\mathrm{d}r} = 0.012 \leq \frac{a}{c^N} \min_{[r_{min}, r_{max}]} \frac{\mathrm{d}p(r)}{\mathrm{d}r} = 0.013a,$$

which can be satisfied for some a in the range $\frac{1}{2} < a < 1$. Therefore, a tracking parameter δ exists such that condition (11) holds.

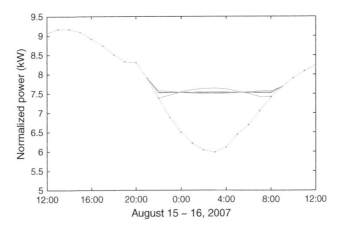

Fig. 2 Convergence of Algorithm 1 for a homogenous PEV population, with $\delta = 0.007$. This violates the condition (11) (© IEEE 2010), reprinted with permission

Figure 1 provides simulation results for the decentralized computation algorithm of Sect. 3, for the homogeneous PEV population of this section. The tracking parameter $\delta = 0.012$ was used for this case. Each line in the figure corresponds to an iterate of the algorithm. We observe that convergence to the NE (shown by the solid flat curve) is achieved in a few cycles. This NE is clearly the globally optimal valley-filling strategy.

The condition on δ established in Sect. 3 for the implementation of NE is sufficient, but not necessary. Figure 2 confirms this. Here,

$$\delta = 0.007 < \frac{1}{2c^N} \max_{[r_{\min}, r_{\max}]} \frac{dp(r)}{dr} = 0.012,$$

yet the system still converges to the same valley-filling solution as in Fig. 1. As δ decreases, however, the process eventually ceases to converge. This can be observed in Fig. 3, which shows the iterations when $\delta = 0.003$. In order to avoid an unreasonably high charging rate during the non-PEV demand valley, we have constrained the charging rate to a maximum of 3 kW. This constraint does not effect the convergence property of the algorithm.

5.2 Heterogeneous PEV Populations

PEV populations are heterogeneous if vehicles do not have identical charging requirements. To examine optimal charging outcomes for heterogeneous populations, we constructed a simplified case with PEVs having one of three charging energy requirements: 10 kWh, 15 kWh or 20 kWh. We further assumed that the number

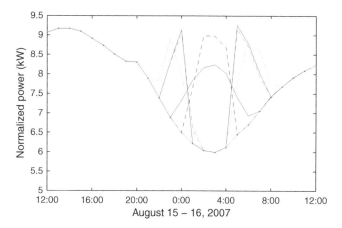

Fig. 3 Nonconvergence of Algorithm 1 for a homogenous PEV population, with $\delta = 0.003$. This significantly violates the condition (11) (© IEEE 2010), reprinted with permission

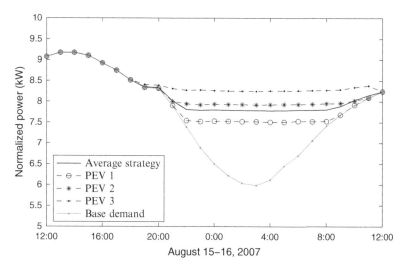

Fig. 4 Convergence of Algorithm 1 for a heterogeneous PEV population (© IEEE 2010), reprinted with permission

of PEVs in each group accounted for about 50, 30, and 20% of the population respectively. We can verify that there exists some a such that $\frac{1}{2} < a < 1$. Figure 4 shows the results for this heterogeneous case, with tracking parameter $\delta = 0.0125$. In particular, the dashed curves with marks show the optimal charging strategies for the 1st, 2nd, and 3rd class of PEVs, and the solid curve provides the average demand value across the entire population. Notice that this curve of average demand is flat between 10 pm and 8 am, where all PEVs are charging, in accordance with the valley-fill property of (12).

6 Conclusions and Future Research

This chapter discussed a class of decentralized charging control problems for large populations of PEVs. These problems are formulated as large-population games on a finite charging interval, and under the right circumstances existence, uniqueness and optimality of the NE can be established. Following algorithm (S1–S4), subject to certain mild conditions, the large-population charging games will converge to a unique NE, which is either globally optimal (for homogeneous populations) or nearly globally optimal (for the heterogeneous case).

We introduced a number of extensions to the basic problem, including the addition of network constraints, user preferences (including a terminal cost), and forecast uncertainty. These extensions are largely open areas of research and are likely to be critically important to supporting large numbers of PEVs in power systems.

References

1. Lemoine D, Kammen D, Farrell A (2008) An innovation and policy agenda for commercially competitive plug-in hybrid electric vehicles. Environ Res Lett 3, 014003
2. Goebel C (2011) The business value of ICT-controlled plug-in electrical vehicle charging. In: Submitted to 32nd International Conference on Information Systems, Shanghai, China, pp 803–810
3. Saber A, Venayagamoorthy G (2010) Efficient utilization of renewable energy sources by gridable vehicles in cyber-physical energy systems. IEEE Syst J 4(3):285–294
4. Waraich R, Galus M, Dobler C, Balmer M, Andersson G, Axhausen K (2009) Plug-in hybrid electric vehicles and smart grid: Investigations based on a micro-simulation. Technical report, Institute for Transport Planning and Systems, ETH Zurich
5. Callaway D, Hiskens I (2011) Achieving controllability of electric loads. Proc IEEE 99(1):184–199
6. Rotering N, Ilic M (2010) Optimal charge control of plug-in hybrid electric vehicles in deregulated electricity markets. EEE Trans Power Syst 26(3):1021–1029
7. Galus M, Andersson G (2008) Demand management of grid connected plug-in hybrid electric vehicles (PHEV). In: Proceedings of the IEEE Energy 2030 Conference, Altanta, GA, pp 1–8
8. Denholm P, Short W (2006) An evaluation of utility system impacts and benefits of optimally dispatched plug-in hybrid electric vehicles. Technical Report NREL/TP-620-40293, National Renewable Energy Laboratory
9. Rahman S, Shrestha G (1993) An investigation into the impact of electric vehicle load on the electric utility distribution system. IEEE Trans Power Deliv 8(2):591–597
10. Koyanagi F, Uriu Y (1997) Modeling power consumption by electric vehicles and its impact on power demand. Electr Eng Jpn 120(4):40–47
11. Koyanagi F, Inuzuka T, Uriu Y, Yokoyama R (1999) Monte Carlo simulation on the demand impact by quick chargers for electric vehicles. In: Proceedings of the IEEE Power Engineering Society Summer Meeting, Istanbul, Turkey, vol 2, pp 1031–1036
12. Ma Z, Callaway D, Hiskens I (2010) Decentralized charging control for large populations of plug-in electric vehicles. In: Proceedings of the 49th IEEE Conference on Decision and Control, Atlanta, GA, pp 206–212
13. Ma Z, Hiskens I, Callaway D (2011) Distributed MPC methods in charging control of large populations of plug-in electric vehicles. In: Proceedings of the IFAC World Congress, Milano, Italy, pp 10493–10498

14. Ma Z, Callaway D, Hiskens I (2011) Decentralized charging control of large populations of plug-in electric vehicles. in revision, IEEE Transactions on Control Systems Technology, to appear

15. Monderer D, Shapley L (1996) Potential games. Game and Econ Behav 14:124–143

16. Altman E, Başar T, Jiménez T, Shimkin N (2001) Routing into two parallel links: Game-theoretic distributed algorithms. J Parallel Distr Comput 61(9):1367–1381

17. Altman E, Boulogne T, El-Azouzi R, Jimenez T, Wynter L (2006) A survey on networking games in telecommunications. Comput Oper Res 33:286–311

18. Christodoulou G, Mirrokni V, Sidiropoulos A (2006) Convergence and approximation in potential games. In: Proceedings of the 23rd Symposium on Theoretical Aspects of Computer Science (STACS), Dresden, Germany, pp 349–360

19. Even-Dar E, Kesselman A, Mansour Y (2001) Convergence time to Nash equilibirum in load balancing. ACM Trans Comput Logic 2(3):111–132

20. Berenbrink P, Friedetzky T, Goldberg L, Goldberg P, Hu Z, Martin R (2006) Distributed selfish load balancing. In: Proceedings of the 17th Annual ACM-SIAM Symposium on Discrete Algorithms, Miami, Florida, pp 354–363

21. Blum A, Even-Dar E, Ligett K (2006) Routing without regret: On convergence to Nash equilibria of regret-minimizing algorithms in routing games. In: Proceedings of the 25th ACM Symposium on Principles of Distributed Computing, Denver, CO, pp 45–52

22. Fischer S, Olbrich L, Vocking B (2008) Approximating Wardrop equilibria with finitely many agents. Distr Comput 21(2):129–139

23. Fischer S, Racke H, Vocking B (2006) Fast convergence to Wardrop equilibria by adaptive sampling methods. In: Proceedings of the 38th Annual ACM Symposium on Theory of Computing, Seattle, WA, pp 653–662

24. Borenstein S, Bushnell J, Wolak F (2002) Measuring market inefficiencies in California's restructured wholesale electricity market. Am Econ Rev 92(5):1376–1405

25. Bushnell J, Mansur E, Saravia C (2008) Vertical arrangements, market structure, and competition: An analysis of restructured US electricity markets. Am Econ Rev 98(1):237–266

26. Hobbs B (2002) Linear complementarity models of Nash-Cournot competition in bilateral and POOLCO power markets. IEEE Trans Power Syst 16(2):194–202

27. Cunningham L, Baldick R, Baughman M (2002) An empirical study of applied game theory: Transmission constrained Cournot behavior. IEEE Trans Power Syst 17(1):166–172

28. Pettersen E (2004) Managing end-user flexibility in electricity markets. Fakultet for samfunnsvitenskap og teknologiledelse, Institutt for industriell økonomi og teknologiledelse, NTNU

29. Philpott A, Pettersen E (2006) Optimizing demand-side bids in day-ahead electricity markets. IEEE Trans Power Syst 21(2):488–498

30. Huang M, Caines P, Malhamé R (2003) Individual and mass behaviour in large population stochastic wireless power control problems: Centralized and Nash equilibrium solutions. In: Proceedings of the 42th IEEE International Conference on Decision and Control, Maui, Hawaii, pp 98–103

31. Huang M, Caines P, Malhamé R (2007) Large-population cost-coupled LQG problems with non-uniform agents: Individual-mass behaviour and decentralized epsilon-Nash equilibria. IEEE Trans Automat Contr 52(9):1560–1571

32. Wardrop J (1952) Some theoretical aspects of road traffic research. In: Proceedings of the Institution of Civil Engineers, Part 2, pp 1: 325–78

Risk Analysis of Coordinated Cyber Attacks on Power Grid

Siddharth Sridhar, Manimaran Govindarasu, and Chen-Ching Liu

Abstract The supervisory control and data acquisition (SCADA) network provides adversaries with an opportunity to perform coordinated cyber attacks on power system equipment as it presents an increased attack surface. Coordinated attacks, when smartly structured, can not only have severe physical impacts, but can also potentially nullify the effect of system redundancy and other defense mechanisms. This chapter proposes a vulnerability assessment framework to quantify risk due to intelligent coordinated attacks, where risk is defined as the product of probability of successful cyber intrusion and resulting power system impact. The cyber network is modeled using Stochastic Petri Nets and the steady-state probability of successful intrusion into a substation is obtained using this. The model employs a SCADA network with firewalls and password protection schemes. The impact on the power system is estimated by load unserved after a successful attack. The New England 39-bus system is used as a test model to run Optimal Power Flow (OPF) simulations to determine load unserved. We conduct experiments creating coordinated attacks from our attack template on the test system and evaluate the risk for every case. Our attack cases include combinations of generation units and transmission lines that form coordinated attack pairs. Our integrated risk evaluation studies provide a methodology to assess risk from different cyber network configurations and substation capabilities. Our studies identify scenarios, where generation capacity, cyber vulnerability, and the topology of the grid together could be used by attackers to cause significant power system impact.

S. Sridhar (✉)
Iowa State University, Ames, IA, USA
e-mail: sridhar@iastate.edu

M. Govindarasu
Department of Electrical and Computer Engineering, Iowa State University, Ames, IA, USA
e-mail: gmani@iastate.edu

C.-C. Liu
Washington State University, Pullman, WA, USA
e-mail: liu@eecs.wsu.edu

A. Chakrabortty and M.D. Ilić (eds.), *Control and Optimization Methods for Electric Smart Grids*, Power Electronics and Power Systems 3,
DOI 10.1007/978-1-4614-1605-0_14, © Springer Science+Business Media, LLC 2012

1 Introduction

Critical Infrastructures are complex physical and cyber-based systems that form the lifeline of modern society, and their reliable and secure operation is of paramount importance to national security and economic vitality. The Commission on Critical Infrastructure Protection (CCIP) has identified telecommunications, electric power systems, natural gas and oil, banking and finance, transportation, water supply systems and emergency services among the critical infrastructure systems [1]. The cyber system forms the backbone of the nation's critical infrastructures, which means that a compromise of the cyber system could have significant impacts on the reliable and safe operations of the physical systems that rely on it. The focus of this chapter is on the cyber security of electric power infrastructure.

The electric power grid, as of today, is a highly automated network. A variety of communication networks are interconnected to the electric grid for the purpose of sensing, monitoring, and control. These communication networks are closely associated with the supervisory control and data acquisition (SCADA) systems for system operation functions and real-time control of the power system.

The SCADA network connects the control center, generating stations, load substations, and other corporate offices. Supervisory control enables the operator at the control center to manipulate settings on remote equipment such as circuit breakers. Data acquisition enables the operator to monitor and obtain data from remote equipment. The SCADA system communicates with *Remote Terminal Units* (RTU) located in substations to obtain monitoring data and issue control commands. RTUs are microprocessor-based devices that in turn communicate with local devices such as digital relays in substations and gather information for the control center. Information gathered includes status indicators for devices such as circuit breakers and analog measurements from measuring devices such as current and potential transformers. Figure 1 is a schematic of the network connection between the control center and devices in a substation.

Modern RTUs are Internet Protocol (IP) accessible [2]. They use an Ethernet interface and support TCP/IP-based protocols. Some of the IP compatible protocols used by RTUs are DNP3.0, ICCP, UCA2.0 and Modbus [3]. These offers several advantages to system operators, most importantly, the communication infrastructure is readily available from Internet Service Providers. The fact that SCADA operations occasionally use the same infrastructure as the Internet exposes them to cyber attacks.

Threats to SCADA are broadly classified into *Internal Threats* and *External Threats* [4]. Internal threats are attacks by personnel currently employed in the organization, who therefore have greater physical or cyber access to SCADA system equipment confidential and critical information. Attacks by an attacker who does not have physical or cyber access rights to the facility are categorized as external threats. External threats that are specifically directed at a system is the focus of this chapter.

A common threat to SCADA systems is from malware introduced into the system by infected portable storage devices and email attachments. Insufficiently protected wireless access points and WiFi-enabled computers with a LAN connection are also

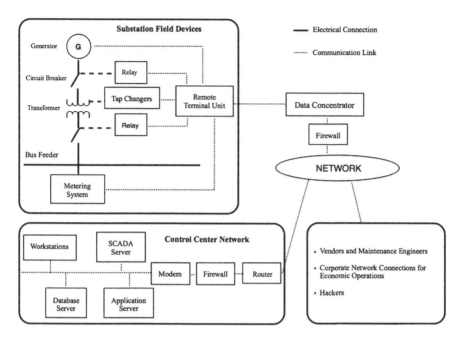

Fig. 1 Schematic of a SCADA Network

potential entry points. Most SCADA networks are connected to corporate networks using technologies such as *Virtual Private Networks* (VPN). Hence, if an attacker has access to the corporate network, he can get into the SCADA network with relatively less difficulty. Therefore, it is very important to ensure that there are no open and unsecured links at the corporate center.

We classify attacks on SCADA into *Intelligent Attacks* and *Brute-Force Attacks*. The above classification is based on the resources available to the attacker, intellectual capability, and level of expertise in power system operation and control. If the attacker has knowledge a priori on the operation of the entire system, he would be able to inflict a much severe attack. Thereby, with this knowledge an attacker would be able to target a substation that is very vital in the power grid. After a successful intrusion into a substation the attacker would have a wide range of attack options that can be executed (e.g., tripping breakers, changing transformer tap settings, capacitor bank settings). If the attacker has expertise in the controls available to him through the SCADA system, he would be able to choose an attack or a combination of attacks to cause maximum damage. On the contrary, an attacker who is not technically skilled will not be able to cause the same impact as an intelligent attacker.

Attacks can also be from a single attacker or a group of attackers working toward a common goal. The former type is called an *isolated attack* and the latter is a *coordinated attack*. Attackers coordinating with one another could attack the power grid at different locations. In combination with the first two classifications, we list the following four different types of attacks on power system SCADA.

1. *Isolated brute-force attacks*: An example for this classification is an attacker tripping a line in the power grid that is carrying only a small fraction of the demand. Power could be rerouted to that segment of the load, without any load being shed.
2. *Isolated intelligent attacks*: Consider a case where an attacker attacks a system at a time when a section of the distribution system is shut down for maintenance and power to a segment of the load is being rerouted. In this case, lines carrying the additional power to the load segment will be stressed. If the attacker could trip such a stressed line, it could potentially lead to unserved load condition or possibly result in a system collapse.
3. *Coordinated brute-force attacks*: An example of the third type of classification is a group of attackers (or just a group of attacks) launching an attack on the system at different locations without a well-defined goal. Such an attack without intelligence could be nulled by power system back-up and security measures.
4. *Coordinated intelligent attacks*: The fourth type of classification involves a clear-cut goal to be attained, knowledge of the topology of the system, and a good understanding of system response to a contingency. This attack pattern can cause severe impacts on the power system because it involves nullifying system defenses that are otherwise effective against an isolated attack. Our work is focused on this classification of attack.

Traditional power system planning techniques have accommodated $(N-1)$ contingencies in their scope. However, the system was not designed to fend against attacks that target system components, and their primary and/or secondary lines of defense. Such coordinated attacks, when carefully structured and executed, can push the system outside the protection zone and cause the above-mentioned impacts. The goal of this chapter is to provide a framework to evaluate risk of coordinated attacks on a power system with substations having a given cyber network configuration.

2 Related Work

Analytical techniques to estimate the impact of cyber attacks on physical systems have gained prominence in the literature in recent years. Research work in this area is not restricted to power system SCADA alone, but has been extended to other domains such as gas pipelines systems.

Authors in [5] propose a novel technique to mathematically model firewall and password protection mechanisms to obtain the probability of a successful attack using Petri Nets. In [6], the author proposes effective metrics to estimate the physical impact of a cyber attack. However, the isolated attack plan in the above works do not take into account system reserve capacities that can compensate for the loss in generation.

In [7], the authors capture a sense of coordinated attacks by attacking different power system components, but do not incorporate system defense mechanisms in their attack plan. In [8], the authors provide a comprehensive review of power system control actions during disturbances.

In [9], the authors propose a tool that, given an industrial network and a description of vulnerabilities, detects possible concerns in the network for the application. In [10], the authors propose the application of a two-layer wireless sensor network to observe the condition of transmission lines in a power system. In [11], the authors develop a model-based design of attack detection techniques to show how such a model of a computer and network system detects cyber attacks. In [12], State Space modeling approach is employed for the representation of coordinated attacks.

In [13], the author presents a view on the interdependency between critical infrastructures. In [14], contingency analysis is performed to quantify the survivability of high consequence systems based on attacker potential. In [15], the authors present statistical data on outage occurrences. The authors of [15] and [16] also identify perceived threats and reasons for stress on the power infrastructure. In [17], the author explains the complexity and challenges faced by the power industry. [18] provides a detailed list of measures that could be employed to enhance SCADA cyber security. Leversage and James [19] presents a three-step process for the calculation of Mean Time to Compromise to efficiently compare different security arrangements. Shaw [4] and Krutz [20] are fine resources that provide insightful details into the operation of SCADA systems and their vulnerabilities. McDonald [21] provides details on operations in an electrical substation.

The authors of [22, 23] propose a security setup called the trust system that enhances SCADA network security with minimal impact on the real-time application the network is monitoring. In [24], the authors develop a test bed imitating the power distribution network and present a set of scenarios that demonstrate the importance of information technology in grid management. In [25], a framework based on defense graphs and influence diagrams is presented to evaluate the security of cyber networks used in power system operation. [26] classifies the power system into four operating states and also enumerates events that trigger transition between two operating states. In [27, 28], the need for an intelligent power system control mechanism is stressed upon and the authors explain the use of devices such as Phasor Measurement Units (PMUs) to satisfy the same purpose. In [29], authors analyze existing key-management protocols to see if they satisfy security requirements of the SCADA system. Additionally, the authors also propose a new scheme that is secure, supports message broadcast, and avoids bottlenecks.

In our work, we make use of separate techniques to analyze the cyber and power network. However, there are a few integrated techniques available in the literature to estimate vulnerability in terms of metrics such as Loss of Load Expectancy (LOLE) and Energy Not Served per Interruption (ENSI) (e.g., [6, 7]). The shortcoming of this approach is that it fails to capture the parameters of cyber network on the value estimated of risk, such as topology of the cyber network, type of security devices, security device settings (e.g. firewall rules), etc. Our approach captures these aspects effectively. Also, using a separate tool for cyber network analysis gives us the flexibility to perform analysis on cyber networks with different configurations and security mechanisms. Moreover, using separate techniques also equips us to evaluate risk precisely, leveraging the existing models in both cyber and physical domains.

3 Coordinated Attack Scenario

Attack on Frequency Control System

The North American power system is functionally divided into *Balancing Authorities Areas* (BAAs). Each BAA, formerly known as Control Area, is maintained by a *Balancing Authority* (BA) – an entity responsible for maintaining generation – load balance within the BAA. Adjacent BAAs are connected by transmission lines (tie-lines) that facilitate power exchange for economic and reliable operation. As long as the load-generation balance within the system is maintained, the frequency remains at its standard value of 60 Hz. However, during contingencies such as the loss of a generating unit, the system experiences frequency excursions. Following such a disturbance, power system frequency drops below the nominal value of 60 Hz. Once this happens, various system defense mechanisms kick in at different time instances in order to restore the system to normal operating conditions. The following is the time-sequence of system defense mechanisms, following the loss of a generator.

1. At $t = 0 +$ s, that is, immediately after the disturbance, generators closest to the tripped generator will contribute to arrest the fall in frequency.
2. After this, the inertia of the other generators in the system contribute to arrest the fall in frequency until $t = 3 - 4$ s.
3. At $t = 4 +$ s, generators equipped with governor control ramp up generation to return the frequency to a steady-state value. However, governor control does not restore frequency to 60 Hz.
4. After the operation of governor control mechanism, the AGC attempts to restore the system frequency to 60 Hz.

An intelligent attacker would utilize the knowledge of this sequence to plan the attack. If the attack plan includes actions to nullify the effect of mitigation strategies at every step along the way, the physical impact caused would be severe. From an attacker's perspective, the attack is most successful when the amount of load unserved is maximum. The following are details that might aid the attacker in creating the attack template.

1. *Geographical location of load dispatch centers* – If the load dispatch center is distant from most generators in the system, it might become difficult to deliver power during contingency situations, due to transmission constraints.

2. *Generators with governor control enabled* – Not all generating stations in North America have active governor control. Hence, targeting generating units with active governor control will affect system response during contingencies.
3. *Largest generating units in the system* – By operational settings, generating units with larger MVA rating will have a greater influence on system recovery. When such units are brought offline, system mitigation is severely hampered. Units with large inertia also contribute to arrest the initial decline in frequency. Targeting such units will have a similar effect on the system.
4. *Time of the day* – An attack performed when the system is under heavy loading conditions will have maximum impact. This is because the rate of decline of frequency is faster under such conditions.

A sophisticated attacker would form an attack vector by carefully choosing targets based on the above criteria. However, the overall impact of the attack depends on the success of compromising each one of the targets. Depending on the number of substations successfully compromised the impact of the attack can vary between zero to significant load shedding.

4 Risk Estimation of Coordinated Cyber Attacks

The process of risk estimation involves two separate sub-operations—determining the probability of a successful intrusion and the resulting physical impact on the system. It is essential to perform separate analysis on the cyber and power network to evaluate these required parameters. Figure 2 is a flowchart that represents the process of risk estimation. The model from [5] is used for cyber network modeling and optimal power flow (OPF) simulations are performed on the test power system to determine load shedding required.

4.1 Cyber Network Modeling and Analysis

The cyber network model employed in this analysis is a network that consists of computers that control components in the power system, Intelligent Electronic Devices (IEDs), substation servers, etc. The network is secured using a firewall protection system. Each computer is secured using a password protection mechanism. It is assumed that a subject who is able to log on to a computer will be able to control physical devices that are controlled by the computer. Also, it is assumed that the attacker is equipped with the technical knowledge required to perform an intrusion. The probability of a successful coordinated attack depends on the probability of successful intrusion into each of the substations. The probability of successful intrusion into a cyber network mainly depends on the following factors:

Fig. 2 Risk modeling procedure

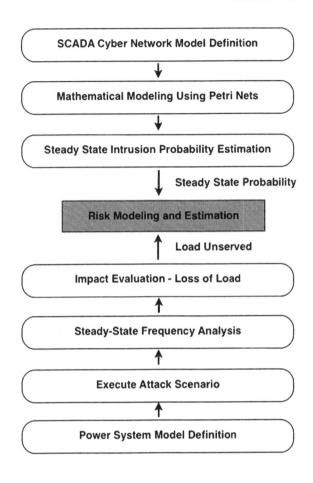

1. The number of access points in the SCADA network (e.g., Wireless access points, remote log-in connections).
2. The strength of the firewall, intrusion detection systems, and other network security schemes.
3. The strength of passwords.
4. The rate of attack on the network.

Hence, for a given network configuration, there exists an associated probability of successful intrusion depending on the above factors. The mathematical modeling tool, *Petri Nets*, has been employed to model the system and obtain the steady-state probability for successful intrusion [30, 31]. The Petri Net for a substation models both, the firewall and password protection mechanisms.

Fig. 3 Firewall Petri Net
model (adopted from [5])

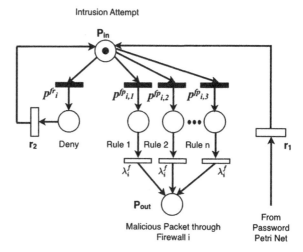

Malicious Packet through
Firewall i

4.1.1 Firewall Modeling

Firewall logs are maintained by system administrators to analyze malicious behavior. Firewall rules at most times are defined based on the type of protocol, IP
address range, port numbers, and sometimes MAC addresses. Any violation of the
rule set will result in the rejection of packets. However, intelligent attackers have
been successful at finding a way past the firewall. The probability of successful
intrusion into a network protected by a firewall is calculated by using information
from firewall logs. Hence, by looking at firewall logs, it is possible to determine
the probability of successful intrusion by identifying illegitimate packets that went
undetected.

The Petri Net model for a firewall, shown in Fig. 3, consists of $n + 1$ immediate
transitions with an associated transition probability, where n represents the number
of rules in the firewall rule set.

The probability of a malicious packet going through a firewall rule policy j at
firewall i is given by

$$P_{i,j}^{fp} = \frac{f_{i,j}^{fp}}{N_{i,j}^{fp}}, \tag{1}$$

where $f_{i,j}^{fp}$ is the frequency of malicious packets through the firewall rule j and
$N_{i,j}^{fp}$ is the total record of firewall rule j in the logs. The probability of n of the $n + 1$
immediate transitions is computed using the above expression.

The probability of packets being rejected at firewall i by rule j is given by

$$P_i^{fr} = \frac{f_{i,j}^{fr}}{N_{i,j}^{fr}}, \tag{2}$$

Fig. 4 Password Petri Net model (adopted from [5])

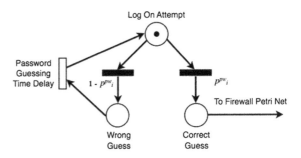

where $f_{i,j}^{fr}$ is the number of rejected packets and $N_{i,j}^{fr}$ represents the total number of packets in the logs. The probability of the deny transition is calculated using the above expression for each one of the firewall rules.

The Petri Net also has n delayed transitions associated with each firewall rule. These transitions have the same rate λ_i^f, which is a function of the firewall instruction execution rate. The transition with rate r_1 represents the *system response time*, which is the time taken by system to respond after processing a log in attempt. The timed transition with rate r_2 is the firewall processing time for denying a packet.

4.1.2 Password Mechanism Modeling

The Petri Net for the password model shown in Fig. 4 consists of two immediate transitions, one each for a correct and an incorrect password guess. The probability P_i^{pw} represents the probability of a successful password guess. This probability is obtained using the expression:

$$P_i^{pw} = \frac{f_i^{pw}}{N_i^{pw}}, \tag{3}$$

where f_i^{pw} is the number of successful intrusion attempts and N_i^{pw} is the number of observed records in the log. The subscript i represents the computer system. The timed transition that connects the place titled *Wrong Guess* and *Log On Attempt* represents the time taken to make a next password guess after a wrong guess.

4.1.3 Cyber Network to Petri Net Equivalent Model Transformation

A given substation network configuration with firewalls and computers can be mathematically modeled using Petri Nets. Every firewall in the network is replaced with its Petri Net equivalent model and every computer is replaced with the password Petri Net model. Figure 5 demonstrates how a sample cyber network is transformed to its Petri Net model. The firewall and computer are replaced with

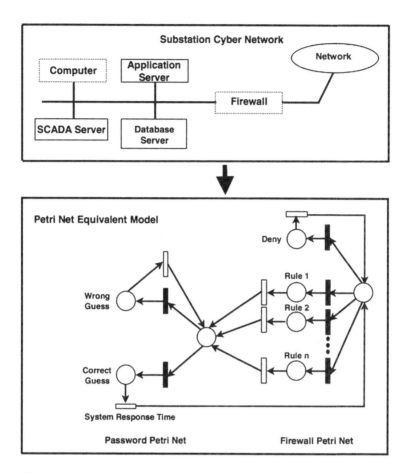

Fig. 5 Transformation from cyber network to Petri Net equivalent

the Firewall and Password Petri Net models, respectively. The resulting Petri Net is called the Petri Net equivalent model of the cyber network. The probability of successful intrusion into the substation computer is obtained from the steady-state probability of the place *Correct Guess* being non-empty.

A coordinated attack pattern consists of two or more attackers working in conjunction. An intrusion type coordinated attack is a success only when intrusions into all the substations in the attack pattern is a success. The probability of a successful coordinated attack is a function of the probabilities of successful intrusion into individual substations in the attack pattern. We assume that the process of intrusion into each substation is an independent event. Therefore, the probability of a successful coordinated attack is equal to the product of the probabilities of individual attacks.

Table 1 Substation intrusion
– steady-state probabilities

Machines	Model A	Model B	Model C
Mac1	0.0333	0.0130	0.0067
Mac2	0.0790	0.0081	0.0100
Mac3	0.0556	0.0123	0.0094
Mac4	0.0714	0.0118	0.0081
Mac5	0.0833	0.0101	0.0091

Fig. 6 Network models A, B and C

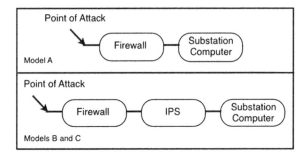

4.1.4 Cyber Network Simulation

The probabilities for successful intrusion into individual machines at three substation cyber network models, namely, A, B, and C were obtained. These are presented in Table 1. The cyber network is transformed to its Petri Net equivalent using the techniques described in Sect. 4. Figure 6 shows the three cyber network models. For an attacker to perform a successful intrusion into a substation computer with network model A, it is assumed that the he has to compromise a firewall and the computer's password protection system. In model B, it is assumed that the attacker would have to contend with a firewall, an *Intrusion Protection System (IPS)* and the password mechanism. Like the firewall, an IPS is also a packet filtering device and is modeled the same way using Petri Nets. In network model C, the substation computer is again protected with a firewall and an IPS, but with more stringent rules for packet filtering.

In network models B and C, the following method was used to model the firewall and IPS in series. The output place of the firewall, P_{out}, becomes the input place of the IPS, P_{in}. Place P_{out} of the IPS becomes the place *Log On Attempt* of the password Petri Net. The *system response time* transition connects the password Petri Net to the firewall Petri Net. The simulation model consists of five rules for both the firewall and the IPS.

Table 1 shows the probability of successful intrusion into machines in all the three cyber network models. The column titled *Machines* presents labels that are given to computers in substations. The numbers in subsequent columns represent the probability of successful intrusion into each of these machines for all the three cyber network models. Machines in the same network model have different intrusion

Table 2 Redispatched generation

Sl	Bus number	P Gen before (MW)	P Gen after (MW)
1	30	350.0	350.0
2	31	595.15	630.0
3	32	591.31	680.69
4	33	577.92	340.79
5	34	577.47	342.01
6	35	583.00	674.50
7	36	579.76	660.00
8	37	567.83	640.00
9	38	928.39	930.00
10	39	997.04	1,100.0

probabilities as a result of varying degree of difficulty in password guessing. This difference between machines in the same substation network is captured as varying probabilities in the password Petri Net model.

4.2 Power System Simulation

The New England 39-bus system, described in the appendix, was used as the test case to run OPF to obtain amount of load to be shed. Simulations were performed to determine the physical impact from both intelligent and dumb coordinated attacks. The following sections provide intelligent and dumb coordinated attack scenarios. The attack impact in each case is measured in terms of the required load shedding (load unserved) for the OPF program to return a feasible solution. The OPF program returns a feasible solution only if the online generation capacity is adequate to satisfy the demand with the given transmission constraints.

4.2.1 Scenario 1 – Coordinated Attack Without Intelligence

A coordinated attack without intelligence is an attack whose primary and secondary attacks have no significant impact on the system. Hence, a dumb coordinated attack would result in either zero or a very small amount of forced load shedding. The following coordinated attack cases are illustrations for dumb coordinated attacks.

Case 1: *Two transmission lines* – Simulation for this case is performed by tripping transmission lines 16–19 and 21–22 in the New England system. If an OPF is run on the system with the remaining online components, a feasible solution is still obtained. This is because of the following factors. When transmission line 16–19 is tripped, the generation units connected to busses 33 and 34 are isolated from the rest of the system. However, the generation resources in the remaining system are adequate to satisfy the demand. Table 2 presents generator operating points before and after line 16–19 is tripped. It is observed that units 4 and 5 step down from 577.92 and 577.47 MW to 340.79 and 342.01 MW, respectively. This is due to the fact that when units 4 and 5 are isolated from the rest of the system, the only load they serve is the one at bus 19. It is also observed that the generating units in the bigger section step up to compensate for the deficiency created by the isolation of units 4 and 5. Further, when transmission line 21–22 is tripped, the power that was delivered through it will be re-routed through line 23–24. The power flowing through 21–24 changes from 409.23 MW with line 21–22 to 1,074.31 MW when it is tripped. Hence, in this case, the system would be able to maintain stable operating conditions by maintaining frequency within acceptable limits.

Case 2: *Generation unit-transmission line combination* – This is a case of a coordinated attack pair that comprises of a generating unit and a line. The primary attack involves the tripping of unit 2. This results in the reduction of system capacity by 630 MW. However, the demand in the system is met by the remaining units in the system. If line 16–17 is tripped as a secondary attack, the system still remains in stable operating conditions as the bulk of the 204.79 MW that was routed through line 16–17 will be re-routed through line 14–4. The power flowing through line 14–4 before line 16–17 is tripped is 288.86 MW and this increases to 432.06 MW after the line is tripped. Therefore, the system will be able to maintain stable operating conditions even after a successful attack.

4.2.2   Scenario 2 – Coordinated Attack With Intelligence

In this section, we provide results of coordinated intelligent attacks on generation units only, transmissions lines only and generation-transmission combinations. As in the case of attack without intelligence, the coordinated attack template involves targeting two power system components at once.

Case 3: *Generation unit and transmission line* – The following scenarios provide examples for this kind of an attack. In the first illustration, the generator connected to bus 31 (unit 2) is tripped and this results in a reduction of system capacity by 630 MW. If the OPF is run on the remaining system, it will return a feasible solution with re-dispatched generation. Hence, without a secondary attack, the primary attack has done no significant damage to the system. As a secondary attack, line 3–4 is tripped. Under these system conditions, the OPF results in an infeasible solution. In other words, with the remaining generation capacity, transmission capacity and line constraints, the generation and load in the system cannot be balanced. Therefore, load shedding has to be performed in order to restore system frequency and regain stability. The following load shedding scheme is assumed. The loads directly served by a tripped generator, that is, the loads that are geographically closest to a tripped unit will be shed first. In this case, a load of 233 MW connected to bus 7 is closest to unit 2. Simulations reveal that a load of 110 MW will have to be shed for the OPF to return a feasible solution. Hence, it is inferred that the above attack on the system will leave 110 MW of the load unserved.

A similar attack case involves the tripping of generation unit 1 (bus 30) and line 16–19. This coordinated attack (*Case 4*) results in a total of 212 MW load unserved. The load in this case was shed from busses 18 and 25.

Table 3 Attack combinations and load unserved

Case	Primary attack	Secondary attack	Load unserved (MW)
1	Line 16–19	Line 21–22	0
2	Unit 2	Line 16–17	0
3	Unit 2	Line 3–4	110
4	Unit 1	Line 16–19	212
5	Line 21–22	Line 23–24	363
6	Unit 1	Unit 5	163

Case 5: Two transmission lines – This type of coordinated attack involves the tripping of two transmission lines in the network. In the New England system, this scenario can be illustrated by tripping lines 21–22 and 24–23. Although, the attack is targeted at transmission lines, it indirectly results in isolating generating units 6 and 7, connected to busses 35 and 36, from the rest of the network. In this case, load is shed from busses 15 and 16 as they are geographically closest to the generators that were isolated from the system. A total of 200 and 163 MW is shed from busses 15 and 16, respectively, to bring the system to stable operating conditions.

Case 6: Two generation units – The New England system has a total of ten generating units and several combinations of generators can be listed for this attack scenario. An attack on generating units 1 and 5, the smallest units in the system, results in a reduction of system capacity by 958 MW. This clearly moves the system into unstable operating conditions as the remaining generation capacity is insufficient to match the system demand and transmission losses. As in the previous cases, load shedding is performed to bring the system back to stable operating conditions. A load of 63 and 100 MW has to be shed from busses 18 and 20, respectively, and the OPF algorithm redispatches generation based on the new demand.

Table 3 compiles the load shed amount for the attacks described in this section. Case 5 has the worst impact as it involves tripping of two transmission lines that isolate two high capacity generation units from the system. Cases 1 and 2 are cases of coordinated attacks without intelligence as they have no impact on the system. The transmission line and generation combinations tripped in these cases are compensated for by other components in the system, resulting in zero load unserved. Case 4, which involves the tripping of the smallest generating unit from the system, still causes 212 MW of load to be shed. Case 6 involves the tripping of the two smallest generating units, unit 1 and 5, from the system and it results in forced load shedding of 163 MW.

4.3   Evaluation of Risk

According to our assumptions, substations housing generating units 1 and 5 in the power system have cyber network configurations Model A and Model C from Table 1, respectively. The coordinated attack is a success only when intrusion into both the substations is a success.

Table 4 Risk estimate

Case	γ	Attack probability	Risk
1	0	0.0010	0
2	0	0.0008	0
3	110	0.0007	0.0770
4	212	0.0004	0.0848
5	363	0.0008	0.2904
6	163	0.0003	0.0489

It is assumed that the process of intrusion into both substations are independent events. This assumption is reasonable as different power companies may use network security solutions and industrial control systems equipment from different vendors. Therefore, the probability of the coordinated attack is equal to the product of probability of individual attacks. Assuming that the breaker connecting generating units to the grid can be controlled from machine *Mac1* in the first substation and *Mac2* in the second substation, the probability of the coordinated attack, $P(Z)$ is

$$P(Z) = P(A) * P(B) = 0.0333 * 0.0100 = 0.0003. \tag{4}$$

Risk is defined as

$$Risk = P(Z) * \gamma, \tag{5}$$

where γ is the physical impact (load unserved). For case 6 coordinated attack on the New England 39-bus system, γ is 163. Using (11), risk is estimated to be

$$Risk = 0.0003 * 163 = 0.0489. \tag{6}$$

Table 4 provides the risk of the system for each of the cases in Table 3.

The above results provide the following inferences. It is interesting to note that the coordinated attack with the highest impact on the system is not the one that targets the generating units directly, but involves two transmission lines. This suggests that an intelligent attacker could exploit critical links in the topology of the power system to cause more damage than directly attacking generating units. Both case 3 and 4 are coordinated attack scenarios involving a generating unit and a transmission line. Although case 4 involves the tripping of the smallest generating unit with capacity 350 MW as opposed to 630 MW from unit 2 in case 3, the impact from case 4 attack is greater than the one from case 3. This suggests that a coordinated attack need not necessarily involve high capacity units to have a significant impact. This also highlights the need for estimating the criticality of a system component in combination with other components in the system.

Case 4 has a higher risk than case 3 despite having a lower probability of success. This is because of the high impact a successful case 4 attack scenario produces. Although case 1 has the highest success probability, along with case 2, it has zero risk. This is due to the fact that a successful case 1 or 2 will result in zero forced load shedding. Hence, dumb coordinated attacks have negligible risk that is either zero or very close to zero. Case 6 coordinated attack, which involves tripping of two generating units, has the least risk. Although the physical impact due to this attack is significant, the probability of attack success is small. This is expected as substations housing generating units are expected to have stringent network security measures and limited access.

5   Conclusion

In this work, we presented a systematic approach to evaluating the risk from coordinated cyber attacks on a power system. Specialized techniques were used to evaluate the probability of successful cyber network intrusion and impact on power system. The resulting risk due to the coordinated attack was estimated. Different combinations of generating units and transmission lines were combined to form coordinated attack pairs. The simulations revealed that important factors that influence physical impact of an attack, among others, are the topology of the power system, the location of the targeted generating substation and other generating substations, and the criticality of the component in combination with other components in the system.

Other forms of coordinated attacks involving other system components could have an equally severe impact. Our future work includes developing a systematic procedure to evaluate the physical impact from coordinated attacks on any system. The aim is to develop an algorithm that returns coordinated attack pairs as an output when provided with a system. The aim is also to make the execution computationally inexpensive without a brute-force approach. Our research also involves developing both, cyber and power system, mitigation techniques for such attacks. Extending the coordinated attack model to systems like water distribution systems and, oil and gas pipeline systems, to evaluate the impact of coordinated attacks is also a part of our plan for future research.

Appendix

The New England 39-bus system, shown in Fig. 7, was used to perform analysis during emergency situations. The 39-bus system has ten generating units divided into three control areas, named 1, 2, and 3. Table 5 shows the amount of power generated and the load demand in each control area. The difference between the total power generated and total load, equal to 47.38 MW, is attributed to transmission losses. Table 6 identifies generators with their capacities and the bus to which they are connected. The column *P Gen* is the amount of power the unit generates under normal conditions. The column *P Limit* represents the maximum generation of that unit and the column *Control Area* shows the control area the generating unit is located in.

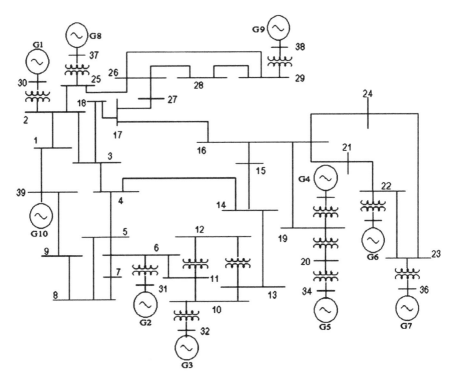

Fig. 7 The New England 39-bus System

Table 5 Control areas – generation and demand

Control area	P Gen (MW)	P Demand (MW)
Control area 1	2,183.5	1,323.5
Control area 2	917.83	1,124
Control area 3	3,246.54	3,853
Total	6,347.87	6,300.5

Table 6 Generator and capacities

Sl	Bus number	P Gen (MW)	P Limit (MW)	Control area
1	30	350.0	350	2
2	31	595.15	630	1
3	32	591.31	750	1
4	33	577.92	732	3
5	34	577.47	608	3
6	35	583.00	750	3
7	36	579.76	660	3
8	37	567.83	640	2
9	38	928.39	930	3
10	39	997.04	1,100	1

Table 7 Load busses and demand

Sl	Bus number	P Demand (MW)	Control area
1	3	322.0	2
2	4	500.0	1
3	7	233.8	1
4	8	572.0	1
4	12	8.5	1
5	15	420.0	3
6	16	329.40	3
7	18	158.0	2
8	20	680.0	3
9	21	274.0	3
10	23	247.5	3
11	24	308.6	3
12	25	224.0	2
13	26	139.0	2
14	27	281.0	2
15	28	206.0	3
16	29	283.5	3
17	31	9.20	1
18	39	1,104.0	3

Table 7 shows the loads that are present in the system. The column *P Demand* is the amount of load in MW units. The bus to which they are connected and the control area is also indicated.

References

1. Presidents Commission on Critical Infrastructure Protection, Critical Foundations: Protecting Americas Infrastructures (1997). [Online]. Available at: http://www.ciao.gov/
2. Motorola SCADA Products: RTU ACE3600
3. Clarke G, Reynders D (2004) Practical modern SCADA protocols: DNP3, 60870.5 and related systems (IDC Technology). September 2004
4. Shaw WT (2008) Cybersecurity for SCADA systems. PennWell Corp., July 28, 2006
5. Ten C-W, Liu C-C, Manimaran G (2008) Vulnerability assessment of cybersecurity for SCADA systems. IEEE Trans Power Syst 23(4):1836–1846
6. Stamp J, McIntyre A, Richardson B (2009) Reliability impacts from cyber attack on electric power systems. In: IEEE PES power systems conference and exposition, PSCE '09, March 15–18, 2009, pp 1–8
7. Salmeron J, Wood K, Baldick R (2004) Analysis of electric grid security under terrorist threat. IEEE Trans Power Syst 19(2):905–912
8. Patel S, Stephan K, Bajpai M, Das R, Domin TJ, Fennell E, Gardell JD, Gibbs I, Henville C, Kerrigan PM, King HJ, Kumar P, Mozina CJ, Reichard M, Uchiyama J, Usman S, Viers D, Wardlow D, Yalla M (2004) Performance of generator protection during major system disturbances. IEEE Trans Power Deliv 19(4):1650–1662
9. Cheminod M, Bertolotti I, Durante L et al (2009) Detecting chains of vulnerabilites in industrial networks. IEEE Trans Ind Inf 5(2):181–193

10. Leon RA, Vittal V, Manimaran G (2007) Application of sensor network for secure electric energy infrastructure. IEEE Trans Power Deliv 22(2):1021–1028
11. Ye N, Giordano J, Feldman J (2001) A process control approach to cyber attack detection. Commun ACM 44(8):76–82
12. Braynov S, Jadliwala M (2003) Representation and analysis of coordinated attacks. In: Proceedings of the 2003 ACM workshop on Formal methods in security engineering (FMSE '03). ACM, New York, NY, USA, pp 43–51
13. Rinaldi SM, Peerenboom JP, Kelly TK (2001) Identifying, understanding and analyzing critical infrastructure interdependencies. IEEE Contr Syst 21(6):11–25
14. McDermott J (2005) Attack-potential-based survivability modeling for high-consequence systems. In: Proceedings of the 3rd IEEE international workshop on information assurance, IWIA '05, pp 119–130
15. Amin M (2003) North America's electricity infrastructure: are we ready for more perfect storms? IEEE Security Privacy 1(5):19–25
16. Goetz E (2002) Cybersecurity for electric power industry. In: Report of investigative research for infrastructure assurance (IRIA), Institute for Security Technology Studies, Dartmouth College, December 2002
17. Amin M (2002) Security challenges for the electricity infrastructure. Computer 35(4):8–10
18. 21 Steps to improve Cyber Security of SCADA Networks. Department of Energy, September 2002
19. Leversage DJ, James E (2008) Estimating a system's mean time to compromise. IEEE Security Privacy 6(1):52–60
20. Krutz RL (2008) Securing SCADA Systems, 1st edn. Wiley, New York
21. McDonald JD (2007) Power substations engineering, 2nd edn. CRC Press, Boca Raton
22. Coates GM, Hopkinson KM, Graham SR, Kurkowski SH (2008) Collaborative, trust-based security mechanisms for a regional utility intranet. IEEE Trans Power Syst 23(3):831–844
23. Coates GM, Hopkinson KM, Graham SR, Kurkowski SH (2010) A trust system architecture for SCADA network security. IEEE Trans Power Deliv 25(1):158–169
24. Dondossola G, Garrone F, Szanto J, Fiorenza G (2007) Emerging information technology scenarios for the control and management of the distribution grid. In: Proceedings of 19th international conference and enhibition on electricity distribution, Vienna, Austria, March 21–24, 2007.
25. Sommestad T, Ekstedt M, Nordstrom L (2009) Modeling security of power communication systems using defense graphs and influence diagrams. IEEE Trans Power Deliv 24(4):1801–1808
26. Huang Y-H, Cardenas AA et al (2009) Understanding the physical and economic consequences of attacks on control systems. Int J Critical Infrastructure Protection 2(3):72–83
27. Kirschen D, Bouffard F (2009) Keep the lights on and the information flowing. IEEE Power Energ Mag 7(1):55–60
28. Giri J, Sun D, Avila-Rosales R (2009) Wanted: a more intelligent grid. IEEE Power and Energy Magazine 7(2):34–40
29. Choi D, Kim H, Won D, Kim S (2009) Advanced key-management architecture for secure SCADA communications. IEEE Trans Power Deliv 24(3):1154–1163
30. Girault C, Valk R (2003) Petri nets for systems engineering. Springer, New York
31. Ajmone Marsan M, Balbo G, Conte G, Donatelli S, Franceschinis G (1995) Modeling with generalized stochastic petri nets, 1st edn. Wiley, New York

Synchronous Measurements in Power Distribution Systems

Daniel A. Haughton and Gerald T. Heydt

Abstract This chapter deals with power distribution engineering measurements and the processing of those measurements. The concept proposed is the use of synchronized measurements. The synchronization is accomplished through the use of a global positioning satellite signal – thus, time stamping measurements. The synchronized measurements are in phasor detail. The use of these measurements in a state estimator is proposed, described, and illustrated. Possible uses of this technology include: control of distribution system components; fault detection and management; energy management.

1 The Present Electric Power Infrastructure and the Smart Grid Initiative

The electric power infrastructure may be subdivided into four general categories: generation, transmission, distribution, and utilization (e.g., system loads). Within each of these categories, significant changes are being proposed or underway. Many of these changes have the potential for system-wide impacts in many areas of design, planning, operations and control of the power system. One of the biggest drivers in all these sectors in the United States is the "Smart Grid" initiative. The salient points related to the modernization of the electric infrastructure are captured by the United States Department of Energy (DoE) "Modern Grid" program [1]:

1. Self-healing from power disturbance events
2. Enabling active participation by consumers in demand response
3. Operating resiliently against physical and cyber attack

D.A. Haughton (✉) • G.T. Heydt
School of Electrical, Computer and Energy Engineering, Arizona State University,
Tempe, AZ 85287, USA
e-mail: Daniel.Haughton@asu.edu; heydt@asu.edu

A. Chakrabortty and M.D. Ilić (eds.), *Control and Optimization Methods for Electric Smart Grids*, Power Electronics and Power Systems 3,
DOI 10.1007/978-1-4614-1605-0_15, © Springer Science+Business Media, LLC 2012

4. Providing power quality for 21st century needs
5. Accommodating all generation and storage options
6. Enabling new products, services, and markets
7. Optimizing assets and operating efficiently

This Smart Grid initiative calls for enhanced levels of sensory information, communications, and controls within the distribution system. It also calls for deployment of energy storage, demand-side-response, and dynamic optimization of grid operations and resources.

Power generation businesses around the world over have seen a significant shift in the political and regulatory landscape. Governments and decision makers are increasingly encouraging the deployment of low-carbon or no-carbon generation sources, particularly from sources considered to be renewable such as wind and solar. Global economic growth continues to fuel the demand for energy. That is, as national economies grow, they continue to consume energy and electricity for powering factories, buildings, houses and transportation. The trends of increasing penetration of intermittent types of renewable resources (wind and solar), and increasing consumption of electricity from developing nations around the world are expected to continue. Technological innovations in the generation, transmission, and distribution subsystems that allow the power system to maintain its integrity are essential to handle high penetration of intermittent sources [2].

Electric transmission networks, generators, and load-serving entities (LSE) in some parts of the United States participate in power markets. Independent system operators or regional transmission organizations generally aim to minimize system operating costs subject to meeting certain system constraints. Through a marketing mechanism driven by auctions, power producing generation resources make sell bids into the market while LSE make buy bids. Ideally, incentivizing system efficiency increases and minimizing costs are both long and short term objectives in power markets; for example, planning system expansion (future growth) may aim to maximize benefit to all market participants. The focus of changes in the transmission system have thus centered on: improved system modeling and analysis tools; data collection; monitoring; new technologies that allow for flexible and reliable operation under a variety of system conditions; and adding capabilities to enhance system operations. The monitoring and control activities that allow transmission system operators to push the system closer to its limits to extract efficiencies from the system are of particular interest in considering the future direction of distribution engineering [3–5].

Electric *distribution* systems may be seen as the next logical point of focus in the evolution of the electric power infrastructure. The electric distribution system has been fairly well modeled and well understood by distribution companies (both independent and within vertically integrated utilities) who serve millions of customers. Distribution systems are often designed in a networked fashion (with isolation switches) but operated as a radial system with main feeders extending from a single distribution substation (with switches between feeders designated *normally-open*). The networked design allows limited reconfigurability where normally-open switches can be closed to change the source of a feeder to another substation. Although the contemporary system designs are well understood, much of the contemporary system has been overdesigned to accommodate load uncertainties, load growth, and help maintain system reliability [4, 5]. The contemporary electric distribution system usually has sensory information collected mostly at the distribution substation; for example, bus voltage magnitudes and line flow measurements. Typically system sensory information or "monitoring" is limited to revenue metering which takes place at each service entrance. Real-time monitoring of feeder voltages, line currents, and customer loads may be performed on an as needed basis.

Electric distribution systems are poised for significant changes as new technologies, different types of loads, and small distributed generation sources are increasingly added to the system. One must keep in mind that the extent, i.e., number of components and investment, of the distribution system is comparable (or may exceed) that in transmission systems. Yet little attention has traditionally been given to enhancing distribution systems. The automation of distribution systems has received attention in recent years, e.g., automating the operation of normally open switches has the potential to significantly improve certain system reliability indices [4–8].

Consideration of the Smart Grid initiatives as previously identified along with the changes seen taking place within the transmission and generation businesses signify a paradigm shift for future distribution systems. Elements within future distribution systems may well become market participants as power marketers and transmission operators look to expand their reach beyond traditional tools (procuring sufficient generation to meet load and using transmission switchyard components to maintain system integrity). Distribution system loads may become just as responsive to changes in price, system state and congestion as generators are in the wholesale power markets [5]. For example, the least-cost operating scenario for an entire region may include reducing consumption by switching off nonnecessary distribution system loads as part of demand-side-response. Networking of distribution system primary feeders and secondary feeders may become favorable in systems with high penetration of distributed generation and renewable sources. To illustrate the foregoing remarks, Fig. 1 shows a pictorial of the transition to the Smart Grid.

2 Emerging Technologies and the Electric Distribution Infrastructure of the Near Future

Future distribution systems are likely to have higher penetrations of new and emerging technologies that are consistent with the objectives identified within the Smart Grid initiative. To facilitate the transition to a "smarter" grid, new technologies, including IT and communications, are required. Technological innovation within the distribution system will likely focus on the development of specific components

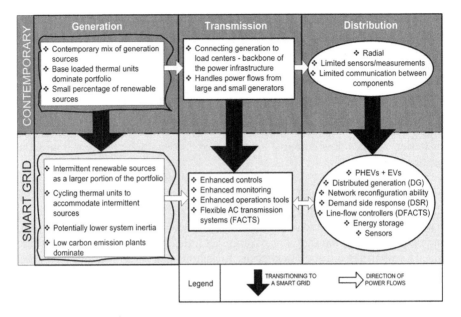

Fig. 1 Transitioning the existing power system toward the Smart Grid

intended to satisfy the Smart Grid vision, control of these components to optimize energy usage across large systems, and developing the communication and IT infrastructure to make this all happen. Critical to the successful integration of new technologies are how they interface with the existing grid, how these devices communicate and interact with each other, and how new technology will enhance system monitoring for both automated and operator assisted control. Examples of Smart Grid technologies include:

- Storage technologies that include thermal storage, compressed air energy storage CAES, flywheels and batteries.
- Distributed generation including biomass, small hydro, and renewable solar, geothermal, and wind.
- Electronic loads requiring AC/DC, AC/DC/DC conversions, such as PHEVs, EVs and others [9].
- Line flow controllers that work on solid state principles.
- Loads that maintain "awareness" and respond to system conditions, including those that participate in demand-response programs.
- Enabling technologies to integrate new components, operate, and optimize these different types of loads, generators and storage elements.
- Wide deployment of sensors and communications infrastructure.

Figure 2 displays how these emerging technologies, including digital control and distribution energy management applications, may be integrated into the distribution system as part of the Smart Grid initiative. A more exhaustive list and further discussion of the next generation of power distribution systems may be found in [5].

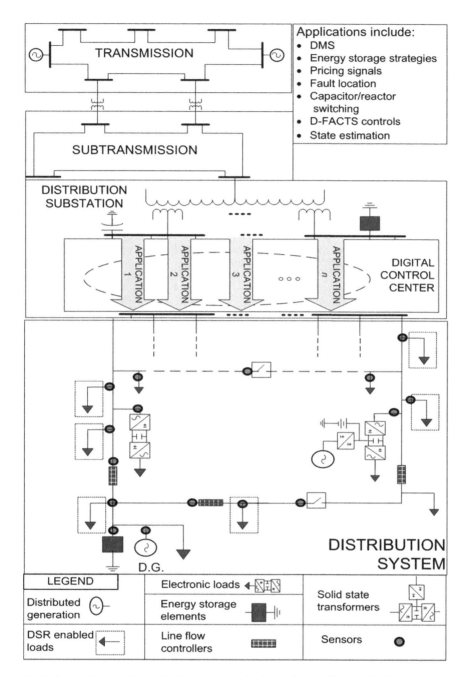

Fig. 2 Integrating emerging technologies, communication, and control into the distribution system of the near future

As previously discussed, the conventional distribution system is built as a network but operated as a radial system. In this distribution system, feeder loads are fairly well known, demand is reasonably well estimated and there is generally an insignificant amount of DG. As a result, system measurements in conventional systems often end at the substation since there were no significant amounts of controllable components, loads, generation resources or energy storage elements. High penetration of the aforementioned Smart Grid technologies poses challenges to the distribution engineer; these challenges encompass the design, protection, monitoring, and operation of the distribution system. A major departure from conventional design and operation is the inclusion of a controllable infrastructure (generators, loads, storage elements, and reconfiguration switches), the communications infrastructure and the system measurements required to accomplish this control. Other issues related to the control of customer loads, demand response, and utilizing energy storage include the designation of responsibility for costs related to equipment maintenance and loss of useful life (i.e., batteries have limited charge/discharge cycles).

3 Synchronous Measurements in Power Distribution Systems

In power distribution engineering, measurements play several roles including: revenue metering; voltage magnitude measurement for shunt capacitor switching; feeder current measurements at the distribution substation; and, in some cases, measurements for the purpose of system asset protection. Quite often such measurements are part of the supervisory control and data acquisition system (SCADA) utilized at all levels throughout the power system [7]. Traditionally, these measurements are asynchronous, i.e., not made at the same time and not triggered by a common timing signal. Part of the widely publicized U.S. DoE "Smart Grid" initiative is the use of measurements throughout the power system at both the transmission and the distribution levels to effectuate control and economic operation.

In order to take full advantage of measurements, especially in control applications, it appears to be appropriate to synchronize the measurements of voltages and currents, thereby making full phasor measurement possible. This approach has been demonstrated in transmission engineering in the form of synchrophasor or phasor measurement units (PMUs) [9–13]. The essence of the idea is the use of an accurate atomic clock and Global Positioning System satellites to allow the estimation of the spatial and temporal coordinates $\{x, y, z, t\}$. Using the time stamp t, it is possible to obtain a least squared error fit of a phasor to a sinusoidal wave, and this yields the positive, negative, and zero sequence components of voltage and current in a three phase measurement. The IEEE Standard C37.118–2005 [15] gives the parameters of a synchrophasor measurement system. The usual data rate is 30 frames per second; however, 60 frames per second can be achieved by many PMU instruments in a 60 Hz system.

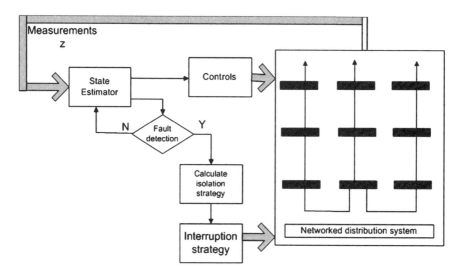

Fig. 3 Pictorial of the concept of a synchrophasor measurement

The concept of a state estimator may also be applied to distribution systems. The state estimates may be used for system visualization, system control, outage detection, and as a supplement to circuit protection/fault management as illustrated in Fig. 3.

Attention turns to the use of synchronized measurements in distribution systems. The motivation is the declining cost of synchrophasor measurement hardware, establishment of the communication infrastructure already under development as part of the advanced metering infrastructure (AMI or smart meters) [3], and the potential of the use of the data obtained for:

- Control of energy flows
- Identification of faults (by location and in time)
- To coordinate controls of solid-state devices in the distribution system (e.g., the future renewable electric energy distribution management, FREEDM, system), including solid state distribution transformer controls [14]
- As a facilitator of energy management systems
- As a facilitator of revenue metering systems (e.g., implementing time of day rates and increasing price transparency within the distribution system)
- Tracking energy flow in a power marketing infrastructure
- Enhancing operator visualization to see the true "state" of the distribution system
- Utilizing this enhanced visualization and operational ability to better plan system expansion, obtain useful life out of system assets and improve overall system efficiencies.

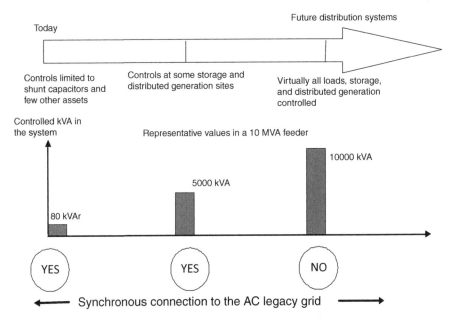

Fig. 4 A time line of controls in electric power distribution systems

Figure 4 is a pictorial showing the evolution of controls in power distribution systems with a conjectured role of synchronous measurements indicated in future systems.

4 State Estimation

In traditional power engineering state estimation, the usual measurements taken are bus voltage magnitudes, real and reactive power (at buses or in lines). In transmission engineering, redundancy of measurements is used to enhance measurement accuracy. Because these measurements are nonlinear functions of the system states, namely the voltages and currents, the problem formulation is nonlinear. To solve the nonlinear state estimation problem, the process is linearized. That is, the basic relationship between the states X and the measurements z is

$$h(X) = z + \eta,$$

where η captures process and measurement noise. The vector-valued nonlinear function $h(X)$ is linearized and the problem formulation proceeds as the solution of the state estimation problem modeled as

$$hx = z + \eta,$$

where h is the process matrix that relates system states to the system model or measurements. The state estimation procedure essentially minimizes the square of the residual, r, between the assumed system model and the set of available measurements

$$r = hx - (z + \eta).$$

The minimization is performed in the least squares sense, that is, the two-norm

$$r'r = (hx - z - \eta)' (hx - z - \eta)$$

is minimized,

$$\frac{\partial}{\partial x}(r'r) = 0.$$

Solving for the system states, the best estimate, $\hat{x}$, is obtained via,

$$\hat{x} = h^+ Z,$$

where h^+ is the pseudoinverse of the process matrix. Note that state estimation is generally performed with an overdetermined set of equations. Consequently, the h matrix is generally not a square matrix, and the pseudoinverse applies.

Note that where synchronous measurements are possible, the states themselves may be metered and time-tagged in real time. That is, the phasor values of the voltages and currents are attained. Therefore, no nonlinear formulation is needed [13]. However, it must be added that there may be circumstances where synchronous measurements are augmented by active power and reactive power measurements – and this state estimation problem is again nonlinear. A full description of the state estimation algorithm and many applications are given in [16,17]. In one formulation, the use of synchronized measurements enables a complex set of measurements with the full system model. Nonsynchronized measurements typically provide magnitude information only, but not phase information as compared to a reference. The full complex state estimation algorithm is constructed with each variable partitioned into a real and imaginary component,

$$h \rightarrow h_r + jh_i$$
$$x \rightarrow x_r + jx_i$$
$$z \rightarrow z_r + jz_i.$$

The residual is then found to be,

$$r = (h_r + jh_i)(x_r + jx_i) - (z_r + jz_i + \eta).$$

The two-norm of the residual vector becomes

$$r^H r = (z_r - h_r x_r + h_i x_i + \eta_r)^2 + (z_i - h_i x_r - h_r x_i + \eta_i)^2.$$

where r^H identifies the Hermitian operation (complex conjugate transpose). The minimization then entails the simultaneous solution of

$$\frac{\partial}{\partial x_r} r^H r = 0 \quad \frac{\partial}{\partial x_i} r^H r = 0.$$

However, the differentiation with respect to a complex variable is generally non-analytic. The Cauchy–Riemann conditions (discussed subsequently) must hold for the differentiation to be valid. To avoid this issue, the partitioning of the system of equations into real and imaginary parts separates the independent real and imaginary variables as follows,

$$\begin{bmatrix} h_r & -h_i \\ h_i & h_r \end{bmatrix} \begin{bmatrix} x_r \\ x_i \end{bmatrix} = \begin{bmatrix} z_r \\ z_i \end{bmatrix} + \begin{bmatrix} \eta_r \\ \eta_i \end{bmatrix}.$$

The solution estimates are then obtained as

$$\begin{bmatrix} \hat{x}_r \\ \hat{x}_i \end{bmatrix} = \begin{bmatrix} h_r & -h_i \\ h_i & h_r \end{bmatrix}^+ \begin{bmatrix} Z_r \\ Z_j \end{bmatrix} + \begin{bmatrix} h_r & -h_i \\ h_i & h_r \end{bmatrix}^+ \begin{bmatrix} \eta_r \\ \eta_i \end{bmatrix}.$$

The Cauchy–Riemann equations for the differentiation with respect to a complex variable require that

$$\frac{\partial \text{Re}(r^H r)}{\partial \text{Re}(x)} = \frac{\partial \text{Im}(r^H r)}{\partial \text{Im}(x)},$$

$$\frac{\partial \text{Re}(r^H r)}{\partial \text{Im}(x)} = \frac{\partial \text{Im}(r^H r)}{\partial \text{Re}(x)},$$

where *Re* refers to the real part and *Im* to the imaginary part of the respective quantity. Note that in the above formulation where the matrix is partitioned, differentiation involves real numbers only.

The process matrix is constructed in a way so as to include the assumed system topology (admittance matrix) and relate individual bus voltage and bus injection measurements to system states. With the ability to obtain system measurements as synchronized quantities, the estimation process can be significantly simplified. The following advantages of the formulation using synchronous measurements are obtained over the conventional transmission system state estimation formulation:

- No need to linearize the state estimation problem
- No iterative solution required
- Faster solution
- Real time control applications envisioned
- Better accuracy from perfectly synchronized measurements

- Potentially lower cost in implementation because of fewer measurements required
- Simplified data processing.

As previously discussed, the number of measurements and the application of those measurements in power distribution primary systems are limited in the contemporary milieu. One objective of the Smart Grid is to take measurements, and to use them to effectuate controls and implement strategies of operation. As an example, renewable resources in a small geographical area could be coordinated effectively and at optimal times through the use of measurements which would actuate controls. An important issue in this approach is the cost effectiveness of the placement of measurements, and the benefits reaped from their use. Another issue is the identification of exactly how many measurements are needed and what accuracy is required to accomplish system control.

Note that as measurement redundancy increases, generally the accuracy of the estimates of the system states increases. Mathematically, this is shown by the sensitivity of state estimates to measurement noise. This sensitivity is somewhat quantified by the singular spectrum of the gain matrix G used in the state estimation formulation. The gain matrix for an unbiased estimator, $h^T h$, may be viewed as *mathematically distant* from a singular matrix, and the distance is determined by the singular value decomposition of $h^T h$. It is well known that the singular spectrum of $h^T h$ determines the condition of the estimate [18–20].

5 Illustrative Example

The following example was developed using the RBTS as published in [21]. Basic assumptions are made for conductor sizes and impedances based on feeder loading and ACSR conductor data sheets provided in [22]. The recast one-line of the distribution system under study is provided in Fig. 5.

The RBTS system model can be incorporated into the state estimation formulation based on synchronized phasor measurements. Two cases illustrate the estimation process and robustness of the estimation formulation to measurement noise. Case I estimates bus voltages in the 84 bus RBTS distribution system assuming that bus voltage measurements are known exactly (i.e., bus voltage measurements agree exactly with power flow study results). This can be accomplished by using the output data of a commercially available power flow software program as the "input measurement vector", z, to the estimator algorithm. Case II presents a more practical illustration where the measurement vector, z, is subject to measurement noise, η, of about 1.0%. Noise in this illustration is assumed to follow a standard normal distribution with unit variance; that is, most of the noise vector entries in this case are within $\pm 1.0\%$ of the expected value of the measurement (results of a power flow study in this case) with decreasing probabilities of larger deviations. Table 1 shows the bus voltage magnitude results of Case I (estimation with no consideration

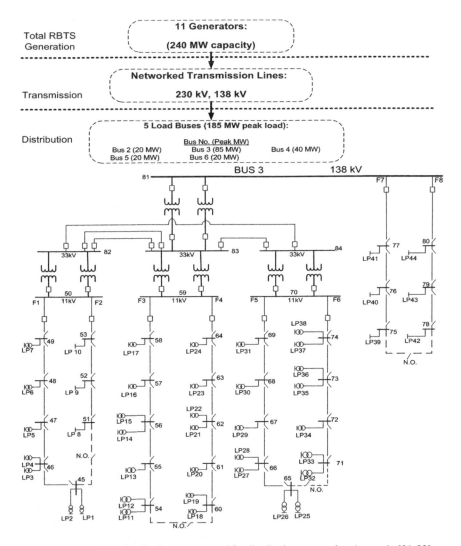

Fig. 5 Redrawn RBTS distribution system used for distribution state estimation study [21, 23]

to noise in measurements) and Case II (1.0% measurement noise considered). Even with measurement noise, the resulting bus voltage magnitudes are well estimated in this case.

Table 1 results indicate that the state estimation algorithm described here provides well-estimated system states – bus voltage magnitudes – for the system studied. Note that all estimated bus voltage magnitudes for Feeders F1 and F2, with measurement noise present, are within 0.75% of the *true* solution. However, limitations exist on the "accuracy" of the state estimates, particularly for feeder laterals. For example, a measurement at a load point bus that is "far away" from

Table 1 Bus voltage magnitudes for a state estimation application to the RBTS, Cases I and II

| | Bus voltage magnitudes | | | | | |
| | Case I | | | | Case II | |
Bus No.	Power flow study result	State estimation solution	Difference	1% noisy signal	State estimation solution	Difference
Feeder F1						
LP1	0.9764	0.9764	0.000	0.9779	0.9778	−0.0014
LP 2	0.9770	0.9770	0.000	0.9786	0.9786	−0.0016
LP 3	0.9797	0.9797	0.000	0.9727	0.9727	0.0070
LP 4	0.9771	0.9771	0.000	0.9720	0.9719	0.0052
LP 5	0.9798	0.9798	0.000	0.9873	0.9873	−0.0075
LP 6	0.9825	0.9825	0.000	0.9816	0.9816	0.0009
LP 7	0.9866	0.9866	0.000	0.9893	0.9894	−0.0028
45	0.9806	0.9808	0.000	0.9807	0.9808	−0.0002
46	0.9815	0.9816	0.000	0.9770	0.9817	−0.0002
47	0.9840	0.9840	0.000	0.9896	0.9850	−0.0010
48	0.9868	0.9867	0.000	0.9884	0.9872	−0.0004
49	0.9907	0.9904	0.000	0.9838	0.9908	−0.0001
50	0.9938	0.9933	0.001	1.003	0.9935	0.0003
Feeder F2						
LP 8	0.9894	0.9895	0.000	0.9900	0.9920	−0.0026
LP 9	0.9902	0.9902	0.000	0.9857	0.9924	−0.0022
LP 10	0.9911	0.9911	0.000	0.9872	0.9925	−0.0014
51	0.9899	0.9901	0.000	0.9920	0.9925	−0.0026
52	0.9906	0.9906	0.000	1.001	0.9927	−0.0021
53	0.9918	0.9917	0.000	0.9998	0.9931	−0.0013

the main feeder provides observability of the distant bus. Measurement noise on this distant bus becomes a more significant determinant of estimated accuracy when compared to estimating states on the main feeders. Figures 6 and 7 provide a graphical illustration of bus voltage magnitude estimates for the power flow solution, estimated solution with no measurement noise, noisy 1.0% and estimated solution with measurement noise. Feeder F1 has "distant" load buses since the load side of a distribution transformer (11/0.415 kV) is being estimated (transformers present large impedances in system models). As a result, feeder load bus state estimates are fairly close to the noisy measurement value (Fig. 6). Feeder F2 load buses are not served through a transformer; consequently, even in the presence of measurement noise, the estimated solution corresponds closely to the *true* solution (Fig. 7).

For the entire 84 bus system Figs. 8 and 9 present, the histograms of estimate error for the cases with no measurement noise and 1.0% measurement noise. Note that error is defined here as the difference between the "true" solution and estimated solution. As previously identified, the estimated solutions are generally well estimated, although the effect of measurement noise is obvious in the different histogram plots.

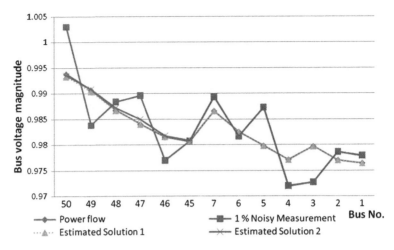

Fig. 6 Feeder F1 bus voltage magnitudes: power flow study result; noisy measurements; estimated magnitude (no measurement noise); estimated magnitude (1.0% measurement noise)

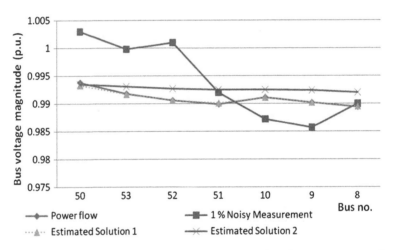

Fig. 7 Feeder F2 bus voltage magnitudes: power flow study result; noisy measurement; estimated magnitude (no measurement noise); estimated magnitude (1.0% measurement noise)

The condition number of the process matrix, $h^T h$, can be found as the ratio of largest singular value to smallest singular value. A large condition number implies that small perturbations to the measurement vector, z, may result in significant error of the solution estimates, $\hat{x}$. Generally, the number and placement of measurements in the system affects the condition of the process matrix. Literature suggests that condition numbers smaller than about 10^{12} provide reasonable estimates in transmission system state estimators [20]. More details about the factors that contribute to ill-conditioning in power systems may be found in [16].

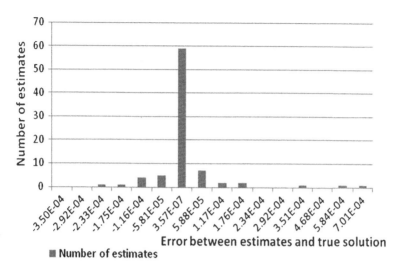

Fig. 8 Histogram of estimation errors for Case I

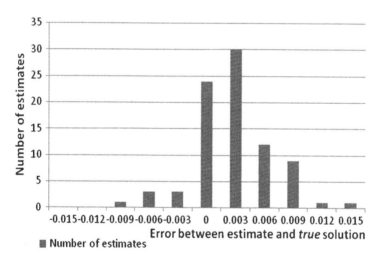

Fig. 9 Histogram of estimation errors for Case II

6 Alternative Applications at the Point of End Use

As stated earlier, distribution engineering may not have received the full attention which it deserves. Similarly, end use may have received even less engineering attention. One exception is the redesign of certain selected consumer appliances such as refrigerators and air conditioners to render them more efficient. In the distribution system of the future, there may be added attention to end-use requirements. The following are offered as potential applications:

Table 2 Innovative concepts for energy distribution and utilization

Concept	Most promising sectors		
	Residential	Commercial	Industrial
Conservation	x	x	
Direct digital control		x	x
Energy storage	x	x	x
Distributed generation	x	x	x
Peak reduction strategies	x		x
Sales of "green" energy	x		

- *Prioritizing loads*: the concept of load prioritization entails the use of an index or rank for every individual load in distribution secondary (end-use) systems. Thus, on a scale of 1 (unimportant) to 10 (important) an air conditioner may have rank 4, while a refrigerator might have rank 9. Ranking would ideally reflect the priority of that load as indicated by the end user. This concept may also be used for load curtailment and load scheduling.
- *High resolution time-of-day rates*: in many venues in the United States, time-of-day rates are in place. These are invariably organized into two time periods corresponding to high demand and low demand. The proposed concept would expand this philosophy to include more time periods that are reflective of changes in system loading. For example, the day might be divided into 12 or 24 periods that range from "high demand" to "low demand". Further, the periods might be adaptive, changing in time, and coordinated with the users' energy management computer that coordinates loads, DG and energy storage elements.
- *Incentivizing the use of more efficient loads*: just as 100 W incandescent lamps have been phased out in some jurisdictions, this concept may be extended to other load types, especially to inefficient loads of older design. In a time multiplexed version of this concept, certain *smart* loads may be switched on/off at predetermined times.
- *Energy storage*: there have been efforts in heating ventilating and air conditioning (HVAC) technologies in which these loads are scheduled optimally for cooling/heating to minimize net energy consumption. The concept might be applied more broadly to accommodate thermal storage in refrigerators; electric automobile battery storage and charging; compressed air energy storage and its scheduling. Also, there may be applications in which water might be pumped to a high holding tank and the water discharged through a turbine for energy recovery.

Table 2 shows some innovative concepts and the milieu in which these concepts might apply.

7 Summary and Conclusions

The principal recommendation of this work is that state estimation in distribution systems is potentially valuable for control, fault management, and energy management. Synchronized phasor measurements appear to be specially suited for this application. The traditional state estimation formulation problem has been modified to accommodate complex (phasor) measurements and this renders the state estimation problem as linear. Examples indicate the value of the concept.

Applied to distribution system control, the linear distribution state estimator may become an effective tool at estimating distribution system bus voltages, bus injections and line flows. Monitoring system nodes, particularly those nodes at which distributed generation, energy storage, significant concentrations of PHEV and EV loads and DSR enabled loads. With effective estimation of system states, decision algorithms or applications directing control actions such as reducing consumption at a node, using energy storage/batteries for regulation purposes, or signaling distributed generators and D-FACTS devices to modify power flows are easily facilitated within the distribution system. How many measurements to include, and where to place them.

References

1. United States Department of Energy Smart grid. http://www.oe.energy.gov/smartgrid.htm. Accessed on October 25, 2011
2. Varaiya PP, Wu FF, Bialek JW (2011) Smart operation of smart grid: risk limiting dispatch. Proc IEEE 99:1, pp. 40–57
3. Garrity TF (2009) Innovation and trends for future electric power systems. IEEE Power Systems Conference, IEEE Clemson SC, 1–8
4. Brown RE (2008) Impact of smart grid on distribution system design. Proceedings of IEEE power energy society general meeting, IEEE, Pittsburgh PA, pp. 1–4
5. Heydt GT (2010) The next generation of power distribution systems. IEEE Trans Smart Grid 1(3):225
6. Momoh JA (2008) Electric power distribution, automation, protection and control, CRC, New York
7. Northcote-Green J, Wilson R (2007) Control and automation of electric power distribution systems, CRC, Boca Raton
8. Chen RL, Sabir S (2001) The benefits of implementing distribution automation and system monitoring in the open electricity market. In: Proceeding of the IEEE Canadian conference on electrical and computer engineering, Toronto, pp. 825–830
9. Roe C, Meisel J, Meliopoulos APS, Evangelos F, Overbye T (2009) Power system level impacts of PHEVs. In: Proceedings 42nd Hawaii international conference on system sciences, University of Hawaii, Kona HI, pp. 1–10
10. North American Synchrophasor Initiative documents. www.naspi.org. Accessed on October 25, 2011
11. Wu J, Zhou J, Wang Z, Zhang D, Xu S (2008) PMU standard of China. In: Proceedings of IEEE Asia Pacific conference on circuits and systems, Macao, pp. 639–641

12. Kyriakides E, Heydt GT (2009) Synchronized measurements in power system operation: international practices and research issues. In: IEEE power energy society general meeting, IEEE Calgary AB, Canada, pp. 1–3
13. Abur A (2009) Impact of phasor measurements on state estimation. In: International conference on electrical and electronics engineering, University of Hawaii, Kona HI, pp. 13–17
14. Bhattacharya S, Zhao T, Wang G, Dutta S, Baek S, Du Y, Parkhideh B, Zhou X, Huang AQ (2010) Design and development of generation-I silicon based solid state transformer. In: Proceedings of 25th annual IEEE applied power electronics conference, IEEE, Palm Springs, CA, pp. 1666–1673
15. IEEE Standard C37.118–2005 (2005) Standard for synchrophasors for power systems. Piscataway, NJ
16. Abur A, Gómez Expósito A (2004) Power system state estimation theory and implementation. Dekker, New York
17. Monticelli A (2000) Electric power system state estimation, Proc IEEE 88(2):262–282
18. Golub GH, Van Loan CF (1996) Matrix computations. John Hopkins University Press, MD
19. Allen MB III, Isaacson EL (1998) Numerical analysis for applied science. Wiley, NY
20. Ebrahimian R, Baldick R (2001) State estimator condition number analysis, IEEE Trans Power Syst 16(2):273–279
21. Billinton R, Kumar S, Chowdhury N, Chu K, Debnath K, Goel L, Khan E, Kos P, Nourbakhsh G, Oteng-Odjei J (1989) A reliability test system for educational purposes – basic data, IEEE Trans Power Syst 4(3):1238–1244
22. Glover JD, Sarma MS, Overbye T (2007) Power system analysis and design. Cenage Learning, KY
23. Billinton R, Jonnavithalu S (1996) A test system for teaching overall power system reliability assessment, IEEE Trans Power Deliv 11(4):1670–1676

The Influence of Time Delays on Decentralized Economic Dispatch by Using Incremental Cost Consensus Algorithm

Ziang Zhang and Mo-Yuen Chow

Abstract In a smart grid, robust energy management algorithms should have the ability to operate correctly in the presence of unreliable communication capabilities, and often in the absence of a central control mechanism. Effective distributed control algorithms could be embedded in distributed controllers to properly allocate electrical power among connected buses autonomously. By selecting the incremental cost of each generation unit as the consensus variable, the incremental cost consensus (ICC) algorithm is able to solve the conventional centralized economic dispatch problem (EDP) in a distributed manner. However, the communication time-delay may cause instability of the system and should be considered during the design process. The mathematical formulation of the ICC algorithm with time-delay is presented in this chapter. Several case studies are also presented to show the system characteristics of the ICC algorithm with time-delay.

1 Introduction

State-of-the-art information and communication technologies (ICT) can provide a much more reliable communication network for the next generation power grid compared with the current power system. Most of the conventional control algorithms and energy management techniques are designed to operate under very limited communication requirements, thus their performance is also limited. With the improvements in communications and the control network, more powerful energy management techniques can be applied to the operation and control of the power system.

Z. Zhang (✉) • M.-Y. Chow
Department of Electrical and Computer Engineering, North Carolina State University, Raleigh, NC 27606, USA
e-mail: zzhang15@ncsu.edu; chow@ncsu.edu

A. Chakrabortty and M.D. Ilić (eds.), *Control and Optimization Methods for Electric Smart Grids*, Power Electronics and Power Systems 3, DOI 10.1007/978-1-4614-1605-0_16, © Springer Science+Business Media, LLC 2012

In the last few years, new operating techniques have been developed for smart grid operations. The conventional optimal power flow (OPF) has been extended to the distribution level [1] and revised under a smart grid environment [2]. The unit commitment (UC) has also been modified for the smart grid in order to push its performance further [3]. The improved particle swarm optimization (IPSO) has been employed to solve the economic dispatch problem (EDP) with a nonconvex cost function [4]. In addition to improving the existing algorithm, new optimization techniques have been applied to smart grid operations, such as risk-limiting dispatch [5] and the jump and shift method [6]. In addition, some new auction-based algorithms, such as the intelligent auction scheme for the smart grid market [7] and flow-gate bidding [8], have been developed.

Unlike the conventional power system, the smart grid will rely heavily on an integrated communication and control network. This type of system is known as a large-scale networked control system (LSNCS), in which the components of the system, such as sensors, controllers, and actuators, are connected by the communication network [9]. A smart grid can be viewed as an LSNCS. The conventional centralized control scheme, however, may encounter severe challenges on applying LSNCS. The control center is required to have a high level of connectivity, which may also impose a substantial computational burden. The centralized control scheme is also more sensitive to failure than distributed control schemes [10]. Moreover, the "plug-and-play" technologies will make the topology of the smart grid vary over time [11]. The power system is inherently distributed, so one of the possible control approaches is to apply distributed control algorithms to power system problems. Figure 1 is a prototype of a future power system. In this multi-agent system (MAS), effective distributed control algorithms could be embedded in distributed controllers to properly allocate electrical power among connected buses autonomously. The agent-based distributed control for a smart grid also has been discussed in [12, 13].

A fundamental problem in distributed control systems is that of having all of the distributed controllers reach a consensus. The consensus problem has been studied extensively in the last several years in the system and control areas [14, 15]. By using the distributed consensus framework, we have proposed an incremental cost consensus (ICC) algorithm as an example to illustrate the use of distributed control on a smart grid in [16]. The ICC algorithm can solve the centralized EDP in a distributed fashion. The extended version of the ICC has been presented in [17], in which the system mismatch can be acquired by an average consensus network. The convergence rate issue and leader selection criterion issue of the algorithm have been discussed in [16, 18]. Since the information is often received with time delays, this chapter focuses on the influence of the time delay on the system convergence rate.

The rest of this chapter organized as follows. In Sect. 2, basic graph theory notation and the consensus algorithm with a time delay are introduced. The problem formulation of the ICC algorithm is established in Sect. 3. The simulation results based on different time-delay scenarios are given in Sect. 4. Finally, some conclusions are drawn in Sect. 5.

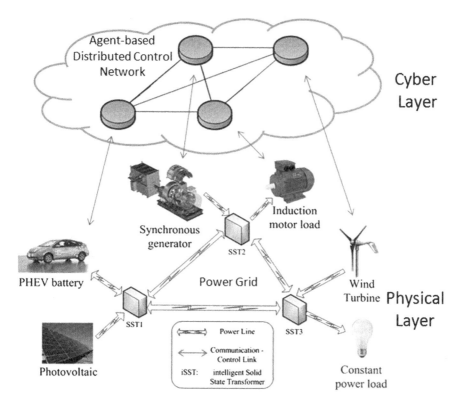

Fig. 1 A prototype of future power system

2   Consensus Algorithm Preliminaries

2.1   Graph Theory Notations

A graph G is used to model the network topology of the system. Graph G is a pair of sets (V, E), where V is a finite nonempty set of elements called "vertices" or "nodes," and E is a set of unordered pairs of distinct vertices called "edges." A simple graph is an unweighted, undirected graph containing no graph loops and no multiple edges [19]. Unless stated otherwise, the unqualified term "graph" usually refers to a simple graph. A graph is connected if there is a path between any distinct pair of nodes. A directed tree is a digraph, where every node except the root has exactly one parent; the root has no parent and has a directed path to every other node. The directed spanning tree of a digraph is a directed tree formed by graph edges that connect all of the nodes on the graph.

An adjacency matrix A of finite graph G of n vertices is the $n \times n$ matrix, where the off-diagonal entry a_{ij} is the number of edges from vertex i to vertex j. In the special case of a finite simple graph, the adjacency matrix is a $(0, 1)$-matrix with zeros on its diagonal. If the graph is undirected, the adjacency matrix is symmetrical. Let matrix $L = [l_{ij}]$ be defined as:

$$l_{ii} = \sum_{i \neq j} a_{ij}, \quad \text{for on-diagonal elements;}$$

$$l_{ij} = -a_{ij}, \quad \text{for off-diagonal elements.}$$

For an undirected graph, L is called the Laplacian matrix, and it has the property of being symmetric positive semidefinite.

2.2 Consensus with Time Delay

Let $x_i \in \mathbb{R}$ denote the state value of node i. The state value of a node might be its physical quantities, such as voltage, output power, or incremental cost. We say that the nodes of a network have reached a consensus if and only if $x_i = x_j$ for all i, j [14]. Assuming each agent has a first-order dynamic, a continuous-time consensus algorithm is given in [14, 15] as:

$$\dot{x}_i(t) = -\sum_{j=1}^{n} a_{ij}(x_i(t) - x_j(t)), \quad i = 1, \ldots, n, \tag{1}$$

where a_{ij} is the (i, j) entry of the adjacency matrix A. The consensus algorithm also can be written in matrix form as

$$\dot{x}(t) = -L_n x(t), \tag{2}$$

where L_n is the $n \times n$ graph Laplacian matrix.

In a real-world network, it takes time for messages to propagate through it. If the message takes time τ to travel from agent i to agent j, τ can be considered as the time delay. Then the consensus algorithm becomes:

$$\dot{x}_i(t) = -\sum_{j=1}^{n} a_{ij}(x_i(t - \tau) - x_j(t - \tau)). \tag{3}$$

The time delay could destabilize the consensus network, which is a sufficient condition for the convergence of (3) introduced in [20]:

$$\tau < \frac{\pi}{2\lambda_n},$$

where λ_n is the largest eigenvalue of graph Laplacian L.

When information takes a fixed time T to travel between nodes, we need to model the consensus network dynamics as a discrete-time dynamic system [15] to facilitate analysis. A discrete-time consensus algorithm is described by:

$$x_i[k+1] = \sum_{j=1}^{n} d_{ij} x_j[k], i = 1, \ldots, n,$$ (4)

where k is the discrete-time index and d_{ij} is the (i, j) entry of the row-stochastic matrix D, which can be defined by the following:

$$d_{ij} = l_{ij} \left/ \sum_{j=1}^{n} |l_{ij}|, i = 1, \ldots, n. \right.$$ (5)

Similarly, as in continuous time, the consensus algorithm with time delay in discrete time [21] is:

$$x_i[k+1] = \sum_{j=1}^{n} d_{ij} x_j[k-\tau].$$ (6)

Since each data packet in a communication network always arrives discretely, the discrete consensus algorithm has been selected for further development.

2.3 Average Consensus Algorithm

The consensus algorithms introduced in the previous section guarantee the consensus can be reached as long as a direct spanning tree exists in the communication network. However, they do not guarantee what the steady-state value is. There is a special case of consensus, called the "average consensus algorithm," in which the final steady-state value is the average of the initial condition of every agent:

$$x_i[k+1] = \sum_{j=1}^{n} w_{ij} x_j[k],$$ (7)

where w_{ij} is the (i, j) entry of the updating matrix W. The paper in [22] shows that the average consensus can be reached asymptotically if the communication topology is balanced and
$$W = I - \mu L,$$ (8)

when $0 < \mu < 1/d_{max}$ and d_{max} is the maximum degree of the graph. Note that an undirected graph is always a balanced graph, thus the two-way communication ability is preferred.

3   Incremental Consensus Algorithm with Time Delay

3.1   Leader–Follower Consensus for Incremental Cost

The objective of economic dispatch is to minimize the fuel cost of the operation. Assume that the generation units have the quadratic cost function:

$$C_i(P_{Gi}) = \alpha_i + \beta_i P_{Gi} + \gamma_i P_{Gi}^2. \tag{9}$$

The objective of the EDP is to minimize the total cost of operations for an n generator system:

$$C_{\text{total}} = \sum_{i=1}^{n} C_i(P_{Gi}). \tag{10}$$

under the power balance constraint:

$$P_{\text{D}} - \sum_{i=1}^{n} P_{Gi} = 0, \tag{11}$$

where P_{Gi} denotes the output power of unit i and P_{D} denotes the total power demand.

Using the consensus algorithm as a basic framework, the EDP can be solved in a distributed manner. The definition of Incremental Cost for each generator is the same as that for the conventional economic dispatch:

$$\text{IC}_i = \frac{\partial C_i(P_{Gi})}{\partial P_{Gi}} = \lambda_i \quad i = 1, 2, \ldots, n. \tag{12}$$

Select IC as the consensus variable, plug in the first-order discrete consensus algorithm (6):

$$\lambda_i[k+1] = \sum_{j=1}^{n} d_{ij} \lambda_j[k - \tau], \quad i = 1, \ldots, n, \tag{13}$$

where d_{ij} is the (i, j) entry of the row-stochastic matrix D_n and τ represents the time delay. Equation (13) is the updating rule for followers.

In order to satisfy the power balance constraint, define ΔP to indicate the mismatch between the total demand and overall power generated:

$$\Delta P = P_{\text{D}} - \sum_{i=1}^{n} P_{Gi}. \tag{14}$$

The update rule for the leader generator becomes:

$$\lambda_i[k+1] = \sum_{j=1}^{n} d_{ij} \lambda_j[k - \tau] + \varepsilon \Delta P, \tag{15}$$

where ε is a positive scalar. We call ε the convergence coefficient and it controls the convergence speed of the leader generator. Equation (15) is the updating rule for the leaders.

3.2 Average Consensus for Power Mismatch

Comparing (13) and (15), it is obvious that the leader has the external information. The information ΔP can be acquired through a distributed average consensus network [17]:

$$\Delta P_i[k+1] = \sum_{j=1}^{n} w_{ij}\Delta P_j[k-\tau], \quad i = 1, \ldots, n, \tag{16}$$

where ΔP_i is the power mismatch at each node and ΔP is the total power mismatch that is equal to the one defined in (14), w_{ij} is the (i, j) entry of the updating matrix W defined by (8).

By using the average consensus algorithm to solve the value of ΔP, every node can acquire the value of the total power mismatch so multiple leaders in the network could improve the convergence rate [17].

The power generation constraint also needs to be considered:

$$\begin{cases} \lambda_i = \lambda_{i\_lower}, \text{ when } P_{Gi} < P_{Gi,min} \\ \lambda_i[k+1] = \sum_{j=1}^{n} d_{ij}\lambda_j[k-\tau], \text{ when } P_{Gi,min} \le P_{Gi} \le P_{Gi,max} \\ \lambda_i = \lambda_{i\_upper}, \text{ when } P_{Gi} > P_{Gi,max}. \end{cases} \tag{17}$$

Equations (13) through (17) are the mathematical representations of the ICC algorithm. Figure 2 is a flowchart that represents the procedure of the ICC algorithm.

4 Simulation Results

In this section, several case studies are discussed. Figure 3 shows the communication topology of a five-unit system. The corresponding graph Laplacian matrix is

$$L = \begin{bmatrix} 3 & -1 & 0 & -1 & -1 \\ -1 & 3 & -1 & 0 & -1 \\ 0 & -1 & 2 & -1 & 0 \\ -1 & 0 & -1 & 2 & 0 \\ -1 & -1 & 0 & 0 & 2 \end{bmatrix}.$$

There is a generation unit and a load in each area. The parameters for the five units are shown in Table 1.

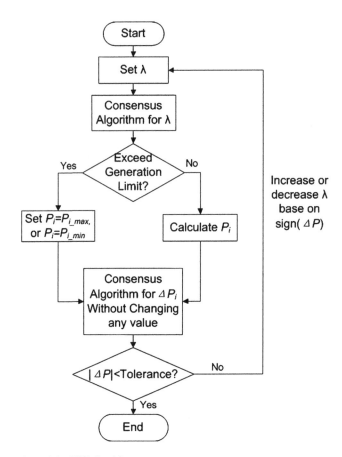

Fig. 2 Flowchart of the ICC algorithm

Fig. 3 Communication topology

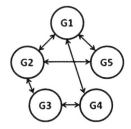

Table 1 Parameters of five-unit system

Unit	α_i	β_i	γ_i
1	561	7.92	0.001562
2	310	7.85	0.00194
3	78	7.8	0.00482
4	561	7.92	0.001562
5	78	7.8	0.00482

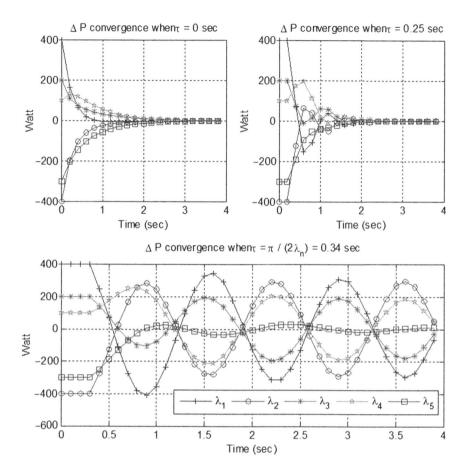

Fig. 4 Convergence of power mismatch under different time delays

4.1 Time Delay on Mismatch Consensus

First, we test the influence of time delay on the low-level mismatch calculation using the average consensus algorithm. Using (16) as the updating rule with a time delay of $\tau = 0, 0.25$, and 0.34, the system responses are shown in Fig. 4. This is a straightforward case, and the characteristics of the system are consistent with the regular average consensus algorithm.

Assume five units each have a different initial ΔP; the goal is to calculate the average value. Based on the Laplacian graph and (4), the upper bound of the time delay for this topology is $\tau = 0.34$ second. Thus, if the time delay is more than 0.34 second, a consensus cannot be reached. As Fig. 4 shows, the time delay can create oscillation in the system. The system has a similar response as a critically damped system when the time delay is equal to the upper bound defined in Sect. 2.2. Further increasing the time delay destabilizes the system.

Since the control decision is based on the system mismatch information, this upper bound of the time delay is the hard limit of the ICC algorithm. It is worth mentioning that if the packet cannot arrive at its neighboring agents before this deadline, the system mismatch cannot be calculated by the average consensus algorithm.

4.2   Time Delay on IC Consensus

After the system mismatch is acquired by the low-level average consensus network, the mismatch information is sent to each agent. If an agent decides to respond to the mismatch signal, then we call this agent a "leader" and it will use (15) as the updating rule. Otherwise, the agent will use (13) as the updating rule and we call it a "follower." The network can have multiple leaders as long as the leader's states are all same.

4.2.1   Single Leader vs. Multi-leader Without Time Delay

In this case study, different numbers of agents have been selected as leader, as shown in the results in Fig. 5. If you select more agents, you can reduce the convergence rate. It is the slowest when only G1 is the leader. After adding G5 as another leader, the mismatch between P_D and P_G reduced more quickly. The system reaches a consensus even faster when you select G3 and G5 as the leaders.

4.2.2   With a Time Delay of $\tau = 0.25$ Second

After adding a 0.25-second delay to the communication network, you will get the simulation results shown in Fig. 6. Similarly, as in the previous case study, the system with the time delay undergoes oscillation. The system with G1 and G5 acting as leaders will have a better performance than the system that only has G1 as leader. However, the performance of the system with G3 and G5 as leaders becomes the worst. Within the given time period, the five units have not reached consensus and the mismatch information has not reached a steady state. This phenomenon may be due to the fact that the Hamming distance between G3 and G5 is larger than the Hamming distance between G1 and G5.

4.2.3   Leader Election Under a Time Delay

The leader election criterion for the ICC algorithm has been discussed in [18]. The basic rule is to use the Eigenvector Centrality as the measurement; pick the node that has the highest Eigenvector Centrality. Here, we extend the conclusion to the

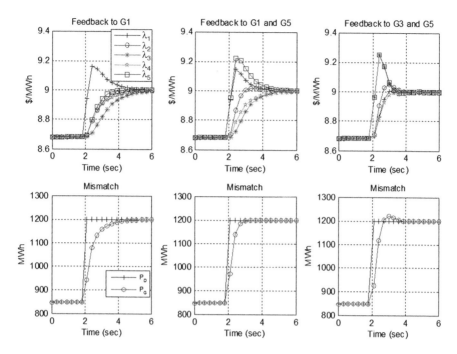

Fig. 5 IC consensus without time-delay

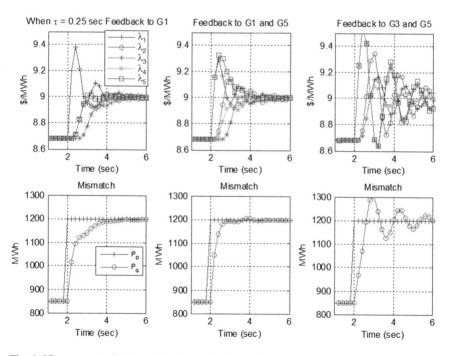

Fig. 6 IC consensus with time delay of $\tau = 0.25$ second

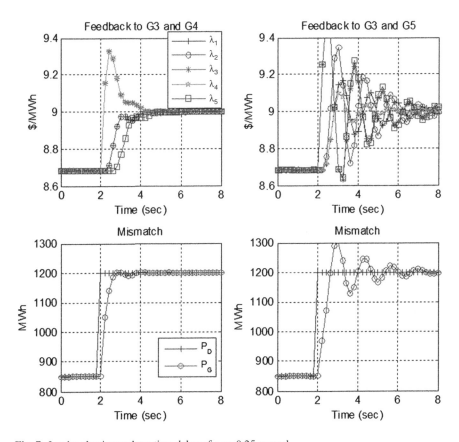

Fig. 7 Leader election under a time delay of $\tau = 0.25$ second

time delay network. The previous case study shows that arbitrarily adding a unit
as leader may not reduce the convergence time under the network with the time
delay. The eigenvector centralities through G1 to G5 are 0.53, 0.53, 0.36, 0.36,
and 0.42, respectively. Based on the simulation results in Fig. 7, the combination
of neighboring units acting as leader has a better performance than the arbitrarily
picked leaders. Even though G5 has high eigenvector centralities, the combination
of G3 and G4 still converges faster than the combination of G3 and G5 because G4
is adjacent to G3.

5 Conclusions

We have explored the use of the consensus algorithm as an effective distributed
control to operate the smart grid. A conventional EDP can be solved in distributed fashion by using the consensus framework. A double consensus-based ICC

algorithm has been discussed to illustrate the use of distributed control on a smart grid. In addition, the sufficient condition of the average consensus algorithm can be applied to the ICC algorithm, which has been extended to taking a communication time delay into consideration.

The influence of a time delay has been discussed through case studies. The low-level average consensus for a power mismatch calculation has the same identical features as the original average consensus problem does. The IC consensus network is more complicated since multiple parameters can affect the convergence rate. We have also found that the same leader configurations have different characteristics depending on whether they are with or without a time delay. For instance, an additional leader can reduce the convergence time when there is no time delay from the communication network. However, it is not always true if the system has a time delay. Moreover, with a time delay in the picture, the centrality indices are no longer the only parameters that can indicate the convergence rate. The relative Hamming distance between the leaders also affects the convergence rate. There is even a chance that arbitrarily chosen leaders may destabilize a stable system.

Acknowledgments These works were partially supported by the National Science Foundation (NSF) under Award Number EEC-08212121.

References

1. Paudyal S, Canizares C, Bhattacharya K (2011) Optimal operation of distribution feeders in smart grids. IEEE Trans Ind Electron 58(10):4495–4503
2. Bruno S, Lamonaca S, Rotondo G et al (2011) Unbalanced three-phase optimal power flow for smart grids. IEEE Trans Ind Electron 58(10):4504–4513
3. Hedman KW, Ferris MC, O'Neill RP et al (2010) Co-optimization of generation unit commitment and transmission switching with N-1 reliability. IEEE Trans Power Syst 25(2):1052–1063
4. Park J-B, Jeong Y-W, Shin J-R et al (2010) An improved particle swarm optimization for nonconvex economic dispatch problems. IEEE Trans Power Syst 25(1):156–166
5. Varaiya PP, Wu FF, Bialek JW (2011) Smart operation of smart grid: risk-limiting dispatch. Proc IEEE 99(1):40–57
6. Chen S, Gooi H (2011) Jump and shift method for multi-objective optimization. IEEE Trans Ind Electron 58(10):4538–4548
7. Ramachandran B, Srivastava SK, Edrington CS et al (2011) An intelligent auction scheme for smart grid market using a hybrid immune algorithm. IEEE Trans Ind Electron 58(10):4603–4612
8. Hedman KW, O'Neill RP, Fisher EB et al (2011) Smart flexible just-in-time transmission and flowgate bidding. IEEE Trans Power Syst 26(1):93–102
9. Gupta RA, Chow M-Y (2010) Networked control system: overview and research trends. IEEE Trans Ind Electron 57(7):2527–2535
10. D'Andrea R, Dullerud GE (2003) Distributed control design for spatially interconnected systems. IEEE Trans Autom Control 48(9):1478–1495
11. Huang AQ, Crow ML, Heydt GT et al (2011) The future renewable electric energy delivery and management (FREEDM) system: the energy internet. Proc IEEE 99(1):133–148
12. Dimeas AL, Hatziargyriou ND (2007) Agent-based control for microgrids. Power engineering society general meeting, IEEE, 24–28 June 2007

13. Suryanarayanan S, Mitra J, Biswas S (2010) A conceptual framework of a hierarchically networked agent-based microgrid architecture. Transmission and distribution conference and exposition. IEEE PES, 19–22 April 2010
14. Olfati-Saber R, Fax JA, Murray RM (2007) Consensus and cooperation in networked multi-agent systems. Proc IEEE 95(1):215–233
15. Ren W, Beard RW (2008) Distributed consensus in multi-vehicle cooperative control. Springer
16. Zhang Z, Chow M-Y (2011) Incremental cost consensus algorithm in a smart grid environment. Power and energy society general meeting, IEEE, 24–29 July 2011
17. Zhang Z, Ying X, Chow M-Y (2011) Decentralizing the economic dispatch problem using a two-level incremental cost consensus algorithm in a smart grid environment. North American power symposium, pp 4–6 August 2011
18. Zhang Z, Chow M-Y (2011) The leader election criterion for decentralized economic dispatch using incremental cost consensus algorithm. IEEE industrial electronics society annual conference, pp 7–20 November 2011, accepted
19. Gibbons A (1985) Algorithmic graph theory. Cambridge University Press, Cambridge
20. Olfati-Saber R, Murray RM (2004) Consensus problems in networks of agents with switching topology and time-delays. IEEE Trans Autom Control 49(9):1520–1533
21. Wang L, Xiao F (2006) A new approach to consensus problems for discrete-time multiagent systems with time-delays. Am Control Conf 2006:14–16
22. Kingston DB, Beard RW (2006) Discrete-time average-consensus under switching network topologies. Am Control Conf 3551–3556

An Adaptive Wide-Area Power System Damping Controller using Synchrophasor Data

Joe H. Chow and Scott G. Ghiocel

Abstract This chapter presents an adaptive wide-area interarea mode damping controller for power systems using synchrophasor data. A key consideration in the control design is the time delay in computing the phasor quantities and the variable communication network latency for controllers to use remote synchrophasor data. The adaptive switching controller comprises several phase compensators, each designed for a specific data latency. Based on the latency of the arriving synchrophasor data, the adaptive controller will select the appropriate compensator to use. The design is illustrated with a two-area power system. Applications to large power systems will be discussed.

1 Introduction

With the deployment of distributed phasor measurement units (PMUs) synchronized by the clock signal from the Global Positioning System (GPS) [1], it becomes possible to use remote signals for the design of power system interarea oscillation damping controllers. These signals, containing both magnitude and phase information, can be used to exhibit more clearly the interarea modes to be damped. One example of a remote signal is the angle difference between two groups of coherent generators [2]. Some of these signals, however, may be remote to the controllers applying the damping torque, and must be transmitted on communication links such as the Internet.

Synchrophasor data have two major sources of latency, namely, a mostly fixed latency for signal processing algorithms to compute the phasors and a variable latency in data transmission on the Internet. A major task in designing power system damping controller is to provide an appropriate phase compensation [3].

J.H. Chow (✉) • S.G. Ghiocel
Rensselaer Polytechnic Institute, Troy, NY, USA
e-mail: chowj@rpi.edu; ghiocs@rpi.edu

A. Chakrabortty and M.D. Ilić (eds.), *Control and Optimization Methods for Electric Smart Grids*, Power Electronics and Power Systems 3,
DOI 10.1007/978-1-4614-1605-0_17, © Springer Science+Business Media, LLC 2012

Signals with latency would require either additional phase-lead compensation, or less phase-lag compensation. In this chapter, we propose an adaptive control where the phase compensator is adjusted according to the variable latency. Such a scheme is practical because synchrophasor data are time-stamped, and thus the latency is known accurately.

The chapter is organized into the following sections. In Sect. 2, we provide an overview of power system electromechanical mode damping controllers, namely power system stabilizers (PSSs) and Flexible AC Transmission System (FACTS) controllers. In Sect. 3, we discuss aspects of using PMU data for damping control. In Sect. 4, we develop the adaptive damping controller to counter the latency of PMU data. Section 5 provides an illustration of the adaptive damping controller for a two-area, four-machine system.

2 Overview of Electromechanical Mode Damping Controllers

In multi-machine power systems with long transmission lines and heavy power transfer, there may be several critical electromechanical modes that are poorly damped [4,5]. PSSs are a relatively inexpensive approach to enhance the damping of these troublesome modes [6]. Traditional PSS design uses a single-input feedback loop, aimed at damping the local mode, that is, the oscillations of a generator versus a stiff equivalent system. The paper [7] systematically investigates the relative merits and disadvantages of the local input signals for damping control, including the machine speed ω, electrical output power P_e, bus frequency f, and terminal voltage magnitude V_T. More recently, dual-input integral-of-accelerating-power PSSs using both ω and P_e have been introduced [8, 9], resulting in much higher damping improvement. Because of the dominant presence of the local mode in the locally available signals, PSSs tend to provide most of their benefits to the local mode. By providing some phase lead at the frequencies of the interarea modes, a PSS may also provide some incremental benefit to the damping of interarea modes. In this approach, a large number of PSSs may have to be used to provide sufficient interarea mode damping benefit. This is the practice of the US Western Power System, in which all generators greater than 30 MVA or belonging a complex having an aggregate capacity greater than 75 MVA are required to have PSSs installed [10]. To make the PSSs more effective in damping interarea modes, signals rich in interarea mode contents need to be communicated to the PSSs.

In addition to PSSs, interarea damping control can be accomplished using transmission network controllers, such as HVDC systems and FACTS controllers, including Static Var Controllers (SVCs), Thyristor-controlled Series Compensators (TCSCs), and Unified Power Flow Controllers (UPFCs) [11]. For an SVC, in addition to the voltage regulation location, a supplemental damping function using a local network signal is used [12]. The paper [12] examines the relative merits of the line current, line active power flow, line current magnitude, bus voltage, and bus frequency as the input signal to the damping controller. Because the SVC is

Fig. 1 Synthesized area
voltages from local voltage
and current measurements

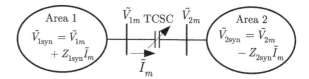

regulating the bus voltage, which is a "nodal" variable, the effective damping control
input signals are "flow" variables such as the line current and active power flow. In
particular, the line current magnitude is recommended, as it is independent of the
power flow direction on the transfer path. The results in [13] extend the SVC design
to include TCSC damping control design, in which the authors develop an analytical
framework based on energy functions.

The use of remote signals for TCSC damping control is investigated in [14] for
a two-area system. The design idea is to use the angular difference $\theta_1 - \theta_2$, where
θ_1 and θ_2 represent the centers of machine angles for Areas 1 and 2, respectively,
as the input to the damping controller. The angles are synthesized using the TCSC
bus voltages and line current flows to compute the angle of an equivalent voltage in
each of the areas (see Fig. 1). In this approach, it is critical that the impedances Z_{1syn}
and Z_{2syn} used to extrapolate the synthesized voltages $\tilde{V}_{1syn}$ and $\tilde{V}_{2syn}$ are selected
properly. The paper [14] investigates the impact on the observability of the interarea
mode as a function of Z_{1syn} and Z_{2syn}. Although the concept is to use remote signals,
the feedback loop is implemented using only local signals. For the control to be
effective, however, it may be necessary to vary Z_{1syn} and Z_{2syn} as the system loading
changes, and if the network configuration is altered, such as in case of a line trip.

3 Using PMU Signals for Damping Control

High-sampling-rate synchronized phasor measurements at various locations
throughout a power grid can provide signals rich in interarea modes. For example,
instead of synthesizing the angle difference between the two areas using locally
measured currents and voltages as shown in Fig. 1, the angles in the areas can
be directly measured by PMUs and communicated to the damping controller. The
practical issues in using the synchrophasor data for damping control include latency,
geographic coverage, and potential data loss.

3.1 PMU Data Latency

Because PMUs are distributed in a power network, it is necessary to have a phasor
data network and communication system to route and collect the distributed data at
locations where they are applied. Figure 2 shows a typical hierarchical architecture,
where the GPS-synchronized time-stamped PMU data measured at a substation is

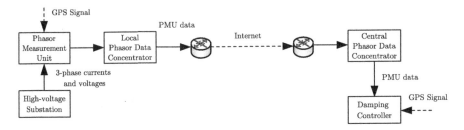

Fig. 2 PMU data communication path

first collected at a local phasor data concentrator (PDC). The PDC then transmits the PMU data via the Internet through a series of routers to a central/regional PDC. In general, the central PDC collects PMU data from many separate local PDCs in order to generate a desired damping controller input signal.

In a phasor data network, each packet of PMU data will be time-stamped. Once all the required PMU data for a particular time instant have been collected, they are sent to the damping controller to generate a damping signal. By comparing the time stamp of the arriving PMU data to the GPS signal, a damping controller can calculate the data latency.

One component of data latency is the time for a data packet to travel in a particular medium. The communication time for a transmission path of 1,000 km of optical fiber is estimated at 5 ms for a single transmission using the UDP protocol and at 15 ms for a retransmission using TCP/IP [15, 16]. A more detailed breakdown for the total time delay T_{ld} in a telecommunication link is provided in [17] as

$$T_{ld} = T_s + T_b + T_p + T_r, \tag{1}$$

where

$$T_s = P_s/D_r \tag{2}$$

is the serial delay, where P_s is the packet size in bits and D_r is the transmission rate of the link in bits/s, T_b is the delay between data packets,

$$T_p = L/v \tag{3}$$

is the propagation delay, where L is the link length in km and v is the propagation speed in the link in km/s, and T_r is the routing delay.

A practical analysis in PMU data latency has been provided in [15] for the Wide-Area Control System (WACS) under construction for the power system in Quebec, Canada. The control system architecture consists of substation PDCs collecting the local measured phasor data and forwarding the data to a central PDC on a dedicated fiber network. The central PDC will then send the relevant phasor data information to be used by PSSs to provide interarea damping torque. An estimate of the latency of the various components of the data communication system of the Quebec power system WACS is shown in Table 1 [15].

Table 1 Estimate of synchrophasor data time delay in Quebec power system		
PMU filter delay	73 ms
Local data concentration	16 ms
2,000 km in optical fiber	10 ms
Central data concentration	10 ms
Total estimated delay	109 ms

According to Table 1, the longest delay is for a PMU to compute synchrophasor data. Ultimately, for interarea damping control, we only need the positive sequence quantities of the phasors, but PMUs are designed to first compute the phasor representation of the individual phases [1]. In order to minimize the impact of noise, a data window of 1–2 cycles (sometimes up to four cycles) of the fundamental frequency is used. Thus, about half of the PMU processing delay may be due to the data window. In another WACS, total PMU data latency was measured to be between 60 and 80 ms, including 40 ms for the PMU data window [18].

As a result, for interarea damping control design, it would be prudent to design for a minimum latency of 100–150 ms. As a rough estimate of the additional complexity in control design, the delay should be quantified as a fraction of the period of the dominant interarea mode to be damped. Most interarea modes involving long distances have frequencies of 0.5 Hz or lower, that is, a period longer than 2 s. From a classical phase compensation design [3], a latency of 150 ms in a 2-s period is equivalent to a phase lag of $0.150/2 \times 360° = 27°$, which although not severe, should be taken into account in the control design. For control architectures not using dedicated optical fiber, the expected PMU data latency will be greater than the one reported in [15].

3.2 Geographical Coverage of PMU Data

Interarea modes are most visible in the speed and angle variables of machines located at the ends of the swing modes. However, most PMUs are located on high-voltage substations and not at the generator terminals. Thus, instead of using machine angles and speeds, bus voltage angles and frequencies can be used. In some systems, if there are PMUs on buses connected to generator substations, then the phasor state estimator technique [19] can be applied to calculate the generator terminal voltages and possibly the internal machine angles as well. However, such computation will increase the latency of the input signal.

The measured network variables will contain both interarea modes and local modes within a slow coherent area. Thus, if there are N_a bus angle variables θ_i and frequency variables f_i available in an area, it would be beneficial to use a weighted sum

$$\theta_a = \sum_{i=1}^{N_a} \alpha_i \theta_i, \quad f_a = \sum_{i=1}^{N_a} \beta_i f_i, \tag{4}$$

where the weights α_i and β_i are selected to eliminate the local mode components in θ_a and f_a. For measurements close to generator buses, the α_i and β_i would be weighting factors set according to the inertias of nearby generators.

When it is desirable to use machine angles and speeds for damping control, which are not measured directly by PMUs, the aggregate machine rotor angle δ_a and speed ω_a can be calculated using the Interarea Model Estimation method [20]. This method requires PMU measurements located along power transfer paths.

3.3   PMU Data Loss

PMU data loss may be due to a PMU not in service, loss of the GPS signal, or network congestion. Most PDCs assemble PMU data according to the time stamp of the data stream [21]. A PDC usually has a time-out function that PMU data not arriving within a specified time T_{to} will be dropped. Thus, late-arriving data are no different from PMU data that do not arrive at all, in a real-time application such as damping control. In [22], an experimental PMU system at the high voltage level in the South Brazil power system reported a 0.01% data loss rate during peak network traffic period. On the other hand, a PMU system at the low voltage level installed at universities throughout Brazil and thus covering a much larger area experienced a data loss rate as high as 14% during periods of network congestion.

To guard against PMU data loss for interarea damping control, the use of multiple signals (4) in the aggregate variables can be helpful. For example, if a measurement θ_i is missing, θ_a can still be computed using the other available angle measurements and an updated set of weights. Thus, the probability of not being able to compute an input signal for feedback control is greatly reduced if the missing data probabilities of the various signals are independent.

4   An Adaptive Damping Controller Using PMU Data

Recently, there is a significant amount of interest in research on power system damping controller design using time-delayed remote signals. For a simplified Brazilian power system [23], the controller needs to use a remote signal to remove a right-half-plane zero, which limits the controller performance. The impact of time delay of the remote signal on the damping control is illustrated via a root-locus analysis. In [17], a WACS using PSSs for damping interarea modes has been proposed. The impact on the interarea mode damping ratio due to synchrophasor data latency is investigated. In [24], the time delay is compensated by a forward rotation of the phasor in the time domain. This is an adaptive process as the time delay may be variable.

In this chapter, we propose an adaptive control algorithm, which compensates for the PMU data latency to improve the interarea mode damping performance.

Fig. 3 Adaptive control scheme

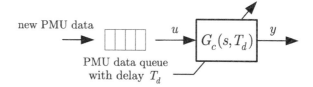

new PMU data

PMU data queue with delay T_d

A schematic of the controller architecture is shown in Fig. 3, which consists of an input data queue and the controller $G_c(s, T_d)$ set according to the latency T_d. There are two important components of this adaptive control.

The first component is a function to continuously monitor the latency of the arriving PMU data by subtracting the time stamp of transmitted PMU data from the actual time obtained from the GPS signal at the controller. There is an ideal minimum latency T_{1b} during periods of light internet traffic. The controller $G_c(s, T_{d1})$ is set for a latency $T_d = T_{d1}$ somewhat higher than T_{1b}. In this low-traffic steady state, the input data queue will mostly be empty except for the time duration between the arrival of a new packet of PMU data and T_{d1}, when it is used and then discarded by the controller. In this period, $G_c(s, T_{d1})$ is fixed, which contains phase compensation for the delay T_{d1} in addition to the appropriate phase compensation without any time delay.

The adaptive part becomes active when for some time $t = t'$, the data latency $T'_d = T_d(t')$ exceeds the latency T_{d1} currently used in the controller $G_c(s, T_{d1})$. The controller will automatically be updated to $G_c(s, T_{di})$, where $T_{di} \geq T'_d$ and will remain unchanged if the incoming PMU data stream has delay time less than T_{di}. The latency monitoring function will continue to track the incoming data stream and determine when it would be appropriate to reduce the delay time T_d in $G_c(s, T_d)$.

The determination of the delay time T_{di} for controller switching is given by the following algorithm.

Adaptive Control Algorithm

Prespecify a set of T_d values: $0 < T_{d1} < T_{d2} < \cdots < T_{dn}$.
At time $t = t_k$, where t is the time at the controller, the time delay T_{di} is used to set the controller.
for $t = t_k + \Delta t$, where Δt is the sampling period of the PMU data if the next data point is already in the input data buffer, or the incremental time of arrival of the next data point if the input data buffer is empty
    if the data delay is larger than the T_{di}, switch to a controller with the lowest latency T_{dj}, which is higher than the data delay
    else if the maximum latency of all the data in the last T_r s is less than T_{dj} where $T_{dj} < T_{di}$, switch in the controller with a lower latency T_{dj}
    else continue with the same controller
end
End algorithm

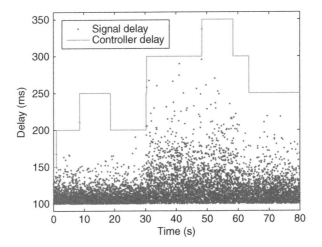

Fig. 4 Adaptive setting of T_d

The latency of an input variable comprising several signals such as the angle difference y is the maximum latency of among all the synchrophasor signals for that time stamp.

A key feature of the Adaptive Control Algorithm is that although switching to a compensator for a higher value of latency T_{di} is immediate because the data point cannot be used otherwise, there is a longer delay T_r for the compensator to switch to a compensator for a lower latency. The objective is to avoid switching constantly between compensators with different latencies, because such rapid switchings may create a sustained or even unstable oscillation, especially during periods of variable internet congestion. Furthermore, excessive switchings of compensators would cause transients in the controller itself. Here, we have used $T_r = 10$ s as the delay to switch to a compensator with a lower latency, which is more than five cycles of the interarea mode oscillation, guaranteeing that the interarea oscillation would have decayed before the compensator is switched.

The adaptive strategy used in this chapter is illustrated in Fig. 4. The PMU data arrival rate at the controller or central data concentrator is modeled as a Poisson process with a minimum delay of 100 ms.

The second component of the adaptive control scheme is the damping controller $G_c(s, T_d)$, which is dependent on the delay T_d and can be designed using a variety of approaches. One such kind of controllers is the classical two-stage lead-lag compensators for PSSs in the form

$$G_c(s, T_d) = K(T_d)\frac{1 + T_1(T_d)s}{1 + T_2(T_d)s}\frac{1 + T_3(T_d)s}{1 + T_4(T_d)s}\frac{T_w s}{1 + T_w s}, \tag{5}$$

where the gain K and the time constants T_i, $i = 1, \ldots, 4$, are functions of T_d. A high-pass filter with a (washout) time constant T_w is also included in (5), such that the controller is inactive for a constant input signal. For any value of latency T_d, we can establish values of K and T_i to provide appropriate controller gain and phase compensation in addition to the damping controller phase compensation when $T_d = 0$. In practice, we can select a set of T_{di} and associate a set of K and T_i for each T_{di}. For speed-input type PSS, which normally requires phase-lead compensation, large time delays may not be readily accommodated by the transfer function (5). Additional lead-lag stages may be needed. Integral-of-accelerating-power type PSSs, using both the speed and electrical power signals as input, require less phase-lead compensation and thus can more readily accommodate longer delays.

Damping controllers for SVC and TCSC, which tend to require phase-lag compensation when $T_d = 0$, can utilize $T_d \neq 0$ as part of the phase-lag compensation, and thus can tolerate longer data delays. An adaptive control design for a TCSC is described in the next section.

5 Design Illustration

The proposed adaptive control design will be illustrated for a TCSC to damp the interarea mode oscillations in a two-area, four-generator system shown in Fig. 5, which is adapted from [25]. Area 1 consists of two coherent generators, Generators 1 and 2, and Area 2 consists of the other two coherent generators, Generators 11 and 12, which are represented by detailed machine and high-response excitation system models. No PSSs are applied to the generators. The generator inertias are changed slightly so that the local modes in the two areas are not identical. The TCSC is located between Buses 201 and 202, on one of the transmission paths connecting the two areas, so that it will be effective in damping the interarea mode. A fourth line connecting Buses 4 and 14 via Bus 999 with relatively high reactance is added. The disturbance used to illustrate the damping controller performance is a short-circuit fault at Bus 999, cleared by removing the Line 4-999. The post-fault system interarea mode would be similar to that of the pre-fault system. The interarea power transfer is at 400 MW. The Power System Toolbox (PST) [26] is used for the linearization and simulation of the nonlinear power system model. Without damping control, the interarea mode of the system is $0.0230 \pm j4.119$, which is unstable. The two local modes are at $-0.6327 \pm j7.0378$ and $-0.5698 \pm j7.2802$.

The damping control design requires using an appropriate input signal. In a WACS, remote signals can also be used, in addition to local signals. Table 2 shows a list of proposed input signals to the damping controller, and the zeros of the transfer function for each input signal that are close to the interarea mode.

Note that in Table 2, V_m denotes the bus voltage magnitude, I_m the line current magnitude, θ the bus voltage angle, δ the machine angle, and ω the machine speed. The number in the subscript of these variables denotes either the bus number or the machine number. Note that V_{m201} and $I_{m(201-202)}$ are local signals, and all the other signals are remotely measured. For an input signal to be effective, none of its zeros

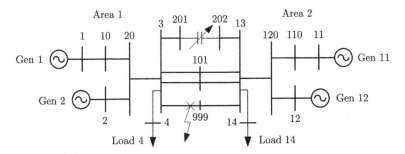

Fig. 5 Four-machine, two-area system

Table 2 Input signals and zeros affecting interarea mode damping control

Input signal	Zeros close to the interarea mode
V_{m201}	$0.379 \pm j2.19$
V_{m101}	None
V_{m13}	$0.126 \pm j5.09$
$I_{m(201-202)}$	$0.0311 \pm j3.80$
$\theta_3 - \theta_{13}$	$-0.0786 \pm j5.63$
$\overline{\theta}_a = 0.5(\theta_1 + \theta_2) - 0.5(\theta_{11} + \theta_{12})$	None
$\overline{\delta}_a = 0.5(\delta_1 + \delta_2) - 0.5(\delta_{11} + \delta_{12})$	$-0.125 + j1.99$
$\overline{\omega}_a = 0.5(\omega_1 + \omega_2) - 0.5(\omega_{11} + \omega_{12})$	None

should be close to the interarea mode. From Table 2, the three effective signals, in descending order of effectiveness based on root-locus analysis, are

$$\overline{\omega}_a, \quad \overline{\theta}_a, \quad V_{m101}. \tag{6}$$

In general, machine speeds are not included in synchrophasor data, and thus the first signal, which is the most effective signal, is not used. We choose the signal $\overline{\theta}_a$ for our design, which represents the difference between the average angles in each area. Note that in the two-area system, the inertias are almost identical and thus the same weights are used. The averaging reduces the presence of the local modes in $\overline{\theta}_a$.

The root-locus plot of the system using the signal $\overline{\theta}_a$ without any phase compensation is shown in Fig. 6a. Note that the angle of departure from the interarea mode is about 255°. This is to be expected as changing the reactance in a power network affects mostly the synchronizing torque, which determines the frequencies of the oscillatory modes. Because the TCSC is located in the power transfer path between the two areas, we expect it to affect only the interarea mode, and not the local modes. The thyristor switching time constant is small, so that it provides very little phase changes. Thus, the root locus of the interarea mode is mostly parallel to the $j\omega$-axis. Hence, for the interarea mode to move in the 180° direction, a phase lag of 75° is needed, assuming there is no latency in the input signal. The root-locus plot with phase-lag compensation is shown in Fig. 6b.

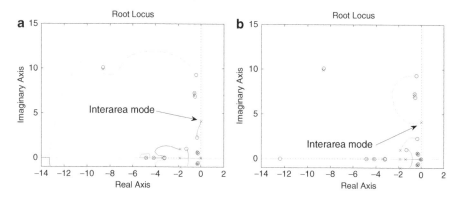

Fig. 6 Root-locus plots: (**a**) no phase compensation, (**b**) with phase-lag compensation

The controller obtained from this analysis is

$$K_0(s) = 13.8 \frac{1 + 0.08087s}{1 + 0.7278s} \frac{T_w s}{1 + T_w s}, \qquad (7)$$

where the time constant of the high-pass filter is $T_w = 10$ s. This controller achieves a damping ratio of 0.268. Note that the phase-lag compensation provided by this single stage is $-53.13°$, less than the optimal level of $-75°$. The performance of this controller for a three-phase short circuit fault on Bus 999 at $t = 0.5$ s, cleared in three cycles by removing Line 4-999, is shown in Fig. 7.

Consider PMU data with a latency of 150 ms, which translates to a phase lag of

$$\frac{150}{1000} \times \frac{4.119}{2\pi} \times 360° = 35.4°, \qquad (8)$$

at the frequency 4.119 rad/s. With the data latency, the phase-lag compensation would reduce to $39.6°$. Figure 8a show the root-locus plot with phase compensation adjusted for latency. One can see that latency does not significantly deteriorate the control design. Note that for the root-locus plots, the latency is modeled by a first-order Padé approximation [27]. Without allowing phase compensation for latency, Fig. 8b shows some deterioration in damping effectiveness.

In designing an adaptive controller, we select latency levels starting at 150 ms, at increments of 50 ms, up to 350 ms. With one stage of phase compensation, the damping controller is implemented as

$$G_c(s, T_d) = K(T_d) \frac{1 + T_1(T_d)s}{1 + T_2(T_d)s} \frac{T_w s}{1 + T_w s}, \qquad (9)$$

where the time constant of the high-pass filter is $T_w = 10$ s. The adaptive parameters $K(T_d)$, $T_1(T_d)$, and $T_2(T_d)$ are given in Table 3.

This adaptive algorithm is applied to the two-area system for the same short-circuit disturbance. The adaptive control performance is shown in Fig. 9. In this

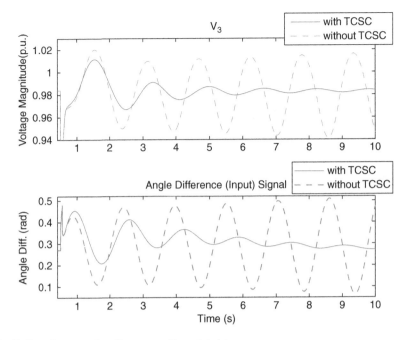

Fig. 7 Damping control performance with no data latency

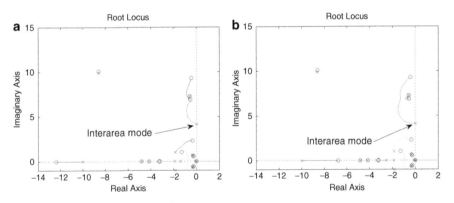

Fig. 8 Root-locus plots: (**a**) no phase compensation for latency, (**b**) phase compensation for latency

simulation, a data buffer function with Poisson arrival probability is created in PST to simulate data latency. The lower plot of Fig. 9a shows the angle difference signal $\bar{\theta}_a$ measured instantaneously at the buses, the time that the signal $\bar{\theta}_a$ actually arrives at the controller, and the $\bar{\theta}_a$ waveform that is used as the damping controller input. In contrast to machine rotor angles which are continuous, bus voltage magnitudes can change abruptly during fault conditions. Note the initial delay for the synchrophasor data $\bar{\theta}_a$ to be picked up by the data buffer. The 150 ms latency compensator is

Table 3 Adaptive phase compensation

Latency (ms)	Controller	$K(T_d)$	$T_1(T_d)$ (ms)	$T_2(T_d)$ (ms)	Phase comp.	Damping ratio
$T_{d1} = 150$	$G_{c150}(s)$	13.7	0.1144	0.5147	$-39.52°$	0.250
$T_{d2} = 200$	$G_{c200}(s)$	10.0	0.1477	0.3987	$-27.35°$	0.264
$T_{d3} = 250$	$G_{c250}(s)$	9.0	0.1861	0.3163	$-15.03°$	0.284
$T_{d4} = 300$	$G_{c300}(s)$	8.3	0.2262	0.2602	$-4.00°$	0.287
$T_{d5} = 350$	$G_{c350}(s)$	7.5	0.2426	0.2426	$0°$	0.267

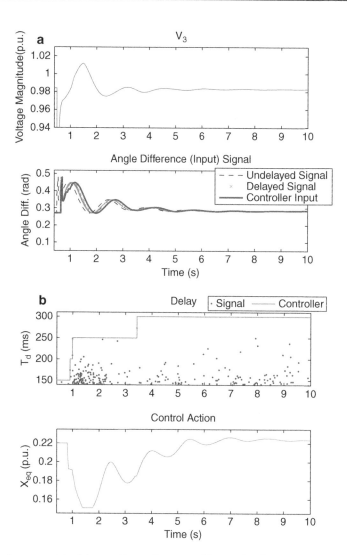

Fig. 9 Performance plots: (**a**) controller performance, (**b**) control action

used at the start. At $t \approx 0.9$ s, the late arrival of a data value causes the 200 ms latency compensator to be switched in. At $t \approx 1$ s, another late arrival causes the 250 ms latency compensator to be switched in. At $t \approx 3.4$ s, the delay is increased to 300 ms, which is used for the remaining portion of the simulation. There are a few data points with latency close to 300 ms but not exceeding it. After switching in the 200, 250, and 300 ms latency compensators, the controller performance is still excellent. To damp the oscillations, the controller drives the effective reactance of the TCSC branch connection Buses 3 and 13 to a minimum, as shown in Fig. 9b. The transients in switching from one compensator to another are visible in the controller output in Fig. 9b.

It should also be noted that using averages of several angles in an area to generate an aggregate angle has the advantage that if one of the signals is lost or does not arrive on time, it is still possible to generate an aggregate angle using the remaining synchrophasor data. For example, if the bus voltage angle measurement θ_{11} of Bus 11 is lost, the angle difference can be readjusted as

$$\hat{\bar{\theta}}_a = 0.5(\theta_1 + \theta_2) - \theta_{12}, \tag{10}$$

in which θ_{12} is now taken as the representative angle of Area 2. Although $\hat{\bar{\theta}}_a$ is less desirable than $\bar{\theta}_a$ because some local mode components will appear in $\hat{\bar{\theta}}_a$, effective interarea damping can still be achieved with $\hat{\bar{\theta}}_a$.

The adaptive control scheme can be applied to PSS and FACTS controllers damp interarea modes of large power systems. An added task is to select those controllers effective for damping certain interarea modes. Then the proper input signals can be selected based on available PMU measurements. An appropriate damping control architecture needs to be established. For phasor networks with several PDCs in the data collection pathway, PMU data latency would expect to be longer and more phase-lead compensation would be required. If a single controller cannot add sufficient damping, then multiple damping controllers working in a decentralized fashion would be required. The basic idea of the Adaptive Control algorithm in terms of the adaptive switching based on PMU data arrival time, and phase compensation for latency are still applicable.

6 Conclusions

In this chapter, we have proposed an adaptive control scheme for developing a power system interarea damping controller to counter the variable PMU data latency. The adaptive control consists of a controller switching algorithm based on the latency of PMU data, and a phase compensation design of the controller for a given set of latency. Latency requires adding phase-lead compensation. In FACTS controller design, a small amount of latency actually reduces the amount of phase-lag compensation required. The adaptive control design is illustrated for a two-area power system.

As a continuation of this work, it would be of interest to apply the adaptive control algorithm to wide-area control using PSSs. It is expected that adding phase-lead to PSSs already providing more than 90° of phase-lead compensation would be quite challenging.

Acknowledgments The research discussed in this chapter is supported in part by a Grant from the Global Climate and Energy Project at Stanford University, and in part by the Power System Research Consortium supported by FirstEnergy, ISO-NE, NYISO, and PJM.

References

1. Phadke AG, Thorp JS (2008) Synchronized phasor measurements and their applications. Springer, New York
2. Chow JH, Peponides G, Kokotovic PV, Avramovic B, Winkelman JR (1982) Time-scale modeling of dynamic networks with applications to power systems. Springer, New York
3. Chow JH, Boukarim GE, Murdoch A (2004) Power system stabilizers as undergraduate control design projects. IEEE Trans Power Syst 19:144–151
4. Kosterev DN, Taylor CW, Mittelstadt WA (1999) Model validation for the August 10, 1966 WSCC system outage. IEEE Trans Power Syst 14:967–979
5. Uhlen K (2010) Wide area monitoring and control activities in Norway and the Nordic power system. NAPSI, Vancouver. Available online: http://www.naspi.org/meetings/workgroup/2010_june/presentations/session_02/uhlen _norway_wide_area_monitoring_20100608.pdf.
6. DeMello FP, Concordia C (1969) Concepts of synchronous machine stability as affected by excitation control. IEEE Trans Power App Syst 88:316–329
7. Larsen EV, Swann DA (1981) Applying power system stabilizers, Part II: performance objectives and tuning concepts. IEEE Trans Power App Syst 100:3025–3033
8. Murdoch A, Venkataraman S, Lawson RA, Pearson WR (1999) Integral of accelerating power type PSS, part 1 - theory, design, and tuning methodology. IEEE Trans Energ Convers 14: 1658–1663
9. Murdoch A, Venkataraman S, Lawson RA (1999) Integral of accelerating power type PSS, part 2 - field testing and performance verification. IEEE Trans Energ Convers 14:1664–1672
10. Western Electricity Coordinating Council Policy Statement on Power System Stabilizers (2002) Available online: http://www.wecc.biz/library/WECC%20Documents/Documents%20 for%20Generators/WECC_PSS-%20Policy-Statement.pdf
11. Hingorani N, Gyugyi L (2000) Understanding FACTS: concepts and technology of flexible AC transmission systems. IEEE Press, New York
12. Larsen EV, Chow JH (1987) SVC control design concepts for system dynamic performance. In: IEEE power engineering society publication 87TH0187-5-PWR. Application of static var systems for system dynamic performance. IEEE PES Winter Meeting, New Orleans, USA, January 1987
13. Noroozian M, Ghandhari M, Andersson G, Gronquist J, Hiskens I (2001) A robust control strategy for shunt and series reactive compensators to damp electromechanical oscillations. IEEE Trans Power Deliv 16:812–817
14. Larsen EV, Sanchez-Gasca JJ, Chow JH (1995) Concepts for design of FACTS controllers to damp power swings. IEEE Trans Power Syst 10:948–956
15. Cyr C, Kamwa I (2010) WACS design at Hydro-Quebec. In: PMU Tutorial, IEEE PES general meeting
16. Kurose JF, Ross KW (2010) Computer networking: a top-down approach, 5th edn. Addison-Wesley, New York

17. Stahlhut JW, Browne TJ, Heydt GT, Vittal V (2008) Latency viewed as a stochastic process and its impact on wide area power system control signals. IEEE Trans Power Syst 23:84–91
18. Shi JH, Li P, Wu XC, Wu JT, Lu C, Zhang Y, Zhao YK, Hu J (2008) Implementation of an adaptive continuous real-time control system based on WAMS. In: Proceedings of CIGRÈ 2nd international conference on monitoring power system dynamics performance, Russia
19. Vanfretti L, Chow JH, Sarawgi S, Fardanesh B (2011) A phasor-data based estimator incorporating phase bias correction. IEEE Trans Power Syst 26:111–119
20. Chow JH, Chakrabortty A, Vanfretti L, Arcak M (2008) Estimation of radial power system transfer path dynamic parameters using synchronized phasor data. IEEE Trans Power Syst 23:564–571
21. IEEE Std C37.188-2005 (2006) IEEE standards for synchrophasors for power systems
22. Decker IC, e Silva AS, Agostino MN, Prioste FB, Mayer BT, Dotta D (2011) Experience and applications of phasor measurements to the Brazilian power system. Euro Trans Electr Power 21:1557–1573
23. Chow JH, Sanchez-Gasca JJ, Ren H, Wang S (2000) Power system damping controller design using multiple input signals. IEEE Contr Syst Mag 20:82–90
24. Chaudhuri NR, Ray S, Majumder R, Chaudhuri B (2010) A new approach to continuous latency compensation with adaptive phasor power oscillation damping controller (POD). IEEE Trans Power Syst 25:939–946
25. Klein M, Rogers GJ, Kundur P (1991) A fundamental study of inter-area oscillations in power systems. IEEE Trans Power Syst 6:914–921
26. Rogers GJ, Chow JH (1995) Hands-on teaching of power system dynamics. IEEE Computer Applications in Power 8(1):12–16
27. Franklin GF, Powell JD, Emami-Naeini A (2010) Feedback control of dynamic systems, 6th edn. Prentice Hall, Upper Saddle River, NJ

A Model Reference Approach for Interarea Modal Damping in Large Power Systems

Aranya Chakrabortty

Abstract In this chapter, we present a set of results on the design of dynamic controllers for electromechanical oscillation damping in large power systems using Synchronized Phasor Measurements. Our approach consists of three steps, namely – (1) *Model Reduction*, where phasor data are used to identify second-order models of the oscillation clusters of the system, (2) *Aggregate Control*, where state-feedback controllers are designed to achieve a desired closed-loop transient response between every pair of clusters, and finally (3) *Control Inversion*, where the aggregate control design is distributed and tuned to actual realistic controllers at the generator terminals until the interarea responses of the full-order power system matches the respective inter-machine responses of the reduced-order system. Although a general optimization framework is needed to formulate these three steps for any n-area power system, we specifically show that model reference control (MRC) can be an excellent choice to solve this damping problem when the power system consists of two dominant areas, or equivalently one dominant interarea mode. Application of MRC to such two-area systems is demonstrated through topological examples inspired by realistic transfer paths in the US grid.

1 Introduction

Model-based control of multi-machine power systems has seen a rich history of nearly fifty years, addressing the fundamental issues of mathematical modeling, stability and robustness of large power networks with a natural progress toward more advanced concepts of component-level dynamics. Discoveries of ideas ranging from coherency and aggregation [1], bifurcation and voltage stability [2], passivity

A. Chakrabortty (✉)
North Carolina State University, Raleigh, NC, USA
e-mail: achakra2@ncsu.edu

A. Chakrabortty and M.D. Ilić (eds.), *Control and Optimization Methods for Electric Smart Grids*, Power Electronics and Power Systems 3,
DOI 10.1007/978-1-4614-1605-0_18, © Springer Science+Business Media, LLC 2012

and energy functions [3, 4], to name a few, have led to a deeper understanding of the intrinsic properties of this large-scale dynamic system, and have also laid the foundations for numerous constructive control designs to regulate its complicated operation. Over the past few years, however, following the US Northeast Blackout of 2003 and similar catastrophic events in the European grid, the research mindset of power system control engineers has steered more toward measurement-based designs. The relevance of this interest has been particularly facilitated by the recent outburst of power system measurement and instrumentation facilities in the form of the Wide-Area Measurement System (WAMS) technology, also referred to these days as the *Synchrophasor* technology [5]. Sophisticated digital recording devices called Phasor Measurement Units or PMUs are currently being installed in accelerating proportions at different points in the North American grid, especially under the smart grid initiatives of the US Department of Energy, to record and export GPS-synchronized, high sampling rate (6–60 samples/s), dynamic power system data. Concerted efforts are being made to develop nationwide "early warning" mechanisms using PMU measurements that will enable power system operators to take timely actions against blackouts and other widespread contingencies. Excellent visualization tools, for example, in the form of Real-Time Dynamics Monitoring System (RTDMS) and US-Wide Frequency Monitoring Network (FNET) [6] are currently being deployed across various corners of the US grid using voltage, current and frequency phasors. This development has been complimented by an equally remarkable progress in data analysis methods and software platforms, some leading examples being the Dynamic System Identification (DSI) software, Prony analysis and Mode Meter [7], Hilbert–Huang transforms [8] and phasor-based state estimation [9].

The majority of research done so far in the Synchrophasor community in the US, however, pertains only to ideas of *monitoring* and *observation*. No rigorous research has yet been done to investigate how Synchrophasors, beyond simply monitoring, can also be used for autonomous *feedback control*. With progress in technology, the *wide-area control* problem, in fact, is becoming more and more prominent in various contexts, one of the most important and complex of which is small-signal stability. It is well known to power system researchers that small-signal instabilities often occur without the knowledge of the grid operators [10], with minor disturbances in the grid growing into devastatingly large events. For example, on August 10, 1996, power oscillations between two groups of *equivalent* generators, one in Alberta and the other in southern California, began oscillating with respect to each other at an uncontrolled rate, resulting in the largest blackout in the US West coast power system. At that time, technology did not permit the diverging oscillations to be observable to the grid operators due to the lack of high-resolution measurements, and the result was an unfortunate disintegration of the entire western interconnection into five separate islands [11]. Today, we have WAMS as the most promising instrumentation technology to detect such instabilities, but the overriding questions that we still need to answer is: Can we design robust, distributed, multivariable controllers using Synchrophasor feedback from selected, spatially distributed nodes (or buses) of a large, interconnected power network to control the interarea power/phase swings between various clusters of the

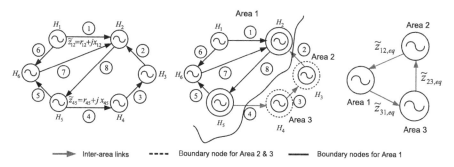

Fig. 1 Linearized Laplacian state variable model

network? – Especially given the fact that these clusters, say *Alberta* and *southern California*, in reality, are some hypothetical aggregations of numerous synchronous machines?

In this chapter, we address this pertinent problem, and present some initial results inspired from the model-reference adaptive control (MRAC) literature, applied to simplistic yet practically relevant topologies of multi-area power systems with a particular focus on two-area systems. The main idea behind our design is a so-called, novel *control inversion* framework, which allows PMU-based linear/nonlinear control designs, developed for reduced-order power systems, to be inverted (or, equivalently *distributed*) to local controllers in actual higher-order systems via suitable optimization methods. The approach, in general, consists of three precise steps, namely:

1. *Model reduction/dynamic equivalencing* – where PMU data are used to identify equivalent models of the oscillation clusters of the entire power system based on the differences in their coupling strengths; for example, Fig. 1 shows a six-machine eight-line power system, where the electrical reactances or the edge weights of lines $\{1, 5, 6, 7, 8\}$ are assumed to be significantly smaller compared to those of lines 2–4, separating the entire network into three coherent clusters or areas, and forcing the system to evolve as an *equivalent* three-machine system over a slow timescale treating all the machines inside Area 1 to be aggregated into one *hypothetical* equivalent machine. The first step in our design involves identification of these equivalent machine models using PMU data available from the terminal buses of each area. Detailed derivations of these measurement-based equivalencing methods have been presented in our recent work [12], and, therefore, will not be our focus in this chapter. Our objective is to design controllers for damping the oscillations between these areas, for which we will simply assume that the area models are available to us by prior identification methods available from [12].

2. *Aggregate control* – where state-feedback controllers are designed to achieve a desired closed-loop transient response between every pair of clusters in the reduced-order system. For the system in Fig. 1 this would mean that a controller, either centralized or distributed, is designed for the three equivalent machines to

achieve a desired dynamic response of the *aggregated* state variables (technically referred to as the *interarea* state responses).

3. *Control inversion* – where the aggregate control design is distributed and tuned back to actual realistic controllers at the generator terminals until the interarea responses of the full-order power system matches the respective inter-machine responses of the reduced-order system. For the six-machine system of Fig. 1, this would mean that the controller designed for the equivalent machine of Area 1 will be mapped back to a set of controllers for the four machines contained in this area, and tuned until the slow modes of this full-order system exhibit similar transient response as the respective closed-loop states in Step 2.

Although a general optimization framework is needed to formulate these three steps for a n-area power system, we specifically show in this chapter that model reference control (MRC) can be an excellent choice to solve this damping problem when the power system consists of two dominant areas, or equivalently one dominant interarea mode. The remainder of the chapter is accordingly organized as follows. Section 2 formulates the wide-area damping problem for a n-machine system as a parametric minimization problem. The discussion is purposely kept at a very generalized level with a conceptual framework of how wide-area damping might be possible for multi-area systems. Section 3 presents a simpler solution to the inversion for two-area power systems via MRC, and illustrates the design with simulations. Section 4 concludes the chapter.

2   Problem Formulation for n-Area Systems

Consider a network of electrical oscillators with n generators (nodes) connected to each other through m tie-lines (edges) with $m \leq n(n - 1)/2$, forming a connected graph with cardinality (n, m), such that atmost one edge exists between any two nodes. This may also be thought of as a power system, although we use the word "power" with reservation as in a real power system generators are not necessarily connected directly but via intermediate buses due to which the network Laplacian becomes extremely complicated, especially for large networks. To avoid this difficulty and in the interest of the specific application discussed in this chapter, we restrict our discussion to networks where each dynamic element i.e., a generator is directly connected to its neighbors. An example of such a network consisting of $n = 6$ generators and $m = 8$ tie-lines is shown in Fig. 1. The arrows along each edge denote the direction of effective power flow. Let the internal voltage phasor of the ith machine be denoted as

$$\tilde{E}_i = E_i \angle \delta_i, \quad i = 1, 2, \ldots, n, \tag{1}$$

where, following synchronous machine theory [4], E_i is constant, δ_i is the angular position of the generator rotor, and $E_i \angle \delta_i$ denotes the polar representation $E_i \varepsilon^{j\delta_i}$ ($j = \sqrt{-1}$). The transmission line connecting the pth and the qth machines is assumed to have an impedance

$$\tilde{z}_{pq} = r_{pq} + jx_{pq}, \tag{2}$$

where 'r' denotes the resistive part and x denotes the reactive part. Here, $p \in \{1, 2, \ldots, n\}$ and $q \in \mathcal{N}_p$, where $\mathcal{N}_p$ is the set of nodes to which the pth node is connected. It follows that the total number of tuples formed by pairing p and q is m. For the rest of the chapter, we will denote the edge connecting the pth and the qth node by e_{pq}. Equation (2) can also be regarded as a complex *weight* of an edge in the network, and implies that $\tilde{z}_{pq} = \tilde{z}_{qp}$. If two nodes do not share a connection, then the impedance corresponding to that non-existing edge is infinite (i.e., open circuit), or equivalently,

$$\tilde{y}_{pq} = \frac{1}{\tilde{z}_{pq}} = \frac{1}{r_{pq} + jx_{pq}} = 0 \quad \forall q \notin \mathcal{N}_p, \tag{3}$$

where $\tilde{y}_{pq}$ is the admittance of e_{pq}. The mechanical inertia of the ith machine is denoted as H_i.

The dynamic electro-mechanical model of the ith generator, neglecting damping, can be written as [4]

$$\dot{\delta}_i = \omega_i - \omega_s, \tag{4}$$

$$2H_i \dot{\omega}_i = P_{mi} - \sum_{k \in \mathcal{N}_i} \left(\frac{E_i^2 r_{ik} - E_i E_k p_{ik} \cos(\delta_{ik} + \alpha_{ik})}{p_{ik}^2} \right), \tag{5}$$

where $\delta_{ik} = \delta_i - \delta_k$, $\omega_s = 120\pi$ is the synchronous speed for a 60 Hz system, ω_i is the rotor angular velocity, P_{mi} is the mechanical power input, $p_{ik} = \sqrt{r_{ik}^2 + x_{ik}^2}$ and $\alpha_{ik} = \tan^{-1}(x_{ik}/r_{ik})$. All quantities are in per unit except for the phase angles which are in radians. We assume that the network structure is known, i.e., the set $\mathcal{N}_i$ for all $i = 1, 2, \ldots, n$ in (4)–(5) is known.

We linearize (4)–(5) about an initial equilibrium $(\delta_{i0}, 0)$, where $0 < \delta_{i0} < 90°$ for all $i = 1, 2, \ldots, n$, and denote the perturbed state variables as

$$\Delta \delta = \mathrm{col}(\Delta \delta_1, \Delta \delta_2, \ldots, \Delta \delta_n), \tag{6}$$

$$\Delta \omega = \mathrm{col}(\Delta \omega_1, \Delta \omega_2, \ldots, \Delta \omega_n). \tag{7}$$

We assume that the control input u enters the system (note: the network graph is connected, by assumption) through the jth node, $j \in \{1, 2, \ldots, n\}$.

$$\begin{bmatrix} \Delta \dot{\delta} \\ \Delta \dot{\omega} \end{bmatrix} = \underbrace{\begin{bmatrix} 0 & I \\ \hline \mathcal{M}^{-1}\mathcal{L} & 0 \end{bmatrix}}_{A} \begin{bmatrix} \Delta \delta \\ \Delta \omega \end{bmatrix} + \underbrace{\begin{bmatrix} 0 \\ \mathcal{E}_j \end{bmatrix}}_{B} u, \tag{8}$$

where I is the n-dimensional identity matrix, $\mathcal{E}_j$ is the jth unit vector with all elements zero except the jth element which is 1, $\mathcal{M} = \mathrm{diag}(M_1, M_2, \ldots, M_n)$, M_i is the inertia of the ith generator, and $\mathcal{L}$ is the $n \times n$ Laplacian matrix with elements:

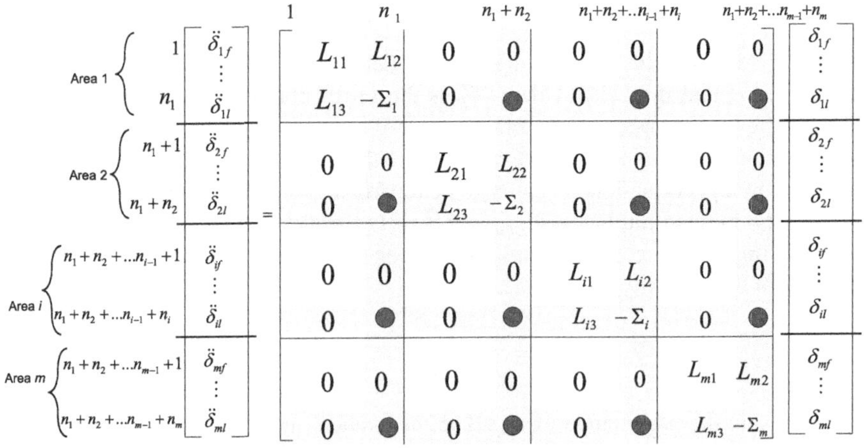

$$
\begin{bmatrix}
\ddot\delta_{1f}\\ \vdots \\ \ddot\delta_{1l}\\
\ddot\delta_{2f}\\ \vdots \\ \ddot\delta_{2l}\\
\ddot\delta_{if}\\ \vdots \\ \ddot\delta_{il}\\
\ddot\delta_{mf}\\ \vdots \\ \ddot\delta_{ml}
\end{bmatrix}
=
\begin{bmatrix}
L_{11} & L_{12} & 0 & 0 & 0 & 0 & 0 & 0\\
L_{13} & -\Sigma_1 & 0 & \bullet & 0 & \bullet & 0 & \bullet\\
0 & 0 & L_{21} & L_{22} & 0 & 0 & 0 & 0\\
0 & \bullet & L_{23} & -\Sigma_2 & 0 & \bullet & 0 & \bullet\\
0 & 0 & 0 & 0 & L_{i1} & L_{i2} & 0 & 0\\
0 & \bullet & 0 & \bullet & L_{i3} & -\Sigma_i & 0 & \bullet\\
0 & 0 & 0 & 0 & 0 & 0 & L_{m1} & L_{m2}\\
0 & \bullet & 0 & \bullet & 0 & \bullet & L_{m3} & -\Sigma_m
\end{bmatrix}
\begin{bmatrix}
\delta_{1f}\\ \vdots \\ \delta_{1l}\\
\delta_{2f}\\ \vdots \\ \delta_{2l}\\
\delta_{if}\\ \vdots \\ \delta_{il}\\
\delta_{mf}\\ \vdots \\ \delta_{ml}
\end{bmatrix}
$$

(Column indices: 1, n_1, n_1+n_2, $n_1+n_2+..n_{i-1}+n_i$, $n_1+n_2+..n_{m-1}+n_m$. Row groupings: Area 1 $\{1,\dots,n_1\}$, Area 2 $\{n_1+1,\dots,n_1+n_2\}$, Area i $\{n_1+n_2+..n_{i-1}+1,\dots,n_1+n_2+..n_{i-1}+n_i\}$, Area m $\{n_1+n_2+..n_{m-1}+1,\dots,n_1+n_2+..n_{m-1}+n_m\}$.)

Fig. 2 Network of six generator nodes and eight tie-line edges

$$\mathscr{L}_{ii} = -\sum_{k\in\mathscr{N}_i}\frac{E_i E_k}{p_{ik}}\sin(\delta_{i0}-\delta_{k0}+\alpha_{ik}), \tag{9}$$

$$\mathscr{L}_{ik} = \frac{E_i E_k}{p_{ik}}\sin(\delta_{i0}-\delta_{k0}+\alpha_{ik}), \quad k\in\mathscr{N}_i, \tag{10}$$

$$\mathscr{L}_{ik} = 0, \quad \text{otherwise} \tag{11}$$

for $i = 1, 2,\ldots,n$. It follows that if $M_i = M_j$, $\forall(i,j)$, then $\mathscr{L} = \mathscr{L}^T$. In general, however, each machine will have distinct inertia as a result of which the symmetry property does not hold. We, therefore, refer to $\mathscr{M}^{-1}\mathscr{L}$ as the unsymmetric Laplacian matrix for the linearized swing model. It is obvious from (8) that the coupling strengths p_{ik} of the links are contained in this matrix, and will decide the separation of areas depending on the differences in the strengths. For example, if the areas are separated by exactly one boundary node in each cluster, then the Laplacian matrix can be written in the form shown in Fig. 2, assuming that the phase angles are stacked according to each area with the last state of each stack being that of the boundary node for that particular area. The dots in the state matrix in Fig. 2, under such a partition, will indicate the *interarea* coupling strengths between any pair of areas, and can be used to reduce the full-order network into a dynamic equivalent system of n-equivalent machines using the parameter identification methods outlined in [12].

A more concrete example is given by the three-area power system shown in Fig. 3, where Areas 1 and 3 consist of two coherent machines each, while Area 2 consists of three coherent machines, and, therefore, reduced to a three-machine equivalent interconnected through an equivalent graph. Since our wide-area control method is based on shaping the closed-loop response of the reduced-order system followed by control inversion, this means that our first task would be to design a distributed excitation control system for each of the three equivalent machines G_{A1},

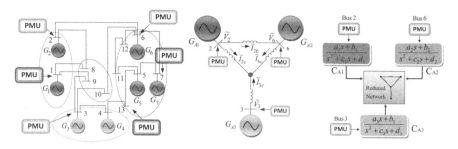

Fig. 3 Network of seven generator nodes divided into three coherent areas

G_{A2}, and G_{A3}, using, for example, classical linear state-feedback designs. None of these controllers, however, can be implemented in practice as G_{A1}, G_{A2}, and G_{A3} do not exist physically. Therefore, keeping the aggregate design as a reference, we next need to design individual excitation for each generator G_1, G_2, ..., G_7 in the 13-bus system of Fig. 3, and tune them optimally until the interarea responses for the 13-bus system coincide with the closed-loop state responses of the three-area system. Our approach for this control distribution is based on various optimization methods, inspired by different engineering applications. For example, one potential direction of our analysis, as described next, will be to distribute the feedback gains of the aggregate controllers via continuous functions (for nonlinear control design) or averaging coefficients (for linear control design), and then to achieve interarea performance matching through optimal tuning of the controller parameters.

Starting from the reduced-order system as in Fig. 3, we consider the swing model of the jth equivalent generator

$$\dot{\delta}_j = \omega_j, \quad M_j \dot{\omega}_j = P_{mj} - D_j \omega_j - P_{ej}. \qquad (12)$$

We assume that the measurements available for feedback for this jth aggregate generator, and their corresponding feedback gains are

$$(y_1^{j_1}, k_{j11}), (y_2^{j_1}, k_{j21}), \ldots, (y_1^{j_2}, k_{j12}), \ldots, (y_m^{j_n}, k_{jmn}),$$

where $y_i^{j_p}$ denotes the ith measured variable by the PMU located at the j_pth bus. The choices for $y_{j_p}^i$, for example, can be voltage and current magnitudes, voltage and current phase angles, bus frequency, and the active power (calculated from the voltage and current phasors) measured by the PMU at Bus j_p. The index j_p corresponds to a bus that is possibly located in close neighborhood of this jth generator in the reduced network. Assigning the mechanical power input $u_j = P_{mj}$ as the control input, an state/output feedback design can then be of the form

$$u_j = f(y_1^{j_1}(t), y_2^{j_1}(t), \ldots, y_m^{j_n}(t), k_{j11}, k_{j21}, \ldots, k_{jmn}), \qquad (13)$$

where $f(\cdot)$ is a smooth function producing a desired closed-loop inter-machine transient response. However, for implementation in the actual 13-bus system u_j

needs to be distributed to each local machine belonging to the jth area. A plausible way of achieving this would be, for example, to construct nonlinear functions $\rho(\cdot)$ mapping each of the feedback gains $(k_{j11}, k_{j21}, \ldots, k_{jmn})$ to each individual machine in the area. The symbol l denotes the total number of machines in the jth area. Stacking the functions $\rho_{mn}^l(\cdot)$ and the gains k_{jmn} for all (j, l, m, n) into vectors $\mathcal{R}$ and $\mathcal{K}$, respectively, the problem that we must, therefore, solve is

$$\min_{\mathcal{R}(\mathcal{K})} ||x_{ij}(t, \mathcal{R}(\mathcal{K})) - \bar{x}_{ij}(t, \mathcal{K})||_2 \quad \text{st.} \quad \mathcal{K} \in \mathcal{K}^* \tag{14}$$

for all (i, j), and for all $t \geq t^* \geq 0$, where: x_{ij} is the interarea state response (phase or frequency) between ith and jth areas in the full-order system, $\bar{x}_{ij}$ is the *designed* inter-machine state response (phase or frequency, respectively) between ith and jth machines in the reduced-order system, and $\mathcal{K}^*$ denotes a constraint set for the feedback gains specifying their allowable upper and lower bounds. Equivalently, we need to search for an optimal function set $\mathcal{R}^*(\mathcal{K})$ such that

$$||x_{ij}(t, \mathcal{R}^*(\mathcal{K})) - \bar{x}_{ij}(t, \mathcal{K})||_2 \leq \epsilon \tag{15}$$

for time $t > t^* > 0$, $\mathcal{K} \in \mathcal{K}^*$, where $\epsilon > 0$ is a chosen tolerance for the performance matching between the two responses.

However, when there are multiple interarea modes in a system, a fundamental concern for solving the optimization problem (15) is the so-called nonuniqueness of the equivalent model with respect to the choice of interarea modes. The idea is briefly shown in Fig. 4. This means that when the phase angle response at any bus in the power system consists of more than one interarea mode, then a specific mode needs to be extracted from the signal via modal decomposition methods such as Prony analysis [7], Eigenvalue Realization Algorithm (ERA) [12], etc., and the impulse response of this extracted mode needs to be used for identifying the equivalent model of the system. However, if a different interarea mode is extracted and used for this identification, then there is no guarantee that model will match with that obtained for the first mode. The mathematical justification behind this is that the transfer function of the full-order power system can be written in the pole residue form as

$$G(s) = \sum_{i \in \mathcal{N}_l} \frac{\sigma_i s + \mu_i}{s^2 + \gamma_i s + \pi_i} + \sum_{j,k \in \mathcal{I}_j \times \mathcal{I}_k} \frac{\sigma_{jk} s + \mu_{jk}}{s^2 + \gamma_{jk} s + \pi_{jk}}, \tag{16}$$

where $\mathcal{N}_l$ is the set of local modes and $\mathcal{N}_i \times \mathcal{N}_j$ is the set of interarea modes operating between the ith and the jth cluster of the system. Since the residue parameter set (σ_{jk}, μ_{jk}) for different pairs of (i, j) are not equal, this implies that the participation of the different interarea modes on the measured signal, as quantified by their participation factors, are different, and, hence, the equivalent model computed based on different pairs of clusters are disparate as well. In terms of our inversion problem, this implies that the optimization (15) needs to be performed for each equivalent interarea model, with an obvious challenge being whether the

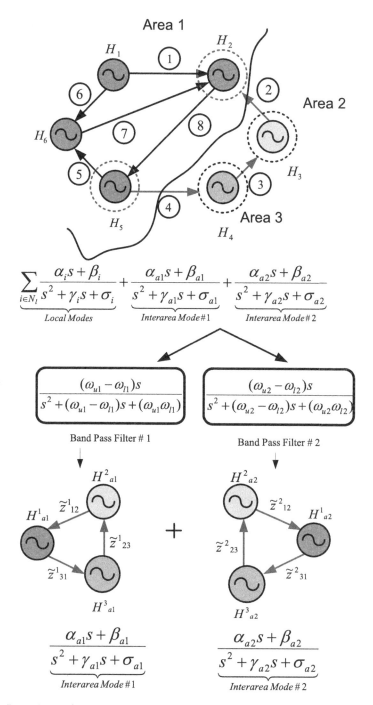

Fig. 4 Dependence of equivalent models on the choice of interarea modes

Fig. 5 Two-cluster oscillator
network

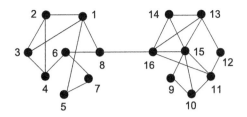

design of a single controller that can damp all interarea modes at their corresponding frequencies, is feasible or not. While the problem needs more introspection, in this chapter we bypass this obstacle by presenting our control designs for a simplistic and yet highly important class of power system, namely a two-area system, i.e., a system with a single dominant interarea mode.

Yet, another important point that needs to be clarified in this context is the need for separating the interarea mode component in a signal from all other local modes, as indicated in (16). One may argue that a state-feedback design to damp a multi-modal PMU signal, in general, will automatically damp its individual components due to each separate mode. However, depending on the *participation factor* [13] of any given mode in the signal the closed-loop damping factor of that mode may be significantly different than that for the overall multi-modal signal. The following example illustrates this fact.

Example 1 Consider the *dumble*-shaped two-area electric network shown in Fig. 5, consisting of 16 nodes, each representing a second-order oscillator in the form of a synchronous generator. The edges of the graph denote the tie-lines between the generators, with each line being characterized by an edge-weight w that denotes the electrical reactance of that line. Nodes inside each area are connected densely to other nodes in that area via strong electrical connections (i.e. small reactances or small values of w), while the two boundary nodes, namely node 8 and 16, are connected to each other by a long transmission line with much larger reactance. The sharp difference between the tie-line reactances internal to each area and that between the areas themselves gives rise to a separation of timescale in the oscillatory responses of all the nodes in the network, as reflected in the eigenvalues of the linearized system. For example, considering the swing dynamics of each node to be of the form

$$2H_i\ddot{\delta}_i = P_{mi} - \sum_{j\in\mathcal{N}_i} \frac{E_i E_j}{x_{ij}} \sin(\delta_i - \delta_j), \quad i = 1, 2, \ldots, 16, \quad (17)$$

where the meaning of the different notations are as in (4) and (5), we get the eigenvalues of the linearized double-integrator dynamic model as listed in Table 1. Referring to (8) these are the eigenvalues of the matrix $\mathcal{M}^{-1}\mathcal{L}$. The actual system matrix A will have 16 pairs of conjugate imaginary eigenvalues that can, respectively, be obtained by taking the square roots of the eigenvalues listed in Table 1, as is obvious from the structure of A given in (8). For the ease of illustration,

Table 1 Open-loop eigenvalues of two-area network (double integrator dynamics)

$\lambda(\mathscr{L})$ # (1 − 8)	0	−0.0012	−3.5552	−2.8885	−2.7026	−2.3496	−2.3320	−2.0957
$\lambda(\mathscr{L})$ # (9 − 16)	−2.0000	−1.7116	−1.5119	−1.0009	−0.9649	−0.7803	−0.6386	−0.4770

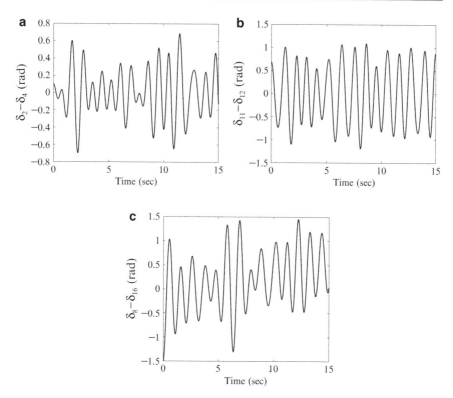

Fig. 6 Phase angle responses of local and boundary machines: (**a**) phase angle between G_2 and G_4, (**b**) phase angle between G_{11} and G_{12}, and (**c**) phase angle between G_8 and G_{16}

here we assume unit inertia for all generators, $P_{mi} = 1$ per unit, $E_i = 1$ per unit for all i, $x_{ij} = 0.1$ per unit for all local tie-lines inside the areas and $x_{8,16} = 10$ per unit, i.e., 100 times more than the local interconnections. It is obvious from the table that $\lambda_1 = 0$ represents the DC mode (since the rows of the Laplacian sum out to zero), $\sqrt{\lambda_2} = \pm j0.00352$ represents the (slow) interarea mode, while all other fourteen pairs of eigenvalues represent the local modes. The open-loop damping factor for all modes is obviously zero as (17) does not include any damping. The relative phase angle response of selected nodes internal to Area 1 is shown in Fig. 6a, that to Area 2 in Fig. 6b, and that for the boundary nodes, namely $\Delta_b \triangleq \delta_8 - \delta_{16}$ is shown in Fig. 6c.

In order to damp the signal Δ_b, we design state-feedback control inputs at a chosen set of generators, say, nodes 1, 5, 8, 10, 12, and 16, via their respective

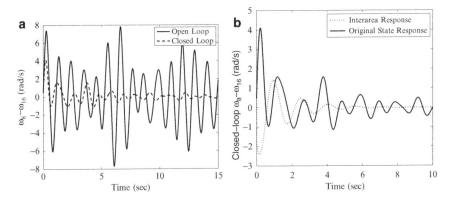

Fig. 7 Comparison of closed-loop damping of state responses and their interarea components: (**a**) open-loop vs closed loop frequency difference between G_8 and G_{16} and (**b**) interarea component of boundary machine frequencies

turbine power inputs (or their excitation system voltages if a third order model is considered) of the form

$$P_{mi} = \bar{P}_{mi} - k_i \dot{\delta}_i, \quad i = 1, 5, 8, 10, 12, 16, \qquad (18)$$

where k_i denotes the speed feedback gain for the ith machine. The gains are chosen so that the closed-loop response of $\Delta_b(t)$ has more than 70% damping, as shown in Fig. 7a. Next, the signal $\Delta_b(t)$ is decomposed into its characteristic modal components using subspace identification method (the algorithm is referred to as the ERA [12]), and the impulse response contributed by the interarea mode is extracted. The damping associated with the interarea mode, however, is found to be significantly less than 50%. Figure 7b shows the actual signal $\Delta_b(t)$ superimposed over the impulse response of its interarea mode, illustrating how the damping of the two signals are significantly different from each other. This example clearly illustrates the fact that if a given amount of damping is intended to be added to a specific interarea mode it may not be enough to design controllers for damping the measured signal by the same exact amount without separating the mode first. Thus, the model reduction approach described in Sect. 2.2 is justified as a feasible way to design damping controllers for any chosen (slow) interarea mode. We next describe details of this control design method for a two-area power system using MRC.

3 Model Reference Control for Two-area Systems

When the full-order transfer function (16) consists of only one interarea mode, then there is only one participation factor present in any measured state from this mode, as a result of which the reduced-order system has a unique topology, and, therefore, admits for a feasible control inversion. In fact, given the simple Laplacian structure of such systems, as already indicated in Sect. 1, MRC can be used as a helpful tool for the inversion problem instead of using the generic optimization framework proposed in Sect. 2. Recalling the swing equation (8), the input and output matrices for the linearized model of the two-area system can be written as $B = \text{col}(0, \mathscr{E})$, $C = [C_1 \ 0]$, where $\mathscr{E}$ is the unit vector with entry 1 at the index corresponding to the control input,[1] and $C_1 \in \mathbb{R}^{1 \times (n_1 + n_2)}$ has all zero entries except for the ith and the jth entries, each belonging to one distinct area, which are $+1$ and -1 (or, vice versa), where i and j are the indices of the nodes whose phase angle difference is being regulated, and n_1 and n_2 are, respectively, the number of nodes in Areas 1 and 2. Denoting $\bar{A} = \mathscr{M}^{-1}\mathscr{L}$, it can be easily shown that for this system,

$$CA^P B = 0, \quad p \geq 0 \text{ is even,} \tag{19}$$

$$CA^{(2r+1)} B = C_1 \bar{A}^r \mathscr{E}, \quad r = 0, 1, 2, \ldots. \tag{20}$$

Furthermore, given that our objective is to track and control the interarea oscillations, we consider the observed output as the phase angle difference between the boundary nodes, and arrange the states as

$$\delta = \text{col}(\delta_{11}, \delta_{12}, \ldots, \delta_{1n_1}, \delta_{21}, \delta_{22} \ldots, \delta_{2n_2}), \tag{21}$$

$$\omega = \text{col}(\omega_{11}, \omega_{12}, \ldots, \omega_{1n_1}, \omega_{21}, \omega_{22}, \ldots, \omega_{2n_2}), \tag{22}$$

where δ_{ij} and ω_{ij} denote the phase and machine speed for the jth machine in the ith area, the total number of machines in Areas 1 and 2 are n_1 and n_2, respectively, and the index pair (n_1, n_2) correspond to the boundary nodes of the two respective areas. To ensure persistency of excitation, we apply the input at any of the boundary nodes so that the control effect will have maximum participation in the interarea mode [12, 14]. For example, considering the input at the boundary node of Area 1, we get

$$C = [C_1 \ 0], \ C_1 = [0 \, 0 \ldots 1 \, 0 \, 0 \ldots -1], \ B = \text{col}(0, \mathscr{E}_{n_1}), \tag{23}$$

where the nonzero entries of C_1 are at the n_1th and n_2th positions, and $\mathscr{E}_{n_1}$ is the unit vector with all entries zero except the n_1th entry which is 1. Rearranging (21)–(22) as

$$\delta_a \triangleq \text{col}(\delta_{11}, \delta_{12}, \ldots, \delta_{1n_1}), \quad \delta_b \triangleq \text{col}(\delta_{21}, \delta_{22} \ldots, \delta_{2n_2}),$$

$$\omega_a \triangleq \text{col}(\omega_{11}, \omega_{12}, \ldots, \omega_{1n_1}), \quad \omega_b \triangleq \text{col}(\omega_{21}, \omega_{22}, \ldots, \omega_{2n_2}),$$

---

[1]For simplicity, we consider only a scalar control input, although the MRC design can be easily extended to multiple control inputs as well.

the state equation then takes the form

$$
\begin{bmatrix} \dot{\delta}_a \\ \dot{\delta}_b \\ \dot{\omega}_a \\ \dot{\omega}_b \end{bmatrix} = \underbrace{\begin{bmatrix} 0 & I \\ \mathscr{L} & 0 \end{bmatrix}}_{A} \begin{bmatrix} \delta_a \\ \delta_b \\ \omega_a \\ \omega_b \end{bmatrix} + \begin{bmatrix} 0 \\ 0 \\ \beta \mathscr{E}_{n1} \\ 0 \end{bmatrix} u, \tag{24}
$$

where $\beta = M_{n_1}^{-1} > 0$ i.e., the reciprocal of the inertia constant of the machine located at the boundary node of Area 1, and the unsymmetric Laplacian matrix $\mathscr{L}$ is the of the form

$$
\mathscr{L} = \begin{bmatrix} \mathscr{L}_1 & \mathscr{L}_2 \\ \mathscr{L}_3 & \mathscr{L}_4 \end{bmatrix}, \quad \mathscr{L}_2 = \begin{bmatrix} 0 & 0 \\ 0 & \gamma_{n_1,n_2} \end{bmatrix}_{n_1 \times (n_1+n_2)}, \tag{25}
$$

$$
\mathscr{L}_3 = \begin{bmatrix} 0 & 0 \\ 0 & \gamma_{n_2,n_1} \end{bmatrix}_{n_2 \times (n_1+n_2)} \tag{26}
$$

with $\gamma_{n_1,n_2} \in \mathbb{R}^{1 \times n_2}$, $\gamma_{n_2,n_1} \in \mathbb{R}^{1 \times n_1}$ given as

$$
\gamma_{n_1,n_2} = \mathrm{col}\left(0,0,\ldots, \frac{E_{n_1} E_{n_2}}{M_{n_1} x_{n_1,n_2}} \cos(\delta_{n_10} - \delta_{n_20}) \right)^{\mathrm{T}},
$$

$$
\gamma_{n_2,n_1} = \mathrm{col}\left(0,0,\ldots, \frac{E_{n_1} E_{n_2}}{M_{n_2} x_{n_1,n_2}} \cos(\delta_{n_10} - \delta_{n_20}) \right)^{\mathrm{T}}.
$$

The RHS of the above expressions follow straight from (5) with all line resistances assumed to be negligible compared to the reactances. The expressions E_{n_j}, M_{n_j} and δ_{n_j0}, respectively, denote the generator voltage, machine inertia and phase angle at pre-disturbance equilibrium for the jth boundary node, $j = 1,2$. x_{n_1,n_2} denotes the reactance or the edge weight of the transmission lines joining the two areas connecting node n_1 with node n_2. Considering (21)–(26), after a few calculations it can be easily shown that (19) and (20) simple reduces to

$$
CB = CAB = CA^2 B := 0, \quad CA^3 B \neq 0 \tag{27}
$$

implying that the relative-degree of the two-area system with the chosen input–output pair is $n^* = 2$. However, we must recall that $\mathscr{L}$ contains both local and interarea couplings implying that the output measurement will contain the modal effects due to $(n_1 + n_2 - 2)$ local modes and one interarea mode of oscillation, say denoted as ν_a. Since our goal is to dampen the interarea mode, the output $y = \delta_{n_1} - \delta_{n_2}$ must be passed through a band-pass filter (BPF) with bandwidth frequency set to ν_a. Typically, such values are learnt a priori by power system operators from offline modal analysis [14]. If the BPF is designed as a relative-degree zero Butterworth filter [15], then the open-loop transfer function of the two-area system retains $n^* = 2$.

The state-space model of the two-machine equivalent of the two-area system, on the other hand, is given as

$$
\underbrace{\begin{bmatrix} \dot{\bar{\delta}}_a \\ \dot{\bar{\delta}}_b \\ \dot{\bar{\omega}}_a \\ \dot{\bar{\omega}}_b \end{bmatrix}}_{\dot{\bar{x}}} = \begin{bmatrix} 0 & I_{2\times2} \\ \mathscr{L} & 0 \end{bmatrix}_{4\times4} \begin{bmatrix} \bar{\delta}_a \\ \bar{\delta}_b \\ \bar{\omega}_a \\ \bar{\omega}_b \end{bmatrix} + \begin{bmatrix} 0 \\ 0 \\ \beta \\ 0 \end{bmatrix} \bar{u}, \tag{28}
$$

$$
\bar{y} = = [1 \ -1; 0 \ 0] \bar{x}, \quad \mathscr{L} = \begin{bmatrix} -\bar{c}_{ab} & \bar{c}_{ab} \\ \bar{c}_{ab} & \bar{c}_{ab} \end{bmatrix}, \tag{29}
$$

with $\bar{c}_{ab} \triangleq E_a E_b \cos(\delta_{a0} - \delta_{b0})/x_{ab}$. The bar sign denotes that the corresponding variables are "equivalents", while subscripts a and b refer to Area 1 and 2, respectively. From (28) and (29), it is clear that the reduced order system has relative degree $\bar{n}^* = 2$ as well.

Summarizing the foregoing analysis, we write the reduced-order, equivalent system as a reference model in the form

$$
\dot{\bar{x}} = \bar{A}\bar{x} + \bar{B}\bar{u}, \quad \bar{y} = \bar{C}\bar{x}, \tag{30}
$$

where a state feedback design of the form $\bar{u} = K\bar{x}(t) \triangleq r(t)$ can be applied to achieve a desired transient response $\bar{y}^*(t)$ of the output.[2] The actual full-order system model, however, has the form

$$
\dot{x} = Ax + Bu, \quad y = Cx, \quad y_m(t) = G(s)[y](t), \tag{31}
$$

where $G(s)$ is a relative-degree zero filter transfer function with Hurwitz zero dynamics [15]. The control objective, therefore, is to design the input u via state feedback such that all closed loop signals are bounded and the plant output $y_m(t)$ tracks $\bar{y}(t)$ asymptotically over time. The following four assumptions hold true for our system:

(A1): Denoting

$$
G_p(s) \triangleq k_p \frac{Z_p(s)}{P_p(s)} = G(s)[C(sI - A)^{-1}Bu](t), \tag{32}
$$

the polynomial $Z(s)$ is stable.

---

[2]Please refer to [12] for the reconstruction of states in the reduced model that will the proposed state feedback design.

(A2): The degree $n = 4$ for $P_p(s)$ is known and fixed.

(A3): Sign of the high frequency gain k_p is known.

(A4): Relative degree of $G_m(s) \triangleq [\bar{C}(sI - \bar{A})^{-1}\bar{B}r](t)$ is same as that of $G_p(s)$, both being equal to 2.

The natural choice for designing the input $u(t)$, therefore, can be a model-reference controller with the following structure [16]:

$$u(t) = \theta_1^T \vartheta_1(t) + \theta_2^T \vartheta_2(t) + \theta_{20} y_m(t) + \theta_3 r(t), \tag{33}$$

where θ's are constant parameters, and

$$\vartheta_1(t) = \frac{q(s)}{\Lambda(s)}[u](t), \quad \vartheta_2(t) = \frac{q(s)}{\Lambda(s)}[y](t), \tag{34}$$

$$q(s) = [1, s, s^2], \quad \vartheta_1, \vartheta_2 \in \mathbb{R}^3, \quad \vartheta_{20} \in \mathbb{R}, \tag{35}$$

and $\Lambda(s)$ is a Hurwitz polynomial of degree 3. Following standard MRC theory, the two filter variables $\vartheta_1(t)$ and $\vartheta_2(t)$, in this case, can be designed simply as

$$\dot{\vartheta}_1(t) = A_\lambda \vartheta_1(t) + B_\lambda u(t), \tag{36}$$

$$\dot{\vartheta}_2(t) = A_\lambda \vartheta_2(t) + B_\lambda y(t), \tag{37}$$

$$A_\lambda = \begin{bmatrix} 0 & 1 & 0 \\ 0 & 0 & 1 \\ -\lambda_0 & -\lambda_1 & -\lambda_2 \end{bmatrix}, \quad B_\lambda = \begin{bmatrix} 0 \\ 0 \\ 1 \end{bmatrix}, \tag{38}$$

where $\Lambda(s) := \lambda_0 + \lambda_1 s + \lambda_2 s^2$. The control parameters in (33) can then be solved for from Diophantyne's equation as in Lemma 5.1 in [16] which states that constants θ_1, θ_2, θ_{20}, and θ_3 exist such that

$$\theta_1^T q(s) P_p(s) + (\theta_2^T q(s) + \theta_{20} \Lambda(s)) k_p Z_p(s)$$
$$= \Lambda(s)(P_p(s) - k_p \theta_3 Z_p(s) P_m(s)), \tag{39}$$

where $P_m(s)$ is the characteristic polynomial of the reference model (30).

As an example, we consider a three-machine cyclic power system, such as the equivalent models shown in Fig. 5, where two generators G_1 and G_2 are connected by a strong link, and, therefore, form one aggregate area, while both machines are connected to a distance generator G_3 via long and, hence, weak tie-lines. The separation of timescales due to the difference in coupling strengths arises due to the three-line reactances and machine inertias, which, for this example, we consider as (all in per unit): $M_1 = 1$, $M_2 = 5$, $M_3 = 10$, $x_{13} = x_{23} = 100 x_{12}$. The three modes (i.e., eigenvalues of the Laplacian matrix, i.e., the state matrix for the double integrator dynamics) of the linearized swing equation are given as $\lambda_1 = 0$, $\lambda_1 = 3.407$, and $\lambda_1 = 0.2307$, clearly indicating the DC mode, the local mode and the interarea mode, respectively. We, therefore, design a second order Butterworth BPF with bandwidth frequency $\omega_s = 0.2307$ (or, alternately with $\omega_s = 3.4$ followed by subtracting the filter output from the original signal), and pass the phase angle

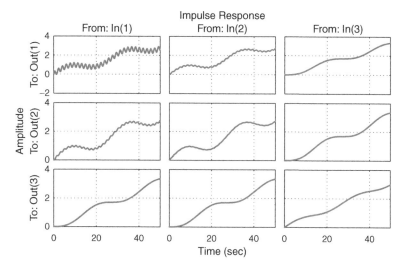

Fig. 8 Phase angle responses for different input–output pairs

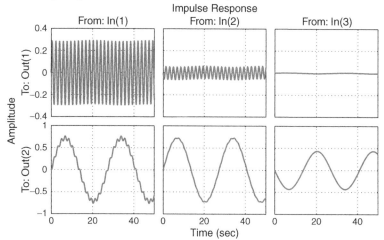

Fig. 9 Phase angle difference for $G_3 - G_1$ and $G_2 - G_1$

difference between G_3 and G_1 through this filter to extract the interarea component. Figure 8 shows the phase angle response of each generator for a unit impulse input applied to each machine separately. Figure 9 shows the phase angles of G_2 and G_3 with respect to G_1, which shows the local and interarea oscillations more explicitly. Figure 10 shows the unimodal output of the BPF superimposed on the actual bimodal phase difference between G_3 and G_1. It can be seen that the interarea response has almost no damping. We next reduce the two-area system into its two-machine equivalent using Interarea Model Estimation [12], and design a state feedback controller to increase the closed-loop damping to 20%. Setting the closed-loop response of the phase angle difference between the two equivalent machines as

Fig. 10 Bimodal phase
response of full-order system
vs unimodal response of
equivalent system

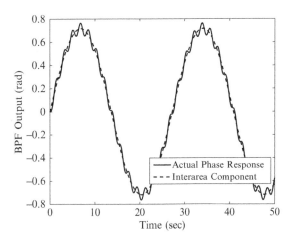

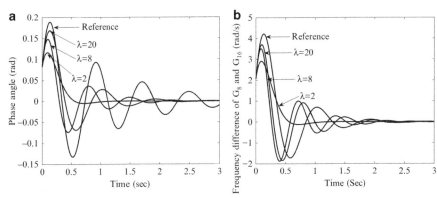

Fig. 11 Interarea damping control via MRC: (**a**) phase angle matching and (**b**) frequency matching

a reference, we next apply the MRC design (33)–(38), and tune λ_0, λ_1, λ_2 so that the filtered output of the full-order system tracks the reference signal as accurately as possible. Figure 11a, b, respectively, show the asymptotic convergence as well as transient matching of the phase angle and frequency for the full-order system response with the reference model over time $t = 0$ to 3 s for three different values of λ_3 fixing $\lambda_1 = \lambda_2 = 1$. Finer matching can be achieved by an iterative choice of the λ's in (38).

4 Conclusions

In this chapter, we presented a set of initial results on the problem of damping interarea oscillations using model-reference state-feedback control designs. The approach consists of model reduction of large power systems into coherent clusters using Synchrophasors, and designing linear state-feedback controllers to achieve a desired damping between the clusters. The final step, thereafter, is to treat the closed-loop response of the reduced system as a reference and employ model-reference control to track this reference for the full-order system. A natural question that may arise is on our proposed approach of *model reduction* preceding control design, i.e., why should one reduce a given network into clusters first, and then invert its closed-loop responses to the actual system? The answer follows from the size and complexity of any realistic power system. The WECC, for example, has roughly 2,000 generators, 11,000 transmission lines, 6,000 transformers, and 6,500 loads [10]. Designing a distributed control mechanism to shape a desired set of *interarea* responses starting from this entire model and using PMU measurements from arbitrary locations would be practically intractable. We believe that the proposed *detour* of reducing such large systems into simpler chunks (even if approximately), and then redistributing their control efforts would give the problem a much more well-defined and less chaotic formulation. Incidentally, two seminal pieces of work done in wide-area damping control so far, namely [17, 18], have also been challenged by similar questions of scalability. Both of these references, however, are experimental case studies addressed toward two very specific interconnections in North America – namely, the north–south Pacific AC Intertie transfer in the WECC [17], and the Hydro-Quebec internal grid [18], as a result of which their methods for solving this problem are more case-specific compared to our method, which is much more generic and network concerted.

References

1. Chow JH, Peponides G, Kokotovic PV, Avramovic B, Winkelman JR (1982) Time-scale modeling of dynamic networks with applications to power systems. Springer, New York
2. Dobson I (1992) Observations on the geometry of saddle node bifurcation and voltage collapse in electric power systems. IEEE Trans Circuit Syst—Part 1 39(3):240–243
3. Willems JL, Willems JC (1970) The application of lyapunov methods to the computation of transient stability regions for multimachine power systems. IEEE Trans Power App Syst 89(5/6): 795–801
4. Pai MA (1989) Energy function analysis for power system stability. Kluwer Academic Publishers, MA
5. Phadke AG, Thorp JS, Adamiak MG (1983) New measurement techniques for tracking voltage phasors, local system frequency, and rate of change of frequency. IEEE Trans Power App Syst 102:1025–1038
6. Liu Y et al (2006) A US-wide power systems frequency monitoring network. In: Proceedings of the IEEE PES general meeting, Montreal, QC, Canada, June 2006
7. Hauer JF, Demeure CJ, Scharf LL (1990) Initial results in prony analysis of power system response signals. IEEE Trans Power Syst 5(1):80–89
8. Messina AR, Vittal V, Ruiz-Vega D, Enriquez-Harper G (2006) Interpretation and visualization of wide-area pmu measurements using hilbert analysis. IEEE Trans Power Syst 21(4): 1760–1771
9. Emami R, Abur A (2009) Reliable placement of synchronized phasor measurements on network branches. In: Proceedings of the IEEE PSCE, Seattle, WA, March 15–19, 2009

10. Dagle JE (2004) Data management issues associated with the August 14, 2003 Blackout Investigation. In: IEEE PES General Meeting, CO, June 2004
11. PNNL EIOC, http://eioc.pnl.gov/research/gridstability.stm, 2011
12. Chakrabortty A, Chow JH, Salazar A (2011) A measurement-based framework for dynamic equivalencing of power systems using wide-area phasor measurements. IEEE Trans Smart Grid 1(2):68–81
13. Pal B, Chaudhuri B (2005) Robust control in power systems. Springer, New York
14. Trudnowski DJ, Dagle JE (1997) Effects of generator and static-load nonlinearities on electromechanical oscillations. IEEE Trans Power Syst 12(3):1283–1289
15. Oppenheim AV, Schafer RW (1989) Discrete-time signal processing. Prentice Hall, Upper Saddle River, NJ
16. Tao G (2003) Adaptive control analysis design and analysis. Wiley, New Jersey
17. Taylor CW, Erickson DC, Martin KE, Wilson RW, Venkatasubramanian V (2005) WACS—wide-area stability and voltage control system: R&D and online demonstration. Proc IEEE 93(5):892–906
18. Kamwa I, Grondin R, Hebert Y (2001) Wide-area measurement based stabilizing control of large power systems—a decentralized/hierarchical approach. IEEE Trans Power Syst 16(1):136–153

Index

A. Chakrabortty and M.D. Ilić (eds.), *Control and Optimization Methods for Electric Smart Grids*, Power Electronics and Power Systems 3, DOI 10.1007/978-1-4614-1605-0, © Springer Science+Business Media, LLC 2012

CPSIA information can be obtained at www.ICGtesting.com
Printed in the USA
LVOW011732050513

332326LV00003B/274/P